THE WORLD'S CLASSICS

HEADLONG HALL AND GRYLL GRANGE

THOMAS LOVE PEACOCK was born in 1785 and grew up in the Thames valley. His early life was unsettled and economically somewhat precarious; he devoted himself to poetry and classical studies but had no formal schooling beyond his earliest teens. His published volumes of poetry caused little stir but impressed Shelley. Peacock became a member of the Shelley circle in 1812 and was closely involved in the poet's personal life thereafter. He published his first novel, *Headlong Hall*, in 1815 and the more ambitious *Melincourt* two years later. *Nightmare Abbey* followed in 1818. In 1819 he began working for the East India Company and the next year married Jane Gryffydh, a parson's daughter, whom he had met whilst living in Wales eight years earlier but with whom he had had no contact in the interim. Over the next fifteen years he became the father of four children and published three more novels, *Maid Marian* (1822), *The Misfortunes of Elphin* (1829), and *Crotchet Castle* (1831), as well as a certain amount of literary journalism and many opera reviews. In the East India Company he was very active and influential in promoting steamship communication between the UK and India and he succeeded James Mill as Chief Examiner of Correspondence in 1836. He published nothing further until after his retirement twenty years later. His second daughter married George Meredith in 1849, but the match proved to be a very unhappy one. Peacock published his 'Memoirs of Shelley' in 1858–60 and his last novel, *Gryll Grange*, in 1860. In spite of domestic misfortune, his later years—spent in scholarly retirement at Lower Halliford, his home since 1823—were serene. He died in 1866.

MICHAEL BARON is a Lecturer in English at Birkbeck College (University of London) and Co-Editor of *English.* He has published essays on Coleridge, Wordsworth, and Yeats and is currently engaged on a book on Wordsworth's poetic style.

MICHAEL SLATER is Reader in English at Birkbeck College (University of London) and was Hon. Editor of *The Dickensian*, 1968–77. He has published, as editor, *Dickens 1970* and Dickens's *Christmas Books* and *Nicholas Nickleby* in the Penguin English Library series, and is the author of *Dickens on America and the Americans* and *Dickens and Women.*

To Robina Barson
and John Grigg

THE WORLD'S CLASSICS

THOMAS LOVE PEACOCK

Headlong Hall and *Gryll Grange*

Edited with introductions by

MICHAEL BARON

AND

MICHAEL SLATER

Oxford New York

OXFORD UNIVERSITY PRESS

1987

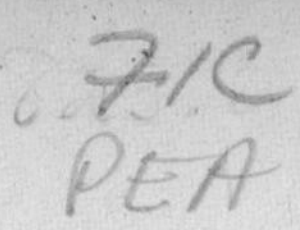

Oxford University Press, Walton Street, Oxford OX2 6DP

Oxford New York Toronto
Delhi Bombay Calcutta Madras Karachi
Petaling Jaya Singapore Hong Kong Tokyo
Nairobi Dar es Salaam Cape Town
Melbourne Auckland

and associated companies in
Beirut Berlin Ibadan Nicosia

Oxford is a trade mark of Oxford University Press

First published as a World's Classics paperback 1987

British Library Cataloguing in Publication Data

Peacock, Thomas Love
Headlong Hall and Gryll Grange.—(The World's classics).
I. Title II. Baron, Michael
III. Slater, Michael
823'.7[F] PR5162.H4
ISBN 0-19-281693-4

Library of Congress Cataloging in Publication Data

Peacock, Thomas Love, 1785–1866.
Headlong Hall; and, Gryll Grange.
(The World's classics) Bibliography: p.
I. Baron, Michael, 1947– . II. Slater, Michael.
III. Peacock, Thomas Love, 1785–1866.
Gryll Grange. 1987. IV. Title.
PR5161.B67 1987 823'.7 87-7849
ISBN 0-19-281693-4 (pbk.)

Typeset by Latimer Trend & Company Ltd, Plymouth
Printed in Great Britain by
Hazell Watson & Viney Ltd.
Aylesbury, Bucks.

CONTENTS

ACKNOWLEDGEMENTS

WE owe a special debt of gratitude to Robina Barson for invaluable help with the Italian Romance material in *Gryll Grange* and with the landscape gardening arguments in *Headlong Hall*; and to our colleague, Dr Roland Mayer, for guidance in tracking Peacock into the by-ways of classical literature and history.

We would also like to thank the following people who have assisted in various ways with the annotation: Professors Harold F. Brooks, J. H. Burns, and Angus Easson, Dr Roy Foster, Jeremy Griffiths, Dr Leslie Parris, Dr Patrick Pollard, William Poole, Dr Richard Ralph, Barbara Rosenbaum, Peter Rowland, Deidre Toomey, Sergio Zoi.

We are very grateful to Hilary Feldman and Diana Godden for their help with the preparation of the manuscript.

We would like to thank the Bodleian Library for permission to reproduce the title-pages from the first editions of *Headlong Hall* and *Gryll Grange* (12 Theta 476 and 250 b 50); also the British Library Board for permission to quote from Add. MSS 36815/6 and 47225, and the Carl and Lily Pforzheimer Foundation, Inc., New York, for permission to quote from MSS TLP 112, 122, and 126 in the Carl H. Pforzheimer Library.

M.B.

M.S.

INTRODUCTION TO *HEADLONG HALL*

DURING the five or six years before *Headlong Hall* was published Peacock's life was as 'Romantic' as anyone could wish. Having survived a year's employment as naval secretary forced upon him by prudent uncles, he drifted into a precarious and largely itinerant life of studying, writing, and repeatedly falling in love. He traced the course of the Thames from its source to Chertsey, wandered in North Wales for a year, visited the Isle of Wight, and followed Shelley to the Lake District, to Edinburgh and back to Bracknell and London. He was often financially desperate, and he contemplated suicide. Three applications for small sums of money were made on his behalf to the Literary Fund; in one of them the applicant, Peacock's friend and publisher Edward Hookham, had 'reason to dread that the fate of Chatterton might be that of Peacock',[1] and early in 1815 Peacock was reduced to writing begging letters to strangers from a sponging-house in Liverpool, after a mysterious affair with a 'rich' heiress. He published in 1812 a revised edition of his poem 'Palmyra' and a new poem 'The Philosophy of Melancholy', but there is no evidence that he made any money out of them. Later in the same year he tried his hand at writing for the stage, which offered the successful author a more rapid and substantial income than verse. At least two of his plays, 'Mirth in the Mountains' (now lost) and 'The Dilettanti', were read by the manager of the Drury Lane company, but neither was performed. He tried the educational market with 'Sir Hornbook' in the summer of 1813, but though it had some success, he apparently made little money out of it. A plan to take in pupils came to nothing. Only in the summer of 1815, when Shelley was able to give him an allowance of £120 per annum, did

[1] Quoted by N. A. Joukovsky in 'Peacock before *Headlong Hall*', *Keats–Shelley Memorial Bulletin*, 36 (1985), 25. My account of Peacock's early life is largely based on Joukovsky's.

Peacock achieve any degree of financial stability. *Headlong Hall* was published at the end of the year.

It is startling to think that 'The Author of "Headlong Hall"' (the name under which Peacock published all his subsequent fiction) was ever compared in high places with Chatterton, who haunted the Romantic mind as a symbol of high-souled despair. Peacock is celebrated, after all, as the creator of Scythrop Glowry, whose resemblance to Shelley was immediately obvious, and of a series of wonderful travesties of Coleridge: Moley Mystic, Flosky, and Skionar. Yet Peacock's fiction is not merely the negative voice of Romanticism. He shared with the poets, particularly Shelley, a fascination with those idealistic states of mind that encompass scepticism as well as enthusiasm. In fact a crucial point in Peacock's early life is his meeting with Shelley late in 1812 and their conversations at Bracknell during the following years. Shelley's house was full of idealisms of various kinds—Godwinism, vegetarianism, egalitarianism, manichaeanism—and a spirit of controversy reigned. Peacock later acknowledged that he sometimes openly laughed at 'the fervour with which opinions utterly inconducive to any practical result were battled for' (*Memoirs of Shelley*, Halliford, viii, 70–1), but his part was probably not as straightforward as that suggests. In 1814 he began 'Ahrimanes', a learned, melancholy, poetic meditation on the good and evil principles in the Zoroastrian religion;[2] but he also wrote, under Shelley's influence, his first literary satire, 'Sir Proteus'. It is much in the vein of Shelley's 'Peter Bell the Third' (written five years later) but spoiled by its heavy-handedness and pedantry. Lampoon was clearly not Peacock's forte (the example in *Gryll Grange*, 'A New Order of Chivalry', is something of an embarrassment), but 'Sir Proteus' at least shows Peacock experimenting with satiric forms at a time when he was also writing a serious poem on a recherché subject. Indeed, the questions raised by Peacock's friendship with Shelley are better approached through consideration of

[2] For an argument that *Headlong Hall* is deeply influenced by Zoroastrian thought see James D. Mulvihill, 'Peacock and Perfectibility in *Headlong Hall*', *Clio*, 13 (1984), 227 ff.

form than by trying to decide whether he took this or that idea seriously.

Marilyn Butler describes *Headlong Hall* as a kind of Socratic dialogue (*Peacock Displayed*, p. 40). This is a good description of the antagonism of Foster and Escot, which is characterized more by set speeches and leading questions than by a free play of intellect and emotion, and Professor Butler reminds us that there is no need to be embarrassed, as many critics in the past have been, by the lack of true characterization in this novel. But this view underestimates Peacock's farce. No reader forgets that the practical outcome of Headlong's enthusiasm for Milestone's plan for improving the estate is the pitching of Cranium into Llyn Peris, from which he is dragged by Escot, who thereby improves his chances of marrying Cranium's daughter Cephalis. The episode is both good farce and an essential part of the plot. So too is the earlier scene in which Escot enters the breakfast room with Cadwallader's skull, innocently causing the 'promiscuous ruin' of plates, cups, and saucers and exciting the 'lively curiosity' of the craniologist. The characteristic quality of *Headlong Hall* is, in fact, its combination of debate and farce:[3] an odd combination, because farce is a closed form (we admire the author's skill in overt contrivance) and debate, in Peacock's hands, an open one. None of the issues the novel raises is intellectually resolved.

Peacock's extant theatrical farces, *The Dilettanti* and *The Three Doctors*, are delightfully silly: there is much elaborate abuse, nose-pulling, wig-snatching, locking people in cupboards and breaking violins and canvases over their heads. We can recognize in them some of the qualities of the novels, especially *Headlong Hall* and *Nightmare Abbey*. *The Three Doctors*, subtitled 'a musical farce', is a series of scenes in which conflicts are summed up in song: duets, trios, even a Gilbertian septet, in which characters express varying attitudes to an absurd situation ('This trigger, if I pull it,|Will

[3] In this it differs widely from Isaac D'Israeli's *Flim Flams* (1797), with which it is sometimes compared, and which is generically nearer to picaresque romance than to farce.

emancipate a bullet . . . ': Halliford, vii. 397–8). Something of this quality is evident in the nine-voice response to Gall's assertion that there is no such thing as good taste left in the world (ch. 10) and, conversely, it lies behind Headlong's repeated attempts to resolve intellectual discord by urging his guests to sing with *one* voice in a glee. What one misses in the plays, however, is the satiric force of the novel's intellectual comedy. Marmaduke Milestone turns up in *The Three Doctors* as a recognizable caricature of Humphry Repton, just as he does in the novel; but he has no *intellectual* antagonist, merely a rival in his suit for Squire Hippy's daughter. The rival lover, O'Fir, has none of the artistic enthusiasms of Patrick O'Prism. For this reason much of the comedy associated with Milestone in the play has little to do with landscape gardening but arises from Peacock's manipulation of the stock characters and situations of farce. A fair sample is the ebullient soliloquy with which Milestone makes his first entry, having been affronted by the boorish Shenkin, Squire Hippy's servant:

That fellow's an uncivilized goat—a mountain-savage—a wild man of the woods. Wants shaving and polishing. As much in need of improvement as the place he inhabits. Great capabilities here. Soon be my own, to clump and level *ad libitum*. Hope the young lady won't prove refractory. Published many books. Sold none. Bad speculation. Present plan much better. Marriage to a fortune cures all evils except itself (Halliford, vii. 404).

The Milestone of *Headlong Hall*, on the other hand, represents parts of a famous controversy which delighted Peacock in 1810. We may suppose that it was not until after his literary experiments in 1813–14 that he was able to find a form, more subtle than farce, in which to explore its ramifications. The controversy about landscape gardening between Humphry Repton (1752–1818), Richard Payne Knight (1750–1824), and Uvedale Price (1747–1829) had its heyday in 1794–5 but revived in 1810 with the publication of Price's *Essays on the Picturesque*, in which he reprinted his and Repton's polemical exchanges. Price's and Knight's initial target was the landscaping practice and influence of Lancelot ('Capability') Brown (1715–83), which is evident in places like the grounds of Blenheim Palace: the house, set in gently undulating parkland, is designed to be viewed across a serpentine river or lake which reflects the house and is crossed by an ornamental bridge. On another side of the house there might be a shrubbery with smooth gravelled walks (a 'pleasure ground') for rainy days or feeble limbs. In *The Landscape* (1794) Knight scorned Brown's 'billiard-table' vistas, preferring a more 'natural'—that is, wilder—landscape. He emphasized, too, a working relationship between house and grounds: an estate which is also a farm and not merely a park. But though it was Brown's ideas that Knight and Price rejected, their attack was directed at Repton, Brown's professional successor, partly because Brown himself had left no written defence of his views. Repton was quite ready to go into print against Brown's detractors, but his position was more ambiguous than that might suggest. On the one hand he endorsed Brown's assumption that the appearance of a country seat should reflect the owner's wealth (this is the point of the joke about 'a stone with distances' in ch. 3). On the other he deplored some of Brown's more 'unnatural' themes, such as an artificial lake on the top of a hill. He was disappointed, even hurt, when Knight attacked him in *The Landscape (Sketches and Hints on Landscape Gardening* (1794), ch. 7). Uvedale Price added his voice to the debate in *An Essay on the Picturesque* (1794), also attacking Brown and his followers, but provoking a new controversy by insisting on a distinction between the beautiful and the picturesque, which Knight would have nothing to do with (cf. the Milestone–O'Prism exchange in ch. 4: 'You will have the goodness to distinguish between the beautiful and the picturesque.' 'Will I . . . Och, but I won't.' Here Milestone takes Knight's view, not Repton's).

It is a bookish debate (of the three, only Repton practised landscape gardening as a profession), but Peacock—who well understood pedantry and the certainties that mere theorizing can lead to—brilliantly released its comic potential by sniffing out the *ad hominem* arguments. The other major debate of the novel—deteriorationism versus perfectibilianism—is more diffuse but has a bearing on the first because both turn upon the ambiguities of the concept of the 'natural': Escot presupposes that the primitive life is more natural, therefore nobler, than sophisticated life. This is a moral agrument chiefly, but the aesthetic sense plays a part in it. Should we applaud the reclamation of land in the Traeth Mawr estuary (a scheme to which Shelley had contributed money and energy in 1812) and the building on its bank of woollen mills and the new township of Tremadoc? Peacock's narrative voice weighs progress against beauty, and his judgement (and Escot's) seems clear:

> all admirers of the magnificence of nature will remember [the estuary] with regret, whatever consolation may be derived from the probable utility of the works which have excluded the waters from their ancient receptacle (ch. 7).

But the deteriorationist issue has other centres, which Peacock found in the work of Lord Monboddo, especially his *Antient Metaphysics* (1779–99) and its sources, chiefly Rousseau's *Discours sur l'origine de l'inégalité parmi les hommes* (1755). Both made extensive use of anthropological research in order to arrive at an idea of what man is like in his 'natural' state. The behaviour of the 'wild man of the woods'—the orang-outan—became a battleground for conflicting theories backed up by absurd stories about the animal's genteel behaviour (many of them retailed in Rees's *Cyclopaedia*, light reading for Mr Panscope in ch. 3). So too did vegetarianism. Both Monboddo and Shelley (in a *Vindication of Natural Diet* (1813)) repeated Rousseau's argument in the *Discours* that man is naturally vegetarian because his stomach and teeth resemble those of the herbivorous orang-outang. Shelley, in fact, was willing to accept strikingly odd notions in support of his vegetarianism, even though he was not a deteriorationist at all.

These are the wilder shores of deteriorationism which Peacock may well have laughed at when he learned about them from Shelley and his manichaean-vegetarian friend J. F. Newton. The sceptical mind is on firmer ground when commenting on the small hypocrisies of the ballroom, on the more blatant abuses of physical and moral health in the factory system, and on the social and monetary pressures in marriage. In these areas, as Marilyn Butler observes (*Peacock Displayed*, p. 46), the deteriorationist is bound to have the better of the argument because it is always easier to be particular, and therefore persuasive, about present evils than about the future improvements. Besides, by 1815 deteriorationism had a powerful new voice in the science of political economy, particularly in Thomas Malthus's *An Essay on the Principle of Population* (1798, greatly enlarged in 1803). Malthus predicted widespread poverty because calculation showed that the population was increasing much more quickly than agricultural productivity. It is the fact that he attempted to *calculate* the future that gave his arguments so much more force than those of the perfectibilians he addressed in his *Essay*: Godwin and Condorcet. They, in comparison, merely speculated from a priori principles. Peacock made extensive satiric reference to Malthus in the figure of Mr Fax in *Melincourt* (1817), but Escot's calculation of the multiplying 'needs' of civilized society has a Malthusian ring.[4]

Rousseau, Monboddo, J. F. Newton, Malthus, and Shelley are Peacock's chief deteriorationist sources. His perfectibilians are Condorcet, Godwin, and—Shelley. In fact Shelley is the reason Peacock was thinking of Godwin at all in 1815. Shelley had expounded uncritically the Godwinian doctrine that there is a logical and necessary connection between our actions and our immediate circumstances (*Queen Mab*, Note to VI, 198), and he followed Godwin in believing that every action is susceptible of purely rational explanation without recourse to the illusory concept of free will (*Political Justice* (1793), iv. 5) or such vague notions as instinct. Virtue consists in the exercise of rational preference, and, as reason spreads (look at the progress of science!), so too will virtue—provided men are educated in the ways of reason. This is the kernel of Godwin's *Political Justice* (1793)

[4] Noted by N. A. Joukovsky in 'A Critical Edition of *Headlong Hall* and *Nightmare Abbey*' (unpubl. D.Phil. thesis, Oxford, 1970).

and the whole story of Condorcet's *Esquisse d'un tableau historique des progres de l'esprit humain* (1795).

Peacock's perfectibilian, Foster, is a pasteboard figure compared with his rival—we know nothing of his inner life—and it is easy to feel that the perfectibilian point of view is less forceful in the novel than the deteriorationist. This is evident when, for example, the philosophers visit the manufactory at Tremadoc. Foster sees a productive Owenite commune; Escot, an inferno where ghastly featured children struggle to keep pace with 'the dizzy and complicated motions of diabolical mechanism'. But Foster does have his moments of satiric force, one of which occurs when he imagines a flower 'falling into a train of theoretical meditation on its original and natural nutriment' and 'work[ing] itself up into a profound abomination' of fertilizers: a *reductio ad absurdum* of Escot's notion that the natural and the original are identical. This is one of the few passages Peacock added in the second edition—presumably to give life to the prefectibilian viewpoint.

There is, in any case, another strand in Peacock's satire of Godwinian rationalism, and it has to do with marriage. 'Marriage', remarks the delightful Sir Telegraph Paxarett in *Melincourt*, 'may sometimes be a stormy lake, but celibacy is almost always a muddy horsepond.' The imagery is as strikingly appropriate to *Headlong Hall* as the sentiment. The penultimate chapter is dominated by Headlong's precipitate marriage-broking: farcical action, but it does have a satiric point because Headlong is a vehicle for Peacock's condemnation of those who do not value marriage, who turn their backs on good-heartedness and the particularities of affection: Panscope and Cranium. Each of these expresses a distinctly coarse attitude to marriage, and Cranium's comment is a classic statement of possessiveness and materialism:

> I simply know ... that if [Cadwallader's skull] were once in my possession, I would not part with it for any acquisition, much less for a wife. I have had one: and, as marriage has been compared to a pill, I can very safely assert that *one is a dose*.

But the more forceful comedy in Cranium's rejection of

marriage proposals arises in his encounter with Headlong, where he becomes an out-and-out necessitarian. Refusing to give credit to Escot for saving his life, he expresses himself in a remarkable parody of Godwin's arguments against free will:

> The whole process of the action was mechanical and necessary. The action of the poker necessitated the ignition of the powder . . . the explosion necessitated my sudden fright, which necessitated my sudden jump . . . The motive or impulse thus adhibited in the person of a drowning man, was as powerful on his material compages as the force of gravitation on mine; and he could no more help jumping into the water than I could help falling into it.

Peacock, then, divides Godwinism into two strands—the spread of reason and virtue on the one hand and the doctrine of necessity on the other—and, broadly speaking, gives the first to Foster and the second to Cranium. Cranium is, of course, principally a phrenologist, but Peacock uses the character in this double fashion partly in order to suggest that phrenology is crudely deterministic, and partly to satirize, in a scene of marriage proposals, a philosophy that had no room for the affections and was infamous for its antagonism to marriage as an institution. It is tempting to suggest that the end of the novel reflects Shelley's relations with Godwin, his prospective father-in-law as well as his philosophical mentor. Having eloped with Godwin's daughter in the summer of 1814, Shelley married her a little over two years later in spite of his earlier Godwinian conviction that marriage was at best merely an empty social convention and at worst institutionalized tyranny. After all, *Nightmare Abbey* offers plenty of evidence that Peacock was happy to satirize Shelley's domestic tangles as well as his intellectual ones.

Peacock's is an art of celebration as well as satire. *Headlong Hall* ends with four marriages; *Gryll Grange* with nine. Of course the comic convention is made fun of in these excesses, yet in both novels marriage is part of a broader social celebration: in *Gryll Grange* the Twelfth Night ball, and in *Headlong Hall* the Christmas ball which marks an era with the 'beau monde of Cambria'. The novel is a series of scenes in which conviviality is disturbed, whether by '*un morne silence*' or by a furious row; by the appearance of a skull at the breakfast table or by an overlong and incomprehensible lecture. At the end images of conviviality persist and give the novel a sense of harmony that is conspicuously lacking in both the intellectual debates and in the perfunctory tying of the plot. (Minor characters simply disappear when their time is up—though we may find some satisfaction in the symbolic absence of asperity and flattery, Gall and Treacle.) Gaster, who now comes into his own, is both *bon vivant* and 'whipper-in' at the marriages, linking social and nuptial celebration. It is true that 'society' is only flimsily represented in the novel—it exists largely in the clannish mind of Miss Brindle-mew Grimalkin Phoebe Tabitha Ap-Headlong—but this is sufficient, with the images of conviviality, to cast a refracting light on both the sullenness of Panscope and Cranium and, more important, on the idealistic misanthropy of Escot. For Peacock's fictions of marriage are an answer to that Romantic strain of thought which sees the 'natural' in the solitary, the pre-social. Peacockian comedy reaches a climax at the very end of the novel in Escot's powerful diatribe on the pressures that lead to bad marriages: the speech is partly ridiculous, being an 'admirable counterpoint' to Escot's *actions*, and partly serious, reflecting back on the exploration of human nature that lies at the heart of the novel.

M.B.

INTRODUCTION TO *GRYLL GRANGE*

Peacock published five more novels during the sixteen years following the appearance of *Headlong Hall*, novels in which satire and romance are mixed in varying proportions. A series of farcical occurrences continues to supply the plot element in these books (his one attempt at melodrama, towards the end of *Melincourt*, is perfunctory in the extreme) but there is a discernible development in Peacock's presentation of his dramatis personae, as Tennyson's friend James Spedding noted in his appreciative survey of Peacock's novelistic output in the *Edinburgh Review* in 1839 (vol. 68, pp. 432–59). Spedding identifies a progression in what he calls 'kindliness' in successive novels: 'the disputants are more in earnest, and less like scoffers in disguise; there is more of natural warmth and life in the characters . . .' It is certainly true that the latest of these books, *Crotchet Castle* (1831), contains two of Peacock's best humorous creations, the Revd Dr Folliott and Lady Clarinda Bossnowl. Whereas Gaster in *Headlong Hall* really is, as his name and Peacock's facetious footnote (p. 3) suggest, little more than a walking, talking stomach, Folliott is a delightful blend of good sense and ripe prejudice, geniality and irascibility, sound learning and pedantry. The beautiful, witty Lady Clarinda puts all Peacock's preceding heroines into the shade as, with a charming mercilessness, she instructs her lovesick swain Captain Fitzchrome about the realities of life for a spirited and intelligent woman in her position:

I am not fit to be a poor man's wife. I cannot take any kind of trouble, or do any one thing that is of any use. Many decent families roast a bit of mutton on a string; but if I displease my father I shall not have as much as will buy the string, to say nothing of the meat; and the bare idea of such cookery gives me the horrors.

Crotchet Castle ends very unsatisfactorily and uncomfortably, however, with Folliott leading a charge to disperse an attack on Chainmail Hall by 'poverty in despair', the starving local peasantry, and Lady Clarinda marrying poor Fitzchrome after all, in the last sentence of the book. The kind of farcically neat yet satisfactorily resonant ending that Peacock had achieved for his two earlier 'novels of talk', *Headlong Hall* and *Nightmare Abbey*, eluded him here; and it may well be that some consciousness of failure in this respect had as much to do with his abandonment of novel-writing for the next twenty-five years or so as the reasons usually adduced, i.e. the death of his most valued reader, his mother, and the demands of his Chief Examinership at India House. Whatever the reason, it was not until the year he retired, 1856, that Peacock gave a public hint—in a Preface to a reprint of *Melincourt*—that a new work on the old lines might be forthcoming. Over his usual novelist's signature, 'The Author of "Headlong Hall"', he wrote:

> Of the disputants whose opinions and public characters (for I never trespassed on private life) were shadowed in some of the persons of the story, almost all have passed from the diurnal scene. Many of the questions, discussed in the dialogues, have more of a general than of a temporary application, and still have their advocates on both sides: and new questions have arisen, which furnish abundant argument for similar conversations, and of which I may yet perhaps, avail myself on some future occasion.

Four more years were to elapse, however, before *Gryll Grange* began appearing monthly in the pages of *Fraser's Magazine*. During this time his publications, all in *Fraser's*, had mainly related to one of the great abiding passions of his life, Greek literature, and to the most important friendship of his life, that with Shelley. An intensely private man, by no means given to public autobiography, Peacock was nevertheless provoked into compiling his fascinating 'Memoirs' of Shelley by the misrepresentations of facts about the poet's life, especially about his relations with his first wife, in some recent publications such as Hogg's *Life of Shelley* and Lady Shelley's *Shelley Memorials*.

Both Greek literature and the luminous figure of Shelley feed into *Gryll Grange* as they had into Peacock's earlier novels. In none of them, however, had Greek been so important as in *Gryll Grange*. Opimian's apposite quotation from Homer is what gains him admission to Falconer's Tower and so sets the story in motion, the production of a modern Aristophanic comedy unites the characters in a communal effort, and Greek supplies a good half of the festive epigraphs for the chapters. As for Shelley, he is certainly recalled in some aspects of young Falconer, the romantic idealist in his Tower, though by no means as fully or as obviously as in Scythrop in *his* tower in *Nightmare Abbey*.

Apart from his essays in *Fraser's*, Peacock's literary output in the 1850s consists of a number of untitled and unfinished stories (some barely started) in which he is evidently working towards the novel that became *Gryll Grange*. One concerns an ancient Roman Epicurean, his daughter and her lover; another, set in 'Boozabowt Abbey', is comic/medieval; and two more, very close in tone to *Gryll Grange*, have modern Thames Valley settings and touch on many things that feature in that novel—such as the three chapels of St Katharine, St Martha and St Anne, St Katharine herself and her history, and the poisoning of the Thames and other results of the 'march of "improvement"'. Enthusiasts for classical literature appear in both of these stories also. All these unfinished novels were printed from the manuscript fragments in the British Library in the Halliford edition of Peacock's works (vol. 8) and further fragments, as well as three chapters of yet another story, set in the Holy Land, about the Knights of St Katharine, are in the Pforzheimer Library, New York.

The publication of *Gryll Grange* seems to have attracted little attention. A rather hostile notice in the *Westminster* found it very much the mixture as before, but the *Saturday Review* and the *Spectator* were more appreciative. Both stressed the uniqueness and originality of Peacock. The *Saturday* found that the volume read 'like a few numbers of *Notes and Queries* jumbled up with a funny love story, and pervaded by a fine Pagan morality', and the *Spectator* welcomed it with an image that Peacock no doubt felt to be entirely appropriate:

The reappearance of the author of 'Headlong Hall' produces a feeling like that with which we await a bottle of wine that we recollect in its 'relish fiery-new' but which we have reason to believe calculated to improve by a twenty years' sojourn in the cellar . . . Mellow is the word which perhaps most adequately expresses the tone of thought which pervades *Gryll Grange*.

James Hannay, writing in the *North British Review* a few months after Peacock's death, declared *Gryll Grange* to be 'quite as fresh as any book of the "Headlong Hall" series, and even more remarkable than the best of them for ingenuity, liveliness of humour, general vigour of wit, and wide reading in literature' (vol. 45, pp. 75–104). Since then, this autumnal flowering of Peacock's genius has not lacked admirers, notably David Garnett in his excellent one-volume edition of the novels (1948) and Marilyn Butler who calls it, in *Peacock Displayed*, 'the richest, most ambitious and complete of his satires' (p. 235). What will surely strike the reader of the present volume most forcibly as he or she passes from *Headlong Hall* to *Gryll Grange* is how profoundly different the later novel is, despite its obvious formal resemblances to the earlier one.

The basic difference is that *Gryll Grange* is not structured around an ongoing intellectual debate like that between the two young philosophers in *Headlong Hall* or between the traditional Toryism of Folliott and the political economy of MacQuedy in *Crotchet Castle*. MacQuedy's successor in *Gryll Grange*, Mr MacBorrowdale, does not appear until chapter 13 and he steadfastly refuses to be drawn into debate, making a clear distinction between the urbane pleasures of post-prandial 'conversation' and the horrors of 'after-dinner lecturers' where

the rest of the good company, or rather the rest which without them would have been good company, was no company. No one could get in a word. They went on in one unvarying stream of monotonous desolating sound. This makes me tremble when a discussion begins. I sit in fear of a lecture.

Mr MacBorrowdale voices this civilized objection to debate at the dinner table, but he is in any case not presented as a tunnel-visioned fanatic in the way MacQuedy was. He has, he confesses in chapter 29, 'talked a great deal of nonsense' about political economy in his time.

It might seem that Lord Curryfin, another figure who does not appear until chapter 13 but who is discussed in chapter 1, is intended for Opimian's 'mighty opposite' in debate. This young man is very much the representative of the bustling modern world of technological and scientific progress, 'the march of mind', competitive examinations, and all the other things that are anathema to Opimian who keeps his feet so firmly on his beloved *vias antiquas*. Curryfin is, moreover, an active member of the 'Pantopragmatic Society' (Peacock's satirical name for the newly-founded National Association for the Advancement of Social Science—see note to p. 137, below), which, for Opimian, is the very epitome of the mischievous fatuity of contemporary legislators and politicians. But, as with MacBorrowdale, Peacock is careful to depict him as an individual rather than as simply a walking collection of *idées fixes*:

being one of those 'over sharp wits whose edges are very soon turned', he did not adhere to any opinion with sufficient earnestness to be on any occasion betrayed into intemperance in maintaining it. So far from this, if he found any unfortunate opinion in a hopeless minority of the company he happened to be in, he was often chivalrous enough to come to its aid, and see what could be said for it.

And this, we feel, is just what he is doing in chapter 19 when he makes some attempt to defend technological progress, Americans, and even competitive examinations against Opimian's broadsides. The reader, in fact, quickly becomes, like Opimian, more interested in Curryfin's emotional life than in his intellectual opinions and, in any case, we eventually discover that he has renounced 'Pantopragmatics', which now seem to him as 'ridiculous' as they do to other *bien pensants* (p. 290).

The kind of debate that seems natural to the world of *Gryll Grange* is an incidental one on scholarly matters rather than a continuing one on intellectual issues. Such is the argument about the Bald Venus of the Romans (ch. 29), the possibility of whose existence Opimian hotly denies. Was Pope's description of the game of Ombre quite correct? What did ancient Greek wine actually taste like—could Alcaeus, Anacreon, and Nonnus have possibly composed the poems they did 'under the inspiration of spirit of turpentine'? Of what complexion was Cleopatra? Did the Vestal Virgins wear their own hair? These are the kind of Sir-Thomas-Browne-like questions that, even though they may sometimes lurk in footnotes, one feels to be the really important ones in *Gryll Grange*, rather than the vast amorphous social issues so passionately canvassed in the world outside—by the Pantopragmatics, for example, or by such desolating orators as those MacBorrowdale has encountered:

> The bore of all bores was the third. His subject had no beginning, middle, nor end. It was education. Never was such a journey through the desert of mind: the Great Sahara of intellect. The very recollection makes me thirsty.

Much of the pleasure and interest of *Gryll Grange* lies in the learned preoccupations of Dr Opimian, fuelled as we feel them to be by his creator's passion for historical accuracy, truth to nature, and the civilization and culture of Ancient Greece. But it is the 'funny love story' that both provides the truly novelistic enjoyment of the book and also conveys to us its deeper meaning. Peacock engages our interest in this aspect of his novel by presenting the four young lovers as very distinct but always appealing personalities. We have on the one hand the gaiety and wit of the poetry-loving Morgana Gryll, and on the other the fascinating reserve combined with glowing physical vitality and frankness of manner of Alice Niphet; on the one hand the brooding sensitivity and aestheticism of the withdrawn Mr Falconer, and on the other the social and physical vigour and the intellectual curiosity of the outgoing Lord Curryfin. The interplay between these four involves Peacock in the kind of psychological and emotional analysis to be found in many Victorian novels (Trollope's presentation of the dilemmas of young lovers offers the closest parallel perhaps). Here, for example, is Curryfin beginning to meditate on his 'perplexities':

He asked himself how it could be, that having begun by making love to Miss Gryll ... he was now evidently in a transition state towards a more absorbing and violent passion for a person who, with all her frankness, was incomprehensible ... He was in a dilemma between Morgana and Melpomene. It had not entered his thoughts that Morgana was in love with him; but he thought it nevertheless very probable that she was in a fair way to become so ... On the other hand he could not divest himself of the idea that Melpomene was in love with him ... (ch. 17).

His relationship with 'Melpomene' (Alice Niphet) develops through shared physical pleasures—playing battledore and shuttlecock or skating or dancing the polka; Algernon Falconer's with Morgana develops through their shared literary enthusiasms—indeed, Boiardo's *Orlando Innamorato* proves to be a vital means of communication in bringing matters to a head between them. Both Morgana and Alice are refreshingly free from the Victorian-heroine stereotype. The most violent action in the book—the burning of Lord Curryfin's 'infallibly safe' sail—is performed by Alice, and Morgana's 'sportiveness' is a constant source of delight as, for example, when she is releasing Curryfin from his commitment towards her:

You offered yourself to me, to have and to hold, for every and aye. Suppose I claim you. Do not look so frightened. You deserve some punishment but that would be too severe. But, to a certain extent, you belong to me, and I claim the right to transfer you. I shall make a present of you to Miss Niphet (ch. 30).

Given such characters as *Crotchet Castle*'s Lady Clarinda and Morgana Gryll here, it is not surprising that critics have tended to see Peacock's influence in the novels of his son-in-law, George Meredith, with their series of high-spirited and unconventional heroines.

It is not only the young lovers, however, that Peacock succeeds in individualizing so well. Most of the older generation, the young's benign 'guardians', are also vividly realized. Mr Gryll may be a pale apology for Squire Headlong or Mr Glowry of Nightmare Abbey (see note to p. 104, below) but Opimian is as much an advance in depth of characterization on Folliott in *Crotchet Castle* as Folliott is on *Headlong Hall*'s Gaster, and his one scene with his splendid wife (Mrs Folliott is a mere off-stage joke in the earlier novel) is so enjoyable that it makes us wish that we saw more of her. The elderly spinster, Miss Ilex, who shows a strong intelligence, cultured tastes that are genuinely hers and not merely fashionable, and a generous sympathy towards the young, notably in her wise counselling of Morgana in chapter 26, is one of the triumphs of the book—again, a world away from the contemporary fictional stereotyping of 'old maids'.

Generally, Peacock would seem to have moved in *Gryll Grange* from one class of comic fiction to another as such classes were defined by himself in an essay on 'French Comic Romances' written a quarter of a century earlier:

> In respect of presenting or embodying opinion, there are two very distinct classes of comic fictions: one in which the characters are abstractions or embodied classifications, and the implied or embodied opinions the main matter of the work; another, in which the characters are individuals and the events and the action those of actual life—the opinions however prominent they may be made, being merely incidental. To the first of these classes belong the fictions of Aristophanes, Petronius Arbiter, Rabelais, Swift, and Voltaire . . .

Peacock's example of the second class is the work of a now forgotten novelist called Pigault le Brun whose 'heroes and heroines are all genuine flesh and blood and invest themselves with the opinions of the time as ordinary mortals do, carrying on the while the realities of every-day life'. Such phrases as this last, and the earlier one about the 'events and the action' being those of 'actual life', must give us pause, however, when seeking to argue that *Gryll Grange* belongs wholly to this second category. The story we find in this novel seems to be more in the mode of Shakespearian Romantic Comedy than in that of realism. Just as friendly discussion in this novel replaces the passionate debate of *Headlong Hall*, so does Romantic Comedy replace farce as the mode of the plot. The dream-like existence of Mr Falconer in his elegantly appointed Tower, attended by his seven 'Vestals' and contemplating his images of an idealized young virgin, seems comparable to that of Orsino at the beginning of *Twelfth Night* rather than to 'the realities of every-day life' in mid-Victorian England, even for a wealthy and reclusive young aesthete; whilst Morgana, sitting at the receipt of suitors in her uncle's hospitable mansion, seems more in the situation of a Portia trusted to make the right choice for herself than of a Hampshire squire's marriageable niece of the mid-century. (The echo of Portia becomes unmistakable in the last chapter when Morgana wittily deplores her unsuccessful suitors as her uncle names them.) And by the time seven bridegrooms have been produced for Falconer's seven handmaidens, the number of marriages giving the book its festive ending (orchestrated by Opimian with 'a peal of Bacchic ordnance' as all the champagne corks are released simultaneously) exceeds even those presided over by Hymen himself at the end of *As You Like It*.

As in Shakespeare's Romantic Comedies we have in *Gryll Grange* a form of literature that combines the presentation and exploration of the emotional life, especially of young love, with romantic and idyllic settings and such fantastic turns of events as Harry Hedgerow's production of six more swains like himself to court Dorothy's sisters at the Tower. Peacock also seems to be doing what Shakespeare often does in these plays—balancing one way of life, or set of values, against another before uniting in marriage his heroes and heroines, together with many of their followers, to form a true 'congenial society'. We see Falconer's private world of Epicurean pleasure (a sort of Tennysonian Palace of Art in which, however, he cannot escape the 'intense pain' of consciousness of the 'mass of poverty and crime' outside his gates); and we hear much about the public world of politics and 'Pantopragmatics' in which Curryfin cuts such a figure, a world in which 'schemes for breeding pestilence' are called 'sanitary improvements', where 'the art of teaching everything, except what will be of use to the recipient' is called 'national education', and so on. Gryll Grange with, within its walls, a circle of friends dining congenially together and sharing other such social pleasures, and, outside, a 'numerous light-rented and well-conditioned tenantry' fattening innumerable pigs serves as Peacock's microcosmic image of the golden world. Like Illyria or Belmont or the Forest of Arden, it is enchanted ground as its châtelaine's name suggests ('... to please [Mr Gryll], she had been called Morgana. He had had some thoughts of calling her Circe, but acquiesced in the name of a sister enchantress ...') and it reclaims Falconer from his self-indulgent 'hermitage' and Curryfin from his brilliant but empty life of lecturing fishermen on fish and other such Pantopragmatical activities. Peacock's meaning is plain: neither in romantic retreat, however aesthetic, nor in the bustle of public life, however well intentioned, does true happiness lie but in family life and private friendships, in the communal life that grows out of these friendships and in the discriminating enjoyment of all the other good things of life—healthful exercise, good food and wine, the beauties of nature, and the pleasures of music, literature, and the other arts.

Peacock had in his time had experience both of romantic retreat, composing his poetry in the wilds of Merionethshire, and, more extensively, of the bustle of public life during nearly forty years' service at India House. The last stage of his existence was that of his retirement at Lower Halliford when, according to his granddaughter Edith Nicolls, 'his life was spent among his books, and in the garden, in which he took great pleasure, and on the river'.[1] *Gryll Grange*, written during this period, may be seen as the septuagenarian author's literary summing-up, his final testament, as it were (further literary work was planned, it seems, but little actually got written before his death). It is undoubtedly the most personal of all his novels. We learn from his cousin, Harriet Love, that the description of Miss Ilex's lover (see note to p. 260, below) is certainly a portrait of the artist as a young man, and Opimian with his comfortable, well-regulated household, his gastronomic expertise, his passion for the Greeks, and his scholarly pursuits clearly reflects the

[1] See 'Biographical Introduction' to vol. 1 of Halliford edition, p. ccvi.

Peacock of later years. All the great writers that, over the years, had done most to (in Arnoldian phrase) prop Peacock's mind—Homer, Aristophanes, the Greek lyricists, Petronius, the egregious Nonnus, Rabelais, Samuel Butler, Voltaire—are distilled into the festive epigraphs and the quotations and allusions that are so marked a feature of the book. But the mellowness and Epicurean poise of *Gryll Grange* must not be seen as the easy result of a long life of unclouded happiness. Peacock's early years seem to have contained more than the average share of emotional upheaval and financial insecurity for a middle-class youth of his time, and later he was, as J. B. Priestley comments (*Thomas Love Peacock* (1966 edn.), pp. 103–4), 'sorely tried by fortune':

> His favourite child died when she was three; his wife became an invalid; his eldest daughter was quickly widowed and then made her tragic marriage with Meredith; his son was unstable and a constant source of anxiety; his youngest daughter lost her two children and died herself not long afterwards; few men have known more misfortunes in their domestic life. The man who wrote *Gryll Grange* . . . must have had a brave heart or a very unfeeling one, and we have ample testimony that Peacock was anything but insensitive.

Knowledge of this poignant biographical background enriches our reading of the novel for it helps us to see it for what it essentially is, a celebration and a life-affirming gesture by an old man who has experienced very different facets of life, enjoyed much, and also suffered much, but who can truly say with his beloved Petronius:

> Always and everywhere I have so lived, that I might consume the passing light, as if it were not to return.[2]

M.S.

[2] See epigraph to ch. 1 of *Gryll Grange*.

NOTE ON THE TEXTS

Headlong Hall was first published in December 1815, although the title-page bears the date 1816. A second edition, revised, appeared in 1816, and a third in 1822 with further revisions, notably the introduction of chapter headings and a table of contents and a thorough revision in the spelling of Mac Laurel's speeches. In 1837 it was published, along with *Nightmare Abbey*, *Maid Marian*, and *Crotchet Castle*, as the 57th volume of Bentley's Standard Novels. The present text, like that of the Halliford edition, is based on the 1837 volume, which, according to its title-page, was corrected by the author. Printing errors have been silently corrected on the authority of the earlier editions, and inconsistencies in the capitalization of titles (e.g. 'the Reverend Doctor') have been removed. In preparing this text the editors have benefited from consulting Nicholas A. Joukovsky's 'A Critical Edition of *Headlong Hall* and *Nightmare Abbey*' (unpublished D.Phil. thesis, Oxford, 1970), the text of which is based on the first edition but incorporates most of the subsequent revisions.

This edition of *Gryll Grange* follows the text of the first volume edition, 1861, for which P. lightly revised the text serialized in *Fraser's Magazine* (vols. 61–2, April–December 1860). Printer's errors, either those left uncorrected from *F.* or those freshly introduced in *1861*, have been silently emended. Some of the running headings in *1861* do not appear in this edition owing to differences of pagination, but all are collected by chapters in the contents pages, as in *1861*.

Substantive variants for both novels are listed in Appendices B and C below. Greek accents and breathing-marks in *Gryll Grange* have been corrected and made consistent; they have not been supplied in *Headlong Hall* since P. did not use them at all in this novel.

The footnotes are P.'s. Editorial explanatory notes, which appear on pp. 337–426, are keyed to the text by catch-phrases. We have decided not to clutter P.'s text with superscript numbers, but it is hoped that the running headings to the explanatory notes will aid reference from text to notes and vice versa.

SELECT BIBLIOGRAPHY

THE standard edition of Peacock's works is the *Halliford Edition of the Works of Thomas Love Peacock*, ed. H. F. B. Brett-Smith and C. E. Jones, 10 vols. (London, 1924–34). David Garnett's *The Novels of Thomas Love Peacock* (London, 1948; rev. edn. 1953) is usefully annotated. *Headlong Hall* and *Nightmare Abbey* have been re-edited by Nicholas A. Joukovsky in 'A Critical Edition of Peacock's *Headlong Hall* and *Nightmare Abbey*' (unpublished D. Phil. thesis, Oxford, 1970). Professor Joukovsky is preparing an edition of Peacock's letters for Oxford University Press. A bibliography of works by or about Peacock is provided by Bill Read, 'Thomas Love Peacock: an Enumerative Bibliography', *Bulletin of Bibliography*, 24 (1963–4).

In the absence of a definitive modern biography the reader is referred to Carl Van Doren's *The Life of Thomas Love Peacock* (London and New York, 1911); to J.-J. Mayoux's *Un Epicurien anglais: Thomas Love Peacock* (Paris, 1933); or to Carl Dawson's critical biography *His Fine Wit* (London, 1970). The best account of Peacock's early life is Nicholas Joukovsky's 'Peacock before *Headlong Hall*', *Keats–Shelley Memorial Bulletin*, 36 (1985). Sir Henry Cole's privately printed *Thomas Love Peacock: Biographical Notes* (BL Cat. no. C.121.b.20) is a useful source of personal reminiscence.

J. L. Madden's *Thomas Love Peacock* (London, 1967) is a short introductory study. Stimulating full-length critical discussions of the novels are J. B. Priestley's *Thomas Love Peacock* (London, 1927); Howard Mills's *Peacock, his Circle and his Age* (Cambridge, 1968); and Marilyn Butler's *Peacock Displayed: A Satirist in his Context* (London, 1979). Bryan Burns's recent study *The Novels of Thomas Love Peacock* (London, 1985) is also useful.

Among shorter essays the following are notable: A. E. Dyson, 'The Wand of Enchantment', in his *The Crazy Fabric* (London, 1966); V. S. Pritchett's 'The Proximity of Wine' (1944), reprinted in his *A Man of Letters* (1985); and Edmund Wilson's 'The Musical Glasses of Thomas Love Peacock', in his *Classics and Commercials* (New York, 1950). There is a selection of critical essays in Lorna Sage (ed.), *Peacock: The Satirical Novels* (Macmillan's Casebook Series, London, 1976).

A CHRONOLOGY OF THOMAS LOVE PEACOCK

Note. This chronology is based on the 'Biographical Introduction' to vol. 1 of the Halliford edition of Peacock's works, ed. H. F. B. Brett-Smith and C. E Jones (1934), and N. A. Joukovsky's 'Peacock before *Headlong Hall*: a New Look at his Early Years' in *The Keats–Shelley Memorial Bulletin*, no. 36 (1985).

1785 Born 18 October at Weymouth, Dorset, only child of Samuel Peacock, a London glass merchant, and Sarah, daughter of Thomas Love, a Master in the Royal Navy, who had retired on a pension to the village of Chertsey, Middlesex, after losing a leg in battle against the French.

1792 Sent to school at Englefield Green near Windsor, where he stayed about six years. P. wrote later (Halliford, viii. 259), 'The master [John Harris Wicks] was not much of a scholar; but he had the art of inspiring his pupils with a love of learning, and he had excellent classical and French assistants.'

1793(?) Death of Samuel Peacock 'in poor circumstances' (Joukovsky, p. 5). Sarah Peacock joins her parents in Chertsey.

1800 Wins a prize from a new magazine, *The Monthly Preceptor, or, Juvenile Library*, for an answer, in heroic couplets, to the question, 'Is History or Biography the More Improving Study?' By this time working as a clerk for the London firm of Ludlow, Fraser & Co., Throgmorton Street, living, with his mother, on the firm's premises. Later P. wrote, 'I passed many of my best years with my mother, taking more pleasure in reading than in society' (Halliford, viii. 259).

1801–4 Writing much occasional verse, the bulk of which he eventually collects together in a manuscript volume (now in the Berg Collection, New York Public Library) including a farcical narrative, 'The Monks of St Mark', subsequently printed as a separate pamphlet.

1805 (Dec.) Publication of *Palmyra, and Other Poems* (vol. dated 1806). Thomas Love dies. P. and his mother return to Chertsey (?)

1807 Becomes friendly with Thomas and Edward Hookham,

sons of Thomas Hookham, a well-known bookseller and publisher 'whose circulating library and reading rooms at 15 Old Bond Street were recognized as "the habitual resort of the *littérateurs* of the day"' (Joukovsky, pp. 17–18); 'deeply in love' with, and engaged to, a young neighbour in Chertsey, Fanny Falkner, but the engagement was broken off 'in an unjustifiable manner' by one of her relatives (Halliford, vol. i, xxxvi–xxxvii).

1808 (May) Appointed secretary to Home Popham, Captain of HMS *Venerable*, anchored off the Downs. The Hookhams send him books and he continues to write poetry as well as prologues for dramatic performances on board.

1809 (2 Apr.) Relinquishes his post aboard the *Venerable* and concentrates on working on a long poem about the Thames; (late May/early June) in order to gather material he traces the course of the river on foot from its source to London.

1810–11 Living a retired life in Wales at Maentwrog (Gwynedd) where he is strongly attracted at different times by two young women, one of them the parson's younger daughter, Jane Gryffydh, aged 20. Parson Gryffydh himself P. describes as 'a little dumpy, drunken, mountain-goat' (Halliford, vol. i, xliii). *The Genius of the Thames* published by the Hookhams during the first half of 1810, to mainly favourable reviews.

1811 (Dec.) P. receives a grant of £21 from the Royal Literary Fund. His grandmother's death (Dec. 1810) had meant the end of her naval widow's pension and the last of the small annuities left to P.'s mother by his father expired at Michaelmas 1811, so that the family had no income.

1812 *The Philosophy of Melancholy: A Poem in Four Parts, with a Mythological Ode* published (Feb.), followed by new edn. of *The Genius of the Thames, Palmyra and Other Poems* (Apr.). P. receives a further grant of £30 from the Literary Fund (May). Introduced to Shelley (who had expressed admiration of his poetry) by Thomas Hookham. Walking tour with a friend, Joseph Gulston, of the Isle of Wight (Sept.). Falls passionately in love with Gulston's 15-year-old cousin, Clarinda Knowles.

1812–13 Working on farces (*Mirth in the Mountains, The Dilettanti,*

The Three Doctors) for Drury Lane Theatre, none of which get produced.

1813 Receives a further grant of £10 from the Literary Fund after Thomas Hookham's report that 'he is reduced to a state of utter distress' (Joukovsky, p. 30). Walking tour in North Wales (summer). Visits Shelley at Bracknell and accompanies him and Harriet on a journey to the Lakes and Edinburgh. *Sir Hornbook*, a 'grammatico-allegorical ballad', for 'The Juvenile Library' published late in the year (vol. dated 1814).

1814 *Sir Proteus: A Satirical Ballad*, attacking Southey, published. Helps Shelley in dealing with lawyers and money-lenders after Shelley's elopement with Mary Godwin and return to England.

1815 Presumed engaged to Marianne de St Croix whose family he had known since his earliest days in London (Joukovsky, pp. 13–14), but becomes involved in an escapade with a mysterious supposed heiress, 'Charlotte', who turns out to have nothing; arrested for debt in Liverpool and bailed out by either Shelley or the Hookhams. Shelley makes him a regular allowance of £120 a year and he settles with his mother near Marlow, Buckinghamshire; accompanies Shelley and Mary on a boating expedition up the Thames. *Headlong Hall* published late in the year (dated 1816).

1816 *The Round Table; or, King Arthur's Feast*, a versified account for children of the Sovereigns of England, published anonymously. Begins work on a new novel, 'Sir Calidor', but never completes it.

1817 *Melincourt* published.

1818 *Rhododaphne: or, The Thessalian Spell. A Poem* published. Shelley and Mary leave for Italy (March). *Nightmare Abbey* published in the autumn. Begins essay 'On Fashionable Literature', eventually abandoned. Begins work on *Maid Marian*, 'a comic Romance of the Twelfth Century'.

1819 (May) Appointed, with James Mill and Edward Strachey, as Assistants to the Examiner of Indian Correspondence at the East India Company's offices in Leadenhall Street. In close touch by mail with Shelley, corrects

	proofs of *Prometheus Unbound* and other work for him. Writes proposing marriage to Jane Gryffydh, unseen and uncommunicated with for eight years (20 Nov.).
1820	(Jan.) 'The Four Ages of Poetry' published in *Ollier's Literary Miscellany*; it provokes Shelley into writing his 'Defence of Poetry' (not published until 1840). Marries Jane Gryffydh in Wales (22 Mar.).
1821	First child, Mary Ellen, born. Holidays in North Wales (Sept.)
1822	*Maid Marian* published and sells better than any of his previous works. Produced as an opera (adapted by J. R. Planché, music by Henry Bishop) at Covent Garden in Dec. Death of Shelley (8 July): P. and Byron are his executors but almost the whole business of settling his estate and protecting the interests of Mary Shelley and her son devolves on P., Byron having refused to act.
1823	Buys two cottages at Lower Halliford in the Thames Valley, converts them into a single house and settles there with his family, including his mother. Second daughter, Margaret, born.
1824(?5)	Son Edward born.
1825–6	Writing 'Paper Money Lyrics' (not published until 1837).
1826	Death of Margaret, which severely affects Jane Peacock: she remains an invalid from this time on. Shortly afterwards a little girl resembling Margaret, Mary Rosewell, daughter of a local villager, is adopted by the Peacocks.
1827	Scathingly reviews Moore's novel *The Epicurean* in *Westminster Review* (Oct.).
1828	Birth of daughter Rosa Jane.
1829	*The Misfortunes of Elphin* published.
1829–40 (approx)	Very active in promoting steam communication with India; gives evidence to Parliamentary Select Committees (1832, 1834, 1837), writes on the subject in *Edinburgh Review* (Jan. 1835), oversees the building of the Company's first steamships, which he calls his 'iron chickens'.
1830	Further contributions to *Westminster*, including attack on Moore's *Byron*.

1830–6 Writing much on opera: over 100 reviews in the *Globe* and *Examiner*, also essays for the *London Review*.

1831 *Crotchet Castle* published. Hailed as 'the wittiest writer in England' by the *Literary Gazette* but viciously attacked in *Fraser's* (Aug.).

1833 Death of his mother ('some time after her death, he remarked to a friend that he had never written with any zeal since', Halliford, vol. i, xix).

1835 'French Comic Romances' contributed to *London Review* (Oct.), followed by 'The Epicier' (Jan., 1836).

1836 Succeeds James Mill as Chief Examiner of Correspondence at India House at a salary of £2,000 p.a.

1837–8 Contributing verse and prose to *Bentley's Miscellany*.

1837 *Headlong Hall*, *Nightmare Abbey*, *Maid Marian*, and *Crotchet Castle* published together as vol. 57 of Bentley's 'Standard Novels' series (substantial, appreciative review by James Spedding in *Edinburgh*, Jan. 1839).

1839 Beginning of warm, lifelong friendship with Byron's friend Sir John Cam Hobhouse, later Lord Broughton; later meets Disraeli and Thackeray on visits to Hobhouse's country house.

1844 (Jan.) Mary Ellen marries Lieut. Edward Nicolls, who is drowned later that year; their daughter Edith born in the autumn.

1849 Mary Ellen marries George Meredith (seven years her junior).

1851 Collaborates with Mary Ellen on an article for *Fraser's*, 'Gastronomy and Civilisation' (published over her initials in Dec.) Death of Jane Peacock. Meredith publishes *Poems* dedicated to P.

1856 Retires from India House. *Melincourt* republished with new Preface.

1857 Death of Rosa Jane (who had married against P.'s wishes).

1858 Mary Ellen elopes with Henry Wallis (returning alone to England in 1859). 'Memoirs of Percy Bysshe Shelley' in *Fraser's* (June). ('Memoirs' Part II followed in *Fraser's* Jan., 1860, 'Unpublished Letters of Shelley' in March

1860, and 'Shelley. Supplementary Notice' in March 1862.)

1859 Review of Müller and Donaldson's *History of Greek Literature* in *Fraser's* (Mar.).

1860 *Gryll Grange* serialized in *Fraser's* (Apr.–Dec.).

1861 Volume publication of *Gryll Grange*, which is appreciatively reviewed by the *Spectator* and the *Saturday Review*. Death of Mary Ellen Meredith.

1862 *Gli'Ingannati, or the Deceived* (prose translation of Italian play) published. P. projects a 'Collection of Miscellanies', writes an autobiographical essay, 'The Last Day of Windsor Forest'.

1866 (23 Jan.) Dies at Lower Halliford, leaving his whole estate to Mary Rosewell, named as sole executrix. His library, auctioned at Sotheby's in June, realizes £503.7*s*.6*d*. (catalogue reproduced in vol. 1 of *Sale Catalogues of Libraries of Eminent Persons*, ed. A. N. L. Munby, 1971).

HEADLONG HALL.

All philosophers, who find
Some favourite system to their mind,
In every point to make it fit,
Will force all nature to submit.

LONDON:
PRINTED FOR T. HOOKHAM, JUN. AND CO.
OLD BOND STREET.

1816.

CONTENTS

HEADLONG HALL

CHAPTER I

THE MAIL

THE ambiguous light of a December morning, peeping through the windows of the Holyhead mail, dispelled the soft visions of the four insides, who had slept, or seemed to sleep, through the first seventy miles of the road, with as much comfort as may be supposed consistent with the jolting of the vehicle, and an occasional admonition to *remember the coachman*, thundered through the open door, accompanied by the gentle breath of Boreas, into the ears of the drowsy traveller.

A lively remark, that *the day was none of the finest*, having elicited a repartee of *quite the contrary*, the various knotty points of meteorology, which usually form the exordium of an English conversation, were successively discussed and exhausted; and, the ice being thus broken, the colloquy rambled to other topics, in the course of which it appeared, to the surprise of every one, that all four, though perfect strangers to each other, were actually bound to the same point, namely, Headlong Hall, the seat of the ancient and honourable family of the Headlongs, of the Vale of Llanberris, in Caernarvonshire. This name may appear at first sight not to be truly Cambrian, like those of the Rices, and Prices, and Morgans, and Owens, and Williamses, and Evanses, and Parrys, and Joneses; but, nevertheless, the Headlongs claim to be not less genuine derivatives from the antique branch of Cadwallader than any of the last named multi-ramified families. They claim, indeed, by one account, superior antiquity to all of them, and even to Cadwallader himself; a tradition having been handed down in Headlong Hall for some few thousand years, that the founder of the family was preserved in the deluge on the summit of Snowdon, and took the name of Rhaiader, which signifies a *waterfall*, in consequence of his having accompanied the water in its descent or diminution, till he found himself comfortably seated on the rocks of Llanberris. But, in later days, when commercial bagsmen began to scour the country, the ambiguity of the sound induced his descendants to drop the suspicious denomination of *Riders*, and translate the word into English; when, not being well pleased with the sound of the *thing*, they substituted that of the *quality*, and accordingly adopted the name *Headlong*, the appropriate epithet of waterfall.

> I cannot tell how the truth may be:
> I say the tale as 't was said to me.

The present representative of this ancient and dignified house, Harry Headlong, Esquire, was, like all other Welsh squires, fond of shooting, hunting, racing, drinking, and other such innocent amusements, *μειζονος δ' αλλου τινος*, as Menander expresses it. But, unlike other Welsh squires, he had actually suffered certain phenomena, called books, to find their way into his house; and, by dint of lounging over them after dinner, on those occasions when he was compelled to take his bottle alone, he became seized with a violent passion to be thought a philosopher and a man of taste; and accordingly set off on an expedition to Oxford, to inquire for other varieties of the same genera, namely, men of taste and philosophers; but, being assured by a learned professor that there were no such things in the University, he proceeded to London, where, after beating up in several booksellers' shops, theatres, exhibition-rooms, and other resorts of literature and taste, he formed as extensive an acquaintance with philosophers and dilettanti as his utmost ambition could desire; and it now became his chief wish to have them all together in Headlong Hall, arguing, over his old Port and Burgundy, the various knotty points which had puzzled his pericranium. He had, therefore, sent them invitations in due form to pass their Christmas at Headlong Hall; which invitations the extensive fame of his kitchen fire had induced the greater part of them to accept; and four of the chosen guests had, from different parts of the metropolis, ensconced themselves in the four corners of the Holyhead mail.

These four persons were, Mr. Foster,* the perfectibilian; Mr. Escot,† the deteriorationist; Mr. Jenkison,‡ the statu-quo-ite; and the Reverend Doctor Gaster§, who, though of course neither a philosopher nor a man of taste, had so won on the Squire's fancy, by a learned dissertation on the art of stuffing a turkey, that he concluded no Christmas party would be complete without him.

The conversation among these illuminati soon became animated; and Mr. Foster, who, we must observe, was a thin gentleman, about thirty years of age, with an aquiline nose, black eyes, white teeth, and black hair—took occasion to panegyrize the vehicle in which they were then travelling, and observed what remarkable improvements had been made in the means of facilitating intercourse between distant parts of the kingdom: he held forth with great energy on the subject of roads and railways, canals and tunnels, manufactures and machinery: 'In short,' said he, 'everything we look on attests the progress of mankind in all the arts of life, and demonstrates their gradual advancement towards a state of unlimited perfection'

Mr. Escot, who was somewhat younger than Mr. Foster, but rather more pale and saturnine in his aspect, here took

* Foster, quasi Φωστηρ,—from φαος and τηρεω, lucem servo, conservo, observo, custodio,—one who watches over and guards the light; a sense in which the word is often used amongst us, when we speak of *fostering* a flame.

† Escot, quasi ες σκοτον, *in tenebras*, scilicet, intuens; one who is always looking into the dark side of the question.

‡ Jenkison: This name may be derived from αιεν εξ ισων, *semper ex æqualibus*—scilicet, mensuris, omnia metiens: one who from equal measures divides and distributes all things; one who from equal measures can always produce arguments on both sides of a question, with so much nicety and exactness, as to keep the said question eternally pending, and the balance of the controversy perpetually in statu quo. By an aphæresis of the α, an elision of the second ε, and an easy and natural mutation of ξ into κ, the derivation of this name proceeds according to the strictest principles of etymology: αιεν εξ ισων—Ιεν εξ ισων—Ιεν εκ ισων—Ιεν 'κ ισων—Ιενκισων—Ienkison—Jenkison.

§ Gaster: scilicet Γαστηρ—Venter,—et præterea nihil.

up the thread of the discourse, observing, that the proposition just advanced seemed to him perfectly contrary to the true state of the case: 'for,' said he, 'these improvements, as you call them, appear to me only so many links in the great chain of corruption, which will soon fetter the whole human race in irreparable slavery and incurable wretchedness: your improvements proceed in a simple ratio, while the factitious wants and unnatural appetites they engender proceed in a compound one; and thus one generation acquires fifty wants, and fifty means of supplying them are invented, which each in its turn engenders two new ones; so that the next generation has a hundred, the next two hundred, the next four hundred, till every human being becomes such a helpless compound of perverted inclinations, that he is altogether at the mercy of external circumstances, loses all independence and singleness of character, and degenerates so rapidly from the primitive dignity of his sylvan origin, that it is scarcely possible to indulge in any other expectation, than that the whole species must at length be exterminated by its own infinite imbecility and vileness.'

'Your opinions,' said Mr. Jenkison, a round-faced little gentleman of about forty-five, 'seem to differ *toto cœlo*. I have often debated the matter in my own mind, *pro* and *con*, and have at length arrived at this conclusion,—that there is not in the human race a tendency either to moral perfectibility or deterioration; but that the quantities of each are so exactly balanced by their reciprocal results, that the species, with respect to the sum of good and evil, knowledge and ignorance, happiness and misery, remains exactly and perpetually *in statu quo*.'

'Surely,' said Mr. Foster, 'you cannot maintain such a proposition in the face of evidence so luminous. Look at the progress of all the arts and sciences,—see chemistry, botany, astronomy——.'

'Surely,' said Mr. Escot, 'experience deposes against you. Look at the rapid growth of corruption, luxury, selfishness——.'

'Really, gentlemen,' said the Reverend Doctor Gaster, after clearing the husk in his throat with two or three hems, 'this is a very sceptical, and, I must say, atheistical conversation, and I should have thought, out of respect to my cloth——.'

Here the coach stopped, and the coachman, opening the door, vociferated—'Breakfast, gentlemen;' a sound which so gladdened the ears of the divine, that the alacrity with which he sprang from the vehicle superinduced a distortion of his ankle, and he was obliged to limp into the inn between Mr. Escot and Mr. Jenkison; the former observing, that he ought to look for nothing but evil, and, therefore, should not be surprised at this little accident; the latter remarking, that the comfort of a good breakfast, and the pain of a sprained ankle, pretty exactly balanced each other.

CHAPTER II

THE SQUIRE. THE BREAKFAST

SQUIRE Headlong, in the mean while, was quadripartite in his locality; that is to say, he was superintending the operations in four scenes of action—namely, the cellar, the library, the picture-gallery, and the dining-room,—preparing for the reception of his philosophical and dilettanti visitors. His myrmidon on this occasion was a little red nosed butler, whom nature seemed to have cast in the genuine mould of an antique Silenus, and who waddled about the house after his master, wiping his forehead and panting for breath, while the latter bounced from room to room like a cracker, and was indefatigable in his requisitions for the proximity of his vinous Achates, whose advice and co-operation he deemed no less necessary in the library than in the cellar. Multitudes of packages had arrived, by land and water, from London, and Liverpool, and Chester, and Manchester, and Birmingham, and various parts of the mountains: books, wine, cheese, globes, mathematical instruments, turkeys, telescopes, hams, tongues, microscopes, quadrants, sextants, fiddles, flutes, tea, sugar, electrical machines, figs, spices, air-pumps, soda-water, chemical apparatus, eggs, French-horns, drawing books, palettes, oils, and colours, bottled ale and porter, scenery for a private theatre, pickles and fish-sauce, patent lamps and chandeliers, barrels of oysters, sofas, chairs, tables, carpets, beds, looking-glasses, pictures, fruits and confections, nuts, oranges, lemons, packages of salt salmon, and jars of Portugal grapes. These, arriving with infinite rapidity, and in inexhaustible succession, had been deposited at random, as the convenience of the moment dictated,—sofas in the cellar, chandeliers in the kitchen, hampers of ale in the drawing-room, and fiddles and fish-sauce in the library. The servants, unpacking all these in furious haste, and flying with them from place to place, according to the tumultuous directions of Squire Headlong and the little fat butler who fumed at his heels, chafed, and crossed, and clashed, and tumbled over one another up stairs and down. All was bustle, uproar, and confusion; yet nothing seemed to advance: while the rage and impetuosity of the Squire continued fermenting to the highest degree of exasperation, which he signified, from time to time, by converting some newly unpacked article, such as a book, a bottle, a ham, or a fiddle, into a missile against the head of some unfortunate servant who did not seem to move in a ratio of velocity corresponding to the intensity of his master's desires.

In this state of eager preparation we shall leave the happy inhabitants of Headlong Hall, and return to the three philosophers and the unfortunate divine, whom we left limping with a sprained ankle into the breakfast-room of the inn; where his two supporters deposited him safely in a large armchair, with his wounded leg comfortably stretched out on another. The morning being extremely cold, he contrived to be seated as near the fire as was consistent with his other object of having a perfect command of the table and its apparatus; which consisted not only of the ordinary comforts of tea and toast, but of a delicious supply of new-laid eggs, and a magnificent round of beef; against which Mr. Escot immediately pointed all the artillery of his eloquence, declaring the use of animal food, conjointly with that of fire, to be one of the principal causes of the present degeneracy of mankind. 'The natural and original man,' said he, 'lived in the woods: the roots and fruits of the earth supplied his simple nutriment: he had few desires, and no diseases. But, when he began to sacrifice victims on the altar of superstition, to pursue the goat and the deer, and, by the pernicious invention of fire, to pervert their flesh into food, luxury, disease, and premature death, were let loose upon the world. Such is clearly the correct interpretation of the fable of Prometheus, which is a symbolical portraiture of that disastrous epoch, when man first applied fire to culinary purposes, and thereby surrendered his liver to the vulture of disease. From that period the stature of mankind has been in a state of gradual diminution, and I have not the least doubt that it will continue to grow *small by degrees, and lamentably less*, till the whole race will vanish imperceptibly from the face of the earth.'

'I cannot agree,' said Mr. Foster, 'in the consequences being so very disastrous. I admit, that in some respects the use of animal food retards, though it cannot materially inhibit, the perfectibility of the species. But the use of fire was indispensably necessary, as Æschylus and Virgil expressly assert, to give being to the various arts of life, which, in their rapid and interminable progress, will finally conduct every individual of the race to the philosophic pinnacle of pure and perfect felicity.'

'In the controversy concerning animal and vegetable food,' said Mr. Jenkison, 'there is much to be said on both sides; and, the question being in equipoise, I content myself with a mixed diet, and make a point of eating whatever is placed before me, provided it be good in its kind.'

In this opinion his two brother philosophers practically coincided, though they both ran down the theory as highly detrimental to the best interests of man.

'I am really astonished,' said the Reverend Doctor Gaster, gracefully picking off the supernal fragments of an egg he had just cracked, and clearing away a space at the top for the reception of a small piece of butter—'I am really astonished, gentlemen, at the very heterodox opinions I have heard you deliver: since nothing can be more obvious than that all animals were created solely and exclusively for the use of man.'

'Even the tiger that devours him?' said Mr. Escot.

'Certainly,' said Doctor Gaster.

'How do you prove it?' said Mr. Escot.

'It requires no proof,' said Doctor Gaster: 'it is a point of doctrine. It is written, therefore it is so.'

'Nothing can be more logical,' said Mr. Jenkison. 'It has been said,' continued he, 'that the ox was expressly made to be eaten by man: it may be said, by a parity of reasoning, that man was expressly made to be eaten by the tiger: but as wild oxen exist where there are no men, and men where there are no tigers, it would seem that in these instances they do not properly answer the ends of their creation.'

'It is a mystery,' said Dr. Gaster.

'Not to launch into the question of final causes,' said Mr. Escot, helping himself at the same time to a slice of beef, 'concerning which I will candidly acknowledge I am as profoundly ignorant as the most dogmatical theologian possibly can be, I just wish to observe, that the pure and peaceful manners which Homer ascribes to the Lotophagi, and which at this day characterise many nations (the Hindoos, for example, who subsist exclusively on the fruits of the earth), depose very strongly in favour of a vegetable regimen.'

'It may be said, on the contrary,' said Mr. Foster, 'that animal food acts on the mind as manure does on flowers, forcing them into a degree of expansion they would not otherwise have attained. If we can imagine a philosophical auricula falling into a train of theoretical meditation on its original and natural nutriment, till it should work itself up into a profound abomination of bullock's blood, sugar-baker's scum, and other *unnatural* ingredients of that rich composition of soil which had brought it to perfection,* and insist on being planted in common earth, it would have all the advantage of natural theory on its side that the most strenuous advocate of the vegetable system could desire; but

* See Emmerton on the Auricula.

it would soon discover the practical error of its retrograde experiment by its lamentable inferiority in strength and beauty to all the auriculas around it. I am afraid, in some instances at least, this analogy holds true with respect to mind. No one will make a comparison, in point of mental power, between the Hindoos and the ancient Greeks.'

'The anatomy of the human stomach,' said Mr. Escot, 'and the formation of the teeth, clearly place man in the class of frugivorous animals.'

'Many anatomists,' said Mr. Foster, 'are of a different opinion, and agree in discerning the characteristics of the carnivorous classes.'

'I am no anatomist,' said Mr. Jenkison, 'and cannot decide where doctors disagree; in the mean time, I conclude that man is omnivorous, and on that conclusion I act.'

'Your conclusion is truly orthodox,' said the Reverend Doctor Gaster: 'indeed, the loaves and fishes are typical of a mixed diet; and the practice of the Church in all ages shows——'

'That it never loses sight of the loaves and fishes,' said Mr. Escot.

'It never loses sight of any point of sound doctrine,' said the Reverend Doctor.

The coachman now informed them their time was elapsed; nor could all the pathetic remonstrances of the reverend divine, who declared he had not half breakfasted, succeed in gaining one minute from the inexorable Jehu.

'You will allow,' said Mr. Foster, as soon as they were again in motion, 'that the wild man of the woods could not transport himself over two hundred miles of forest, with as much facility as one of these vehicles transports you and me through the heart of this cultivated country.'

'I am certain,' said Mr. Escot, 'that a wild man can travel an immense distance without fatigue; but what is the advantage of locomotion? The wild man is happy in one spot, and there he remains: the civilised man is wretched in every place he happens to be in, and then congratulates himself on being accommodated with a machine, that will whirl him to another, where he will be just as miserable as ever.'

We shall now leave the mail-coach to find its way to Capel Cerig, the nearest point of the Holyhead road to the dwelling of Squire Headlong.

CHAPTER III

THE ARRIVALS

In the midst of that scene of confusion thrice confounded, in which we left the inhabitants of Headlong Hall, arrived the lovely Caprioletta Headlong, the Squire's sister (whom he had sent for, from the residence of her maiden aunt at Caernarvon, to do the honours of his house), beaming like light on chaos, to arrange disorder and harmonise discord. The tempestuous spirit of her brother became instantaneously as smooth as the surface of the lake of Llanberris; and the little fat butler 'plessed Cot, and St. Tafit, and the peautiful tamsel,' for being permitted to move about the house in his natural pace. In less than twenty-four hours after her arrival, everything was disposed in its proper station, and the Squire began to be all impatience for the appearance of his promised guests.

The first visitor with whom he had the felicity of shaking hands was Marmaduke Milestone, Esquire, who arrived with a portfolio under his arm. Mr. Milestone* was a

* Mr. Knight, in a note to the Landscape, having taken the liberty of laughing at a notable device of a celebrated *improver*, for giving greatness of character to a place, and showing an undivided extent of property, by placing the family arms on the neighbouring *milestones*, the improver retorted on him with a charge of misquotation, misrepresentation, and malice prepense. Mr. Knight, in the preface to the second edition of his poem, quotes the improver's words:—'The market-house, or other public edifice, or even a *mere stone with distances*, may bear the arms of the family:' and adds:— 'By a *mere stone with distances*, the author of the Landscape certainly thought he meant a *milestone*: but, if he did not, any other interpretation which he may think more advantageous to himself shall readily be adopted, as it will equally answer the purpose of the quotation.' The improver, however, did not condescend to explain what he really meant by a *mere stone with distances*, though he strenuously maintained that he did *not* mean a *milestone*. His idea, therefore, stands on record, invested with all the sublimity that obscurity can confer.

picturesque landscape gardener of the first celebrity, who was not without hopes of persuading Squire Headlong to put his romantic pleasure-grounds under a process of improvement, promising himself a signal triumph for his incomparable art in the difficult and, therefore, glorious achievement of polishing and trimming the rocks of Llanberris.

Next arrived a post-chaise from the inn at Capel Cerig, containing the Reverend Doctor Gaster. It appeared, that, when the mail-coach deposited its valuable cargo, early on the second morning, at the inn at Capel Cerig, there was only one post-chaise to be had; it was therefore determined that the Reverend Doctor and the luggage should proceed in the chaise, and that the three philosophers should walk. When the reverend gentleman first seated himself in the chaise, the windows were down all round; but he allowed it to drive off under the idea that he could easily pull them up. This task, however, he had considerable difficulty in accomplishing, and when he had succeeded, it availed him little; for the frames and glasses had long since discontinued their ancient familiarity. He had, however, no alternative but to proceed, and to comfort himself, as he went, with some choice quotations from the book of Job. The road led along the edges of tremendous chasms, with torrents dashing in the bottom; so that, if his teeth had not chattered with cold, they would have done so with fear. The Squire shook him heartily by the hand, and congratulated him on his safe arrival at Headlong Hall. The Doctor returned the squeeze, and assured him that the congratulation was by no means misapplied.

Next came the three philosophers, highly delighted with their walk, and full of rapturous exclamations on the sublime beauties of the scenery.

The Doctor shrugged up his shoulders, and confessed he preferred the scenery of Putney and Kew, where a man could go comfortably to sleep in his chaise, without being in momentary terror of being hurled headlong down a precipice.

Mr. Milestone observed, that there were great capabilities in the scenery, but it wanted shaving and polishing. If he could but have it under his care for a single twelvemonth, he assured them no one would be able to know it again.

Mr. Jenkison thought the scenery was just what it ought to be, and required no alteration.

Mr. Foster thought it could be improved, but doubted if that effect would be produced by the system of Mr. Milestone.

Mr. Escot did not think that any human being could improve it, but had no doubt of its having changed very considerably for the worse, since the days when the now barren rocks were covered with the immense forest of Snowdon, which must have contained a very fine race of wild men, not less than ten feet high.

The next arrival was that of Mr. Cranium, and his lovely daughter Miss Cephalis Cranium, who flew to the arms of her dear friend Caprioletta, with all that warmth of friendship which young ladies usually assume towards each other in the presence of young gentlemen.*

Miss Cephalis blushed like a carnation at the sight of Mr. Escot, and Mr. Escot glowed like a corn-poppy at the sight of Miss Cephalis. It was at least obvious to all observers, that he could imagine the possibility of one change for the better, even in this terrestrial theatre of universal deterioration.

Mr. Cranium's eyes wandered from Mr. Escot to his daughter, and from his daughter to Mr. Escot; and his complexion, in the course of the scrutiny, underwent several variations, from the dark red of the piony to the deep blue of the convolvulus.

Mr. Escot had formerly been the received lover of Miss Cephalis, till he incurred the indignation of her father by laughing at a very profound craniological dissertation which the old gentleman delivered; nor had Mr. Escot yet discovered the means of mollifying his wrath.

Mr. Cranium carried in his own hands a bag, the contents of which were too precious to be intrusted to any one but himself; and earnestly entreated to be shown to the chamber appropriated for his reception, that he might deposit his

* 'Il est constant qu'elles se baisent de meilleur cœur, et se caressent avec plus de grace devant les hommes, fieres d'aiguiser impunément leur convoitise par l'image des faveurs qu'elles savent leur faire envier.'—ROUSSEAU, *Emile*, liv. 5.

treasure in safety. The little butler was accordingly summoned to conduct him to his *cubiculum*.

Next arrived a post-chaise, carrying four insides, whose extreme thinness enabled them to travel thus economically without experiencing the slightest inconvenience. These four personages were, two very profound critics, Mr. Gall and Mr. Treacle, who followed the trade of reviewers, but occasionally indulged themselves in the composition of bad poetry; and two very multitudinous versifiers, Mr. Nightshade and Mr. Mac Laurel, who followed the trade of poetry, but occasionally indulged themselves in the composition of bad criticism. Mr. Nightshade and Mr. Mac Laurel were the two senior lieutenants of a very formidable corps of critics, of whom Timothy Treacle, Esquire, was captain, and Geoffrey Gall, Esquire, generalissimo.

The last arrivals were Mr. Cornelius Chromatic, the most profound and scientific of all amateurs of the fiddle, with his two blooming daughters, Miss Tenorina and Miss Graziosa; Sir Patrick O'Prism, a dilettante painter of high renown, and his maiden aunt, Miss Philomela Poppyseed, an indefatigable compounder of novels, written for the express purpose of supporting every species of superstition and prejudice; and Mr. Panscope, the chemical, botanical, geological, astronomical, mathematical, metaphysical, meteorological, anatomical, physiological, galvanistical, musical, pictorial, bibliographical, critical philosopher, who had run through the whole circle of the sciences, and understood them all equally well.

Mr. Milestone was impatient to take a walk round the grounds, that he might examine how far the system of clumping and levelling could be carried advantageously into effect. The ladies retired to enjoy each other's society in the first happy moments of meeting: the Reverend Doctor Gaster sat by the library fire, in profound meditation over a volume of the '*Almanach des Gourmands*:' Mr. Panscope sat in the opposite corner with a volume of Rees's Cyclopædia: Mr. Cranium was busy up stairs: Mr. Chromatic retreated to the music-room, where he fiddled through a book of solos before the ringing of the first dinner-bell. The remainder of the party supported Mr. Milestone's proposition; and, accordingly, Squire Headlong and Mr. Milestone leading the van, they commenced their perambulation.

CHAPTER IV

THE GROUNDS

'I PERCEIVE,' said Mr. Milestone, after they had walked a few paces, 'these grounds have never been touched by the finger of taste.'

'The place is quite a wilderness,' said Squire Headlong: 'for, during the latter part of my father's life, while I was *finishing* my *education*, he troubled himself about nothing but the cellar, and suffered everything else to go to rack and ruin. A mere wilderness, as you see, even now in December; but in summer a complete nursery of briers, a forest of thistles, a plantation of nettles, without any live stock but goats, that have eaten up all the bark of the trees. Here you see is the pedestal of a statue, with only half a leg and four toes remaining: there were many here once. When I was a boy, I used to sit every day on the shoulders of Hercules: what became of *him* I have never been able to ascertain. Neptune has been lying these seven years in the dust-hole; Atlas had his head knocked off to fit him for propping a shed; and only the day before yesterday we fished Bacchus out of the horsepond.'

'My dear sir,' said Mr. Milestone, 'accord me your permission to wave the wand of enchantment over your grounds. The rocks shall be blown up, the trees shall be cut down, the wilderness and all its goats shall vanish like mist. Pagodas and Chinese bridges, gravel walks and shrubberies, bowling-greens, canals, and clumps of larch, shall rise upon its ruins. One age, sir, has brought to light the treasures of ancient learning; a second has penetrated into the depths of metaphysics; a third has brought to perfection the science of astronomy; but it was reserved for the exclusive genius of the present times, to invent the noble art of picturesque gardening, which has given, as it were, a new tint to the complexion of nature, and a new outline to the physiognomy of the universe!'

'Give me leave,' said Sir Patrick O'Prism, 'to take an exception to that same. Your system of levelling, and trimming, and clipping, and docking, and clumping, and polishing, and cropping, and shaving, destroys all the beautiful intricacies of natural luxuriance, and all the graduated harmonies of light and shade, melting into one another, as you see them on that rock over yonder. I never saw one of your improved places, as you call them, and which are nothing but big bowling-greens, like sheets of green paper, with a parcel of round clumps scattered over them, like so many spots of ink, flicked at random out a pen,* and a solitary animal here and there looking as if it were lost, that I did not think it was for all the world like Hounslow Heath, thinly sprinkled over with bushes and highwaymen.'

'Sir,' said Mr. Milestone, 'you will have the goodness to make a distinction between the picturesque and the beautiful.'

'Will I?' said Sir Patrick, 'och! but I won't. For what is beautiful? That which pleases the eye. And what pleases the eye? Tints variously broken and blended. Now, tints variously broken and blended constitute the picturesque.'

'Allow me,' said Mr. Gall. 'I distinguish the picturesque and the beautiful, and I add to them, in the laying out of grounds, a third and distinct character, which I call *unexpectedness*.'

'Pray, sir,' said Mr. Milestone, 'by what name do you distinguish this character, when a person walks round the grounds for the second time?'†

Mr. Gall bit his lips, and inwardly vowed to revenge himself on Milestone, by cutting up his next publication.

A long controversy now ensued concerning the picturesque and the beautiful, highly edifying to Squire Headlong.

The three philosophers stopped, as they wound round a

* See Price on the Picturesque.

† See Knight on Taste, and the Edinburgh Review, No. XIV.

projecting point of rock, to contemplate a little boat which was gliding over the tranquil surface of the lake below.

'The blessings of civilisation,' said Mr. Foster, 'extend themselves to the meanest individuals of the community. That boatman, singing as he sails along, is, I have no doubt, a very happy, and, comparatively to the men of his his class some centuries back, a very enlightened and intelligent man.'

'As a partisan of the system of the moral perfectibility of the human race,' said Mr. Escot,—who was always for considering things on a large scale, and whose thoughts immediately wandered from the lake to the ocean, from the little boat to a ship of the line,—'you will probably be able to point out to me the degree of improvement that you suppose to have taken place in the character of a sailor, from the days when Jason sailed through the Cyanean Symplegades, or Noah moored his ark on the summit of Ararat.

'If you talk to me,' said Mr. Foster, 'of mythological personages, of course I cannot meet you on fair grounds.'

'We will begin, if you please, then,' said Mr. Escot, 'no further back than the battle of Salamis; and I will ask you if you think the mariners of England are, in any one respect, morally or intellectually, superior to those who then preserved the liberties of Greece, under the direction of Themistocles?'

'I will venture to assert,' said Mr. Foster, 'that, considered merely as sailors, which is the only fair mode of judging them, they are as far superior to the Athenians, as the structure of our ships is superior to that of theirs. Would not one English seventy-four, think you, have been sufficient to have sunk, burned, and put to flight, all the Persian and Grecian vessels in that memorable bay? Contemplate the progress of naval architecture, and the slow, but immense, succession of concatenated intelligence, by which it has gradually attained its present stage of perfectibility. In this, as in all other branches of art and science, every generation possesses all the knowledge of the preceding, and adds to it its own discoveries in a progression to which there seems no limit. The skill requisite to direct these immense machines is proportionate to their magnitude and complicated mechanism; and, therefore, the English sailor, considered merely as a sailor, is vastly superior to the ancient Greek.'

'You make a distinction, of course,' said Mr. Escot, 'between scientific and moral perfectibility?'

'I conceive,' said Mr. Foster, 'that men are virtuous in proportion as they are enlightened; and that, as every generation increases in knowledge, it also increases in virtue.'

'I wish it were so,' said Mr. Escot; 'but to me the very reverse appears to be the fact. The progress of knowledge is not general: it is confined to a chosen few of every age. How far these are better than their neighbours, we may examine by and bye. The mass of mankind is composed of beasts of burden, mere clods, and tools of their superiors. By enlarging and complicating your machines, you degrade, not exalt, the human animals you employ to direct them. When the boatswain of a seventy-four pipes all hands to the main tack, and flourishes his rope's end over the shoulders of the poor fellows who are tugging at the ropes, do you perceive so dignified, so gratifying a picture, as Ulysses exhorting his dear friends, his *ΕΡΙΗΡΕΣ ΈΤΑΙΡΟΙ*, to ply their oars with energy? You will say, Ulysses was a fabulous character. But the economy of his vessel is drawn from nature. Every man on board has a character and a will of his own. He talks to them, argues with them, convinces them; and they obey him, because they love him, and know the reason of his orders. Now, as I have said before, all singleness of character is lost. We divide men into herds like cattle: an individual man, if you strip him of all that is extraneous to himself, is the most wretched and contemptible creature on the face of the earth. The sciences advance. True. A few years of study puts a modern mathematician in possession of more than Newton knew, and leaves him at leisure to add new discoveries of his own. Agreed. But does this make him a Newton? Does it put him in possession of that range of intellect, that grasp of mind, from which the discoveries of Newton sprang? It is mental power that I look for: if you can demonstrate the increase of that, I will give up the field. Energy—independence—individuality—disinterested virtue—

active benevolence—self-oblivion—universal philanthropy—these are the qualities I desire to find, and of which I contend that every succeeding age produces fewer examples. I repeat it; there is scarcely such a thing to be found as a single individual man: a few classes compose the whole frame of society, and when you know one of a class you know the whole of it. Give me the wild man of the woods; the original, unthinking, unscientific, unlogical savage: in him there is at least some good; but, in a civilised, sophisticated, cold-blooded, mechanical, calculating slave of Mammon and the world, there is none—absolutely none. Sir, if I fall into a river, an unsophisticated man will jump in and bring me out; but a philosopher will look on with the utmost calmness, and consider me in the light of a projectile, and, making a calculation of the degree of force with which I have impinged the surface, the resistance of the fluid, the velocity of the current, and the depth of the water in that particular place, he will ascertain with the greatest nicety in what part of the mud at the bottom I may probably be found, at any given distance of time from the moment of my first immersion.'

Mr Foster was preparing to reply, when the first dinner-bell rang, and he immediately commenced a precipitate return towards the house; followed by his two companions, who both admitted that he was now leading the way to at least a temporary period of physical amelioration: 'but, alas!' added Mr. Escot, after a moment's reflection, 'Epulæ NOCUERE repostæ!'*

CHAPTER V

THE DINNER

THE sun was now terminating his diurnal course, and the lights were glittering on the festal board. When the ladies had retired, and the Burgundy had taken two or three tours of the table, the following conversation took place:—

* Protracted banquets have been copious sources of evil.

SQUIRE HEADLONG

Push about the bottle. Mr. Escot, it stands with you. No heeltaps. As to skylight, liberty-hall.

MR. MAC LAUREL

Really, Squire Headlong, this is the vara nactar itsel. Ye hae saretainly descovered the tarrestrial paradise, but it flows wi' a better leecor than milk an' honey.

THE REVEREND DOCTOR GASTER

Hem! Mr. Mac Laurel! there is a degree of profaneness in that observation, which I should not have looked for in so staunch a supporter of church and state. Milk and honey was the pure food of the antediluvian patriarchs, who knew not the use of the grape, happily for them.—(*Tossing off a bumper of Burgundy.*)

MR. ESCOT

Happily, indeed! The first inhabitants of the world knew not the use either of wine or animal food; it is, therefore, by no means incredible that they lived to the age of several centuries, free from war, and commerce, and arbitrary government, and every other species of desolating wickedness. But man was then a very different animal to what he now is: he had not the faculty of speech; he was not encumbered with clothes; he lived in the open air; his first step out of which, as Hamlet truly observes, is *into his grave.** His first dwellings, of course, were the hollows of trees and rocks. In process of time he began to build: thence grew villages; thence grew cities. Luxury, oppression, poverty, misery, and disease kept pace with the progress of his pretended improvements, till, from a free, strong, healthy, peaceful animal, he has become a weak, distempered, cruel, carnivorous slave.

THE REVEREND DOCTOR GASTER

Your doctrine is orthodox, in so far as you assert that the original man was not encumbered with clothes, and that he

* See Lord Monboddo's Ancient Metaphysics.

lived in the open air; but, as to the faculty of speech, that, it is certain, he had, for the authority of Moses——

MR. ESCOT

Of course, sir, I do not presume to dissent from the very exalted authority of that most enlightened astronomer and profound cosmogonist, who had, moreover, the advantage of being inspired; but when I indulge myself with a ramble in the fields of speculation, and attempt to deduce what is probable and rational from the sources of analysis, experience, and comparison, I confess I am too often apt to lose sight of the doctrines of that great fountain of theological and geological philosophy.

SQUIRE HEADLONG

Push about the bottle.

MR. FOSTER

Do you suppose the mere animal life of a wild man, living on acorns, and sleeping on the ground, comparable in felicity to that of a Newton, ranging through unlimited space, and penetrating into the arcana of universal motion—to that of a Locke, unravelling the labyrinth of mind—to that of a Lavoisier, detecting the minutest combinations of matter, and reducing all nature to its elements—to that of a Shakspeare, piercing and developing the springs of passion—or of a Milton, identifying himself, as it were, with the beings of an invisible world?

MR. ESCOT

You suppose extreme cases: but, on the score of happiness, what comparison can you make between the tranquil being of the wild man of the woods and the wretched and turbulent existence of Milton, the victim of persecution, poverty, blindness, and neglect? The records of literature demonstrate that Happiness and Intelligence are seldom sisters. Even if it were otherwise, it would prove nothing. The many are always sacrificed to the few. Where one man advances, hundreds retrograde; and the balance is always in favour of universal deterioration.

MR. FOSTER

Virtue is independent of external circumstances. The exalted understanding looks into the truth of things, and, in its own peaceful contemplations, rises superior to the world. No philosopher would resign his mental acquisitions for the purchase of any terrestrial good.

MR. ESCOT

In other words, no man whatever would resign his identity, which is nothing more than the consciousness of his perceptions, as the price of any acquisition. But every man, without exception, would willingly effect a very material change in his relative situation to other individuals. Unluckily for the rest of your argument, the understanding of literary people is for the most part *exalted*, as you express it, not so much by the love of truth and virtue, as by arrogance and self-sufficiency; and there is, perhaps, less disinterestedness, less liberality, less general benevolence, and more envy, hatred, and uncharitableness among them, than among any other description of men.

(*The eye of Mr. Escot, as he pronounced these words, rested very innocently and unintentionally on Mr. Gall.*)

MR. GALL

You allude, sir, I presume, to my review.

MR. ESCOT

Pardon me, sir. You will be convinced it is impossible I can allude to your review, when I assure you that I have never read a single page of it.

MR. GALL, MR. TREACLE, MR. NIGHTSHADE, AND MR. MAC LAUREL

Never read our review!!!!

MR. ESCOT

Never. I look on periodical criticism in general to be a species of shop, where panegyric and defamation are sold, wholesale, retail, and for exportation. I am not inclined to be a purchaser of these commodities, or to encourage a trade which I consider pregnant with mischief.

MR. MAC LAUREL

I can readily conceive, sir, ye wou'd na wullinly encoorage ony dealer in panegeeric: but, frae the manner in which ye speak o' the first creetics an' scholars o' the age, I shou'd think ye wou'd hae a leetle mair predilaction for deefamation.

MR. ESCOT

I have no predilection, sir, for defamation. I make a point of speaking the truth on all occasions; and it seldom happens that the truth can be spoken without some stricken deer pronouncing it a libel.

MR. NIGHTSHADE

You are perhaps, sir, an enemy to literature in general?

MR. ESCOT

If I were, sir, I should be a better friend to periodical critics.

SQUIRE HEADLONG

Buz!

MR. TREACLE

May I simply take the liberty to inquire into the basis of your objection?

MR. ESCOT

I conceive that periodical criticism disseminates superficial knowledge, and its perpetual adjunct, vanity; that it checks in the youthful mind the habit of thinking for itself; that it delivers partial opinions, and thereby misleads the judgment; that it is never conducted with a view to the general interests of literature, but to serve the interested ends of individuals, and the miserable purposes of party.

MR. MAC LAUREL

Ye ken, sir, a mon mun leeve.

MR. ESCOT

While he can live honourably, naturally, justly, certainly: no longer.

MR. MAC LAUREL

Every mon, sir, leeves according to his ain notions of honour an' justice: there is a wee defference amang the learned wi' respact to the defineetion o' the terms.

MR. ESCOT

I believe it is generally admitted, that one of the ingredients of justice is disinterestedness.

MR. MAC LAUREL

It is na admetted, sir, amang the pheelosophers of Edinbroo', that there is ony sic thing as desenterestedness in the warld, or that a mon can care for onything sae much as his ain sel: for ye mun observe, sir, every mon has his ain parteecular feelings of what is gude, an' beautifu', an' consentaneous to his ain indiveedual nature, an' desires to see every thing aboot him in that parteecular state which is maist conformable to his ain notions o' the moral an' poleetical fetness o' things. Twa men, sir, shall purchase a piece o' grund atween 'em, and ae mon shall cover his half wi' a park——

MR. MILESTONE

Beautifully laid out in lawns and clumps, with a belt of trees at the circumference, and an artificial lake in the centre.

MR. MAC LAUREL

Exactly, sir: an' shall keep it a' for his ain sel: an' the other mon shall divide his half into leetle farms of twa or three acres——

MR. ESCOT

Like those of the Roman republic, and build a cottage on each of them, and cover his land with a simple, innocent, and smiling population, who shall owe, not only their happiness, but their existence, to his benevolence.

MR. MAC LAUREL

Exactly, sir: an' ye will ca' the first mon selfish, an' the second desenterested; but the pheelosophical truth is semply this, that the ane is pleased wi' looking at trees, an' the other wi' seeing people happy an' comfortable. It is aunly a matter of indiveedual feeling. A paisant saves a mon's life for the same reason that a hero or a footpad cuts his thrapple: an' a pheelosopher delevers a mon frae a preson, for the same reason that a tailor or a prime menester puts him into it: because it is conformable to his ain parteecular feelings o' the moral an' poleetical fetness o' things.

SQUIRE HEADLONG

Wake the Reverend Doctor. Doctor, the bottle stands with you.

THE REVEREND DOCTOR GASTER

It is an error of which I am seldom guilty.

MR. MAC LAUREL

Noo, ye ken, sir, every mon is the centre of his ain system, an' endaivours as much as possible to adapt every thing aroond him to his ain parteecular views.

MR. ESCOT

Thus, sir, I presume, it suits the particular views of a poet, at one time to take the part of the people against their oppressors, and at another, to take the part of the oppressors against the people.

MR. MAC LAUREL

Ye mun alloo, sir, that poetry is a sort of ware or commodity, that is brought into the public market wi' a' other descreptions of merchandise, an' that a mon is pairfectly justified in getting the best price he can for his article. Noo, there are three reasons for taking the part o' the people: the first is, when general leeberty an' public happiness are conformable to your ain parteecular feelings o' the moral an' poleetical fetness o' things: the second is, when they happen to be, as it were, in a state of exceetabeelity, an' ye think ye can get a gude price for your commodity, by flingin' in a leetle seasoning o' pheelanthropy an' republican speerit: the third is, when ye think ye can bully the menestry into gieing ye a place or a pansion to hau'd your din, an' in that case, ye point an attack against them within the pale o' the law; an' if they tak nae heed o' ye, ye open a stronger fire; an' the less heed they tak, the mair ye bawl; an' the mair factious ye grow, always within the pale o' the law, till they send a plenipotentiary to treat wi' ye for yoursel, an' then the mair popular ye happen to be, the better price ye fetch.

SQUIRE HEADLONG

Off with your heeltaps.

MR. CRANIUM

I perfectly agree with Mr. Mac Laurel in his definition of self-love and disinterestedness: every man's actions are determined by his peculiar views, and those views are determined by the organization of his skull. A man in whom the organ of benevolence is not developed, cannot be benevolent: he, in whom it is so, cannot be otherwise. The organ of self-love is prodigiously developed in the greater number of subjects that have fallen under my observation.

MR. ESCOT

Much less, I presume, among savage than civilised men, who, *constant only to the love of self, and consistent only in their aim to deceive, are always actuated by the hope of personal advantage, or by the dread of personal punishment.**

* Drummond's Academical Questions.

MR. CRANIUM

Very probably.

MR. ESCOT

You have, of course, found very copious specimens of the organs of hypocrisy, destruction, and avarice.

MR. CRANIUM

Secretiveness, destructiveness, and covetiveness. You may add, if you please, that of constructiveness.

MR. ESCOT

Meaning, I presume, the organ of building; which I contend to be not a natural organ of the *featherless biped*.

MR. CRANIUM

Pardon me: it is here.—(*As he said these words, he produced a skull from his pocket, and placed it on the table, to the great surprise of the company.*)—This was the skull of Sir Christopher Wren. You observe this protuberance—(*The skull was handed round the table.*)

MR. ESCOT

I contend that the original unsophisticated man was by no means constructive. He lived in the open air, under a tree.

THE REVEREND DOCTOR GASTER

The tree of life. Unquestionably. Till he had tasted the forbidden fruit.

MR. JENKISON

At which period, probably, the organ of constructiveness was added to his anatomy, as a punishment for his transgression.

MR. ESCOT

There could not have been a more severe one, since the propensity which has led him to building cities has proved the greatest curse of his existence.

SQUIRE HEADLONG—(*taking the skull*)

Memento mori. Come, a bumper of Burgundy.

MR. NIGHTSHADE

A very classical application, Squire Headlong. The Romans were in the practice of adhibiting skulls at their banquets, and sometimes little skeletons of silver, as a silent admonition to the guests to enjoy life while it lasted.

THE REVEREND DOCTOR GASTER

Sound doctrine, Mr. Nightshade.

MR. ESCOT

I question its soundness. The use of vinous spirit has a tremendous influence in the deterioration of the human race.

MR. FOSTER

I fear, indeed, it operates as a considerable check to the progress of the species towards moral and intellectual perfection. Yet many great men have been of opinion that it exalts the imagination, fires the genius, accelerates the flow of ideas, and imparts to dispositions naturally cold and deliberative that enthusiastic sublimation which is the source of greatness and energy.

MR. NIGHTSHADE

*Laudibus arguitur vini vinosus Homerus.**

MR. JENKISON

I conceive the use of wine to be always pernicious in excess, but often useful in moderation: it certainly kills some, but it saves the lives of others: I find that an occasional glass, taken with judgment and caution, has a very salutary effect in maintaining that equilibrium of the system, which it is always my aim to preserve; and this calm and temperate use

* Homer is proved to have been a lover of wine by the praises he bestows upon it.

of wine was, no doubt, what Homer meant to inculcate, when he said:

*Παρ δε δεπας οινοιο, πιειν ὁτε θυμος ανωγοι.**

SQUIRE HEADLONG

Good. Pass the bottle.

(*Un morne silence.*)

Sir Christopher does not seem to have raised our spirits. Chromatic, favour us with a specimen of your vocal powers. Something in point.

Mr. Chromatic, without further preface, immediately struck up the following

SONG

In his last binn SIR PETER lies,
Who knew not what it was to frown:
Death took him mellow, by surprise,
And in his cellar stopped him down.
Through all our land we could not boast
A knight more gay, more prompt than he,
To rise and fill a bumper toast,
And pass it round with THREE TIMES THREE.

None better knew the feast to sway,
Or keep Mirth's boat in better trim;
For Nature had but little clay
Like that of which she moulded him.
The meanest guest that graced his board
Was there the freest of the free,
His bumper toast when PETER poured,
And passed it round with THREE TIMES THREE.

He kept at true good humour's mark
The social flow of pleasure's tide:
He never made a brow look dark,
Nor caused a tear, but when he died.
No sorrow round his tomb should dwell:
More pleased his gay old ghost would be,
For funeral song, and passing bell,
To hear no sound but THREE TIMES THREE.

(*Hammering of knuckles and glasses, and shouts of Bravo!*)

* A cup of wine at hand, to drink as inclination prompts.

MR. PANSCOPE

(*Suddenly emerging from a deep reverie.*)

I have heard, with the most profound attention, every thing which the gentleman on the other side of the table has thought proper to advance on the subject of human deterioration; and I must take the liberty to remark, that it augurs a very considerable degree of presumption in any individual, to set himself up against the *authority* of so many great men, as may be marshalled in metaphysical phalanx under the opposite banners of the controversy; such as Aristotle, Plato, the scholiast on Aristophanes, St. Chrysostom, St. Jerome, St. Athanasius, Orpheus, Pindar, Simonides, Gronovius, Hemsterhusius, Longinus, Sir Isaac Newton, Thomas Paine, Doctor Paley, the King of Prussia, the King of Poland, Cicero, Monsier Gautier, Hippocrates, Machiavelli, Milton, Colley Cibber, Bojardo, Gregory Nazianzenus, Locke, D'Alembert, Boccaccio, Daniel Defoe, Erasmus, Doctor Smollett, Zimmermann, Solomon, Confucius, Zoroaster, and Thomas a-Kempis.

MR. ESCOT

I presume, sir, you are one of those who value an *authority* more than a reason.

MR. PANSCOPE

The *authority*, sir, of all these great men, whose works, as well as the whole of the Encyclopædia Britannica, the entire series of the Monthly Review, the complete set of the Variorum Classics, and the Memoirs of the Academy of Inscriptions, I have read through from beginning to end, deposes, with irrefragable refutation, against your ratiocinative speculations, wherein you seem desirous, by the futile process of analytical dialectics, to subvert the pyramidal structure o. synthetically deduced opinions, which have withstood the secular revolutions of physiological disquisition, and which I maintain to be transcendentally self-evident, categorically certain, and syllogistically demonstrable.

SQUIRE HEADLONG

Bravo! Pass the bottle. The very best speech that ever was made.

MR. ESCOT

It has only the slight disadvantage of being unintelligible.

MR. PANSCOPE

I am not obliged, sir, as Dr. Johnson observed on a similar occasion, to furnish you with an understanding.

MR. ESCOT

I fear, sir, you would have some difficulty in furnishing me with such an article from your own stock.

MR. PANSCOPE

'Sdeath, sir, do you question my understanding?

MR. ESCOT

I only question, sir, where I expect a reply; which, from things that have no existence, I am not visionary enough to anticipate.

MR. PANSCOPE

I beg leave to observe, sir, that my language was perfectly perspicuous, and etymologically corret; and, I conceive, I have demonstrated what I shall now take the liberty to say in plain terms, that all your opinions are extremely absurd.

MR. ESCOT

I should be sorry, sir, to advance any opinion that you would not think absurd.

MR. PANSCOPE

Death and fury, sir———

MR. ESCOT

Say no more, sir. That apology is quite sufficient.

MR. PANSCOPE

Apology, sir?

MR. ESCOT

Even so, sir. You have lost your temper, which I consider equivalent to a confession that you have the worst of the argument.

MR. PANSCOPE

Lightning and devils! sir———

SQUIRE HEADLONG

No civil war!—Temperance, in the name of Bacchus!—A glee! a glee! *Music has charms to bend the knotted oak.* Sir Patrick, you'll join?

SIR PATRICK O'PRISM

Troth, with all my heart: for, by my soul, I'm bothered completely.

SQUIRE HEADLONG

Agreed, then: you, and I, and Chromatic. Bumpers!—bumpers! Come, strike up.

Squire Headlong, Mr. Chromatic, and Sir Patrick O'Prism, each holding a bumper, immediately vociferated the following

GLEE

A heeltap! a heeltap! I never could bear it!
So fill me a bumper, a bumper of claret!
Let the bottle pass freely, don't shirk it nor spare it,
For a heeltap! a heeltap! I never could bear it!

No skylight! no twilight! while Bacchus rules o'er us:
No thinking! no shrinking! all drinking in chorus:
Let us moisten our clay, since 't is thirsty and porous:
No thinking! no shrinking! all drinking in chorus!

GRAND CHORUS

By Squire Headlong, Mr. Chromatic, Sir Patrick O'Prism, Mr. Panscope, Mr. Jenkison, Mr. Gall, Mr. Treacle, Mr. Nightshade, Mr. Mac Laurel, Mr. Cranium, Mr. Milestone, and the Reverend Doctor Gaster.

A heeltap! a heeltap! I never could bear it!
So fill me a bumper, a bumper of claret!
Let the bottle pass freely, don't shirk it nor spare it,
For a heeltap! a heeltap! I never could bear it!

'ΟΜΑΔΟΣ ΚΑΙ ΔΟΥΠΟΣ ΟΡΩΡΕΙ !

The little butler now waddled in with a summons from the ladies to tea and coffee. The Squire was unwilling to leave his Burgundy. Mr. Escot strenuously urged the necessity of immediate adjournment, observing, that the longer they continued drinking the worse they should be. Mr. Foster seconded the motion, declaring the transition from the bottle to female society to be an indisputable amelioration of the state of the sensitive man. Mr. Jenkison allowed the Squire and his two brother philosophers to settle the point between them, concluding that he was just as well in one place as another. The question of adjournment was then put, and carried by a large majority.

CHAPTER VI

THE EVENING

MR. PANSCOPE, highly irritated by the cool contempt with which Mr. Escot had treated him, sate sipping his coffee and meditating revenge. He was not long in discovering the passion of his antagonist for the beautiful Cephalis, for whom he had himself a species of predilection; and it was also obvious to him, that there was some lurking anger in the mind of her father, unfavourable to the hopes of his rival. The stimulus of revenge, superadded to that of preconceived inclination, determined him, after due deliberation, to *cut out*

Mr. Escot in the young lady's favour. The practicability of this design he did not trouble himself to investigate; for the havoc he had made in the hearts of some silly girls, who were extremely vulnerable to flattery, and who, not understanding a word he said, considered him a *prodigious clever man*, had impressed him with an unhesitating idea of his own irresistibility. He had not only the requisites already specified for fascinating female vanity, he could likewise fiddle with tolerable dexterity, though by no means so *quick* as Mr. Chromatic (for our readers are of course aware that rapidity of execution, not delicacy of expression, constitutes the scientific perfection of modern music), and could warble a fashionable love-ditty with considerable affectation of feeling: besides this, he was always extremely well dressed, and was heir-apparent to an estate of ten thousand a-year. The influence which the latter consideration might have on the minds of the majority of his female acquaintance, whose morals had been formed by the novels of such writers as Miss Philomela Poppyseed, did not once enter into his calculation of his own personal attractions. Relying, therefore, on past success, he determined *to appeal to his fortune*, and already, in imagination, considered himself sole lord and master of the affections of the beautiful Cephalis.

Mr. Escot and Mr. Foster were the only two of the party who had entered the library (to which the ladies had retired, and which was interior to the music-room) in a state of perfect sobriety. Mr. Escot had placed himself next to the beautiful Cephalis: Mr. Cranium had laid aside much of the terror of his frown; the short craniological conversation, which had passed between him and Mr. Escot, had softened his heart in his favour; and the copious libations of Burgundy in which he had indulged had smoothed his brow into unusual serenity.

Mr. Foster placed himself near the lovely Caprioletta, whose artless and innocent conversation had already made an impression on his susceptible spirit.

The Reverend Doctor Gaster seated himself in the corner of a sofa near Miss Philomela Poppyseed. Miss Philomela detailed to him the plan of a very moral and aristocratical novel she was preparing for the press, and continued holding forth, with her eyes half shut, till a long-drawn nasal tone from the reverend divine compelled her suddenly to open them in all the indignation of surprise. The cessation of the hum of her voice awakened the reverend gentleman, who, lifting up first one eyelid, then the other, articulated, or rather murmured, 'Admirably planned, indeed!'

'I have not quite finished, sir,' said Miss Philomela, bridling. 'Will you have the goodness to inform me where I left off?'

The Doctor hummed a while, and at length answered: 'I think you had just laid it down as a position, that a thousand a-year is an indispensable ingredient in the passion of love, and that no man, who is not so far gifted by *nature*, can reasonably presume to feel that passion himself, or be correctly the object of it with a well-educated female.'

'That, sir,' said Miss Philomela, highly incensed, 'is the fundamental principle which I lay down in the first chapter, and which the whole four volumes, of which I detailed to you the outline, are intended to set in a strong practical light.'

'Bless me!' said the Doctor, 'what a nap I must have had!'

Miss Philomela flung away to the side of her dear friends Gall and Treacle, under whose fostering patronage she had been puffed into an extensive reputation, much to the advantage of the young ladies of the age, whom she taught to consider themselves as a sort of commodity, to be put up at public auction, and knocked down to the highest bidder. Mr. Nightshade and Mr. Mac Laurel joined the trio; and it was secretly resolved, that Miss Philomela should furnish them with a portion of her manuscripts, and that Messieurs Gall and Co. should devote the following morning to cutting and drying a critique on a work calculated to prove so extensively beneficial, that Mr. Gall protested he really *envied* the writer.

While this amiable and enlightened quintetto were busily employed in flattering one another, Mr. Cranium retired to complete the preparations he had begun in the morning for a lecture, with which he intended, on some future evening, to favour the company: Sir Patrick O'Prism walked out into the grounds to study the effect of moonlight on the snow-clad mountains: Mr. Foster and Mr. Escot continued to make love, and Mr. Panscope to digest his plan of attack on the heart of Miss Cephalis: Mr. Jenkison sate by the fire, reading *Much Ado about Nothing*: the Reverend Doctor Gaster was still enjoying the benefit of Miss Philomela's opiate, and serenading the company from his solitary corner: Mr. Chromatic was reading music, and occasionally humming a note: and Mr. Milestone had produced his portfolio for the edification and amusement of Miss Tenorina, Miss Graziosa, and Squire Headlong, to whom he was pointing out the various beauties of his plan for Lord Littlebrain's park.

MR. MILESTONE

This, you perceive, is the natural state of one part of the grounds. Here is a wood, never yet touched by the finger of taste; thick, intricate, and gloomy. Here is a little stream, dashing from stone to stone, and overshadowed with these untrimmed boughs.

MISS TENORINA

The sweet romantic spot! How beautifully the birds must sing there on a summer evening!

MISS GRAZIOSA

Dear sister! how can you endure the horrid thicket?

MR. MILESTONE

You are right, Miss Graziosa: your taste is correct—perfectly *en règle*. Now, here is the same place corrected—trimmed—polished—decorated—adorned. Here sweeps a plantation, in that beautiful regular curve: there winds a gravel walk: here are parts of the old wood, left in these majestic circular clumps, disposed at equal distances with wonderful symmetry: there are some single shrubs scattered in elegant profusion: here a Portugal laurel, there a juniper; here a lauristinus, there a spruce fir; here a larch, there a lilac; here a rhododendron, there an arbutus. The stream, you see, is become a canal: the banks are perfectly smooth and green, sloping to the water's edge: and there is Lord Littlebrain, rowing in an elegant boat.

SQUIRE HEADLONG

Magical, faith!

MR. MILESTONE

Here is another part of the grounds in its natural state. Here is a large rock, with the mountain-ash rooted in its fissures, overgrown, as you see, with ivy and moss; and from this part of it bursts a little fountain, that runs bubbling down its rugged sides.

MISS TENORINA

O how beautiful! How I should love the melody of that miniature cascade!

MR. MILESTONE

Beautiful, Miss Tenorina! Hideous. Base, common, and popular. Such a thing as you may see anywhere, in wild and mountainous districts. Now, observe the metamorphosis. Here is the same rock, cut into the shape of a giant. In one hand he holds a horn, through which that little fountain is thrown to a prodigious elevation. In the other is a ponderous stone, so exactly balanced as to be apparently ready to fall on the head of any person who may happen to be beneath:* and there is Lord Littlebrain walking under it.

SQUIRE HEADLONG

Miraculous, by Mahomet!

MR. MILESTONE

This is the summit of a hill, covered, as you perceive, with wood, and with those mossy stones scattered at random under the trees.

* See Knight on Taste.

MISS TENORINA

What a delightful spot to read in, on a summer's day! The air must be so pure, and the wind must sound so divinely in the tops of those old pines!

MR. MILESTONE

Bad taste, Miss Tenorina. Bad taste, I assure you. Here is the spot improved. The trees are cut down: the stones are cleared away: this is an octagonal pavilion, exactly on the centre of the summit: and there you see Lord Littlebrain, on the top of the pavilion, enjoying the prospect with a telescope.

SQUIRE HEADLONG

Glorious, egad!

MR. MILESTONE

Here is a rugged mountainous road, leading through impervious shades: the ass and the four goats characterise a wild uncultured scene. Here, as you perceive, it is totally changed into a beautiful gravel-road, gracefully curving through a belt of limes: and there is Lord Littlebrain driving four-in-hand.

SQUIRE HEADLONG

Egregious, by Jupiter!

MR. MILESTONE

Here is Littlebrain Castle, a Gothic, moss-grown structure, half-bosomed in trees. Near the casement of that turret is an owl peeping from the ivy.

SQUIRE HEADLONG

And devilish wise he looks.

MR. MILESTONE

Here is the new house, without a tree near it, standing in the midst of an undulating lawn: a white, polished, angular building, reflected to a nicety in this waveless lake: and there you see Lord Littlebrain looking out of the window.

SQUIRE HEADLONG

And devilish wise he looks too. You shall cut me a giant before you go.

MR. MILESTONE

Good. I'll order down my little corps of pioneers.

During this conversation, a hot dispute had arisen between Messieurs Gall and Nightshade; the latter pertinaciously insisting on having his new poem reviewed by Treacle, who he knew would extol it most loftily, and not by Gall, whose sarcastic commendation he held in superlative horror. The remonstrances of Squire Headlong silenced the disputants, but did not mollify the inflexible Gall, nor appease the irritated Nightshade, who secretly resolved that, on his return to London, he would beat his drum in Grub Street, form a mastigophoric corps of his own, and hoist the standard of determined opposition against this critical Napoleon.

Sir Patrick O'Prism now entered, and, after some rapturous exclamations on the effect of the mountain-moonlight, entreated that one of the young ladies would favour him with a song. Miss Tenorina and Miss Graziosa now enchanted the company with some very scientific compositions, which, as usual, excited admiration and astonishment in every one, without a single particle of genuine pleasure. The beautiful Cephalis being then summoned to take her station at the harp, sang with feeling and simplicity the following air:—

LOVE AND OPPORTUNITY

Oh! who art thou, so swiftly flying?
    My name is Love, the child replied:
Swifter I pass than south-winds sighing,
    Or streams, through summer vales that glide.
And who art thou, his flight pursuing?
    'T is cold Neglect whom now you see:
The little god you there are viewing,
    Will die, if once he's touched by me.

*Oh! who art thou so fast proceeding,
Ne'er glancing back thine eyes of flame?
Marked but by few, through earth I'm speeding,
And Opportunity's my name.
What form is that, which scowls beside thee?
Repentance is the form you see:
Learn then, the fate may yet betide thee:
She seizes them who seize not me.

The little butler now appeared with a summons to supper, shortly after which the party dispersed for the night.

CHAPTER VII

THE WALK

It was an old custom in Headlong Hall to have breakfast ready at eight, and continue it till two; that the various guests might rise at their own hour, breakfast when they came down, and employ the morning as they thought proper; the Squire only expecting that they should punctually assemble at dinner. During the whole of this period, the little butler stood sentinel at a side-table near the fire, copiously furnished with all the apparatus of tea, coffee, chocolate, milk, cream, eggs, rolls, toast, muffins, bread, butter, potted beef, cold fowl and partridge, ham, tongue, and anchovy. The Reverend Doctor Gaster found himself rather *queasy* in the morning, therefore preferred breakfasting in bed, on a mug of buttered ale and an anchovy toast. The three philosophers made their appearance at eight, and enjoyed *les prémices des dépouilles*. Mr. Foster proposed that, as it was a fine frosty morning, and they were all good pedestrians, they should take a walk to Tremadoc, to see the improvements carrying on in that vicinity. This being readily acceded to, they began their walk.

After their departure, appeared Squire Headlong and Mr. Milestone, who agreed, over their muffin and partridge, to walk together to a ruined tower, within the precincts of the

* This stanza is imitated from Machiavelli's *Capitolo dell' Occasione*.

Squire's grounds, which Mr. Milestone thought he could improve.

The other guests dropped in by one's and two's, and made their respective arrangements for the morning. Mr. Panscope took a little ramble with Mr. Cranium, in the course of which, the former professed a great enthusiasm for the science of craniology, and a great deal of love for the beautiful Cephalis, adding a few words about his expectations: the old gentleman was unable to withstand this triple battery, and it was accordingly determined—after the manner of the heroic age, in which it was deemed superfluous to consult the opinions and feelings of the lady, as to the manner in which she should be disposed of—that the lovely Miss Cranium should be made the happy bride of the accomplished Mr. Panscope. We shall leave them for the present to settle preliminaries, while we accompany the three philosophers in their walk to Tremadoc.

The vale contracted as they advanced, and, when they had passed the termination of the lake, their road wound along a narrow and romantic pass, through the middle of which an impetuous torrent dashed over vast fragments of stone. The pass was bordered on both sides by perpendicular rocks, broken into the wildest forms of fantastic magnificence.

'These are, indeed,' said Mr. Escot, '*confracti mundi rudera*:* yet they must be feeble images of the valleys of the Andes, where the philosophic eye may contemplate, in their utmost extent, the effects of that tremendous convulsion which destroyed the perpendicularity of the poles, and inundated this globe with that torrent of physical evil, from which the greater torrent of moral evil has issued, that will continue to roll on, with an expansive power and an accelerated impetus, till the whole human race shall be swept away in its vortex.'

'The precession of the equinoxes,' said Mr. Foster, 'will gradually ameliorate the physical state of our planet, till the ecliptic shall again coincide with the equator, and the equal

* Fragments of a demolished world.

diffusion of light and heat over the whole surface of the earth typify the equal and happy existence of man, who will then have attained the final step of pure and perfect intelligence.'

'It is by no means clear,' said Mr. Jenkison, 'that the axis of the earth was ever perpendicular to the plane of its orbit, or that it ever will be so. Explosion and convulsion are necessary to the maintenance of either hypothesis: for La Place has demonstrated, that the precession of the equinoxes is only a secular equation of a very long period, which, of course, proves nothing either on one side or the other.'

They now emerged, by a winding ascent, from the vale of Llanberris, and after some little time arrived at Bedd Gelert. Proceeding through the sublimely romantic pass of Aberglaslynn, their road led along the edge of Traeth Mawr, a vast arm of the sea, which they then beheld in all the magnificence of the flowing tide. Another five miles brought them to the embankment, which has since been completed, and which, by connecting the two counties of Meirionnydd and Caernarvon, excludes the sea from an extensive tract. The embankment, which was carried on at the same time from both the opposite coasts, was then very nearly meeting in the centre. They walked to the extremity of that part of it which was thrown out from the Caernarvonshire shore. The tide was now ebbing: it had filled the vast basin within, forming a lake about five miles in length and more than one in breadth. As they looked upwards with their backs to the open sea, they beheld a scene which no other in this country can parallel, and which the admirers of the magnificence of nature will ever remember with regret, whatever consolation may be derived from the probable utility of the works which have excluded the waters from their ancient receptacle. Vast rocks and precipices, intersected with little torrents, formed the barrier on the left: on the right, the triple summit of Moëlwyn reared its majestic boundary: in the depth was that sea of mountains, the wild and stormy outline of the Snowdonian chain, with the giant Wyddfa towering in the midst. The mountain-frame remains unchanged, unchangeable; but the liquid mirror it enclosed is gone.

The tide ebbed with rapidity: the waters within, retained by the embankment, poured through its two points an impetuous cataract, curling and boiling in innumerable eddies, and making a tumultuous melody admirably in unison with the surrounding scene. The three philosophers looked on in silence; and at length unwillingly turned away and proceeded to the little town of Tremadoc, which is built on land recovered in a similar manner from the sea. After inspecting the manufactories, and refreshing themselves at the inn on a cold saddle of mutton and a bottle of sherry, they retraced their steps towards Headlong Hall, commenting as they went on the various objects they had seen.

MR. ESCOT

I regret that time did not allow us to see the caves on the sea-shore. There is one of which the depth is said to be unknown. There is a tradition in the country, that an adventurous fiddler once resolved to explore it; that he entered, and never returned; but that the subterranean sound of a fiddle was heard at a farm-house seven miles inland. It is, therefore, concluded that he lost his way in the labyrinth of caverns, supposed to exist under the rocky soil of this part of the country.

MR. JENKISON

A supposition that must always remain in force, unless a second fiddler, equally adventurous and more successful, should return with an accurate report of the true state of the fact.

MR. FOSTER

What think you of the little colony we have just been inspecting; a city, as it were, in its cradle?

MR. ESCOT

With all the weakness of infancy, and all the vices of maturer age. I confess, the sight of those manufactories, which have suddenly sprung up, like fungous excrescences, in the bosom of these wild and desolate scenes, impressed me with as much horror and amazement as the sudden appearance of the stocking manufactory struck into the mind of Rousseau, when, in a lonely valley of the Alps, he had just congratulated himself on finding a spot where man had never been.

MR. FOSTER

The manufacturing system is not yet purified from some evils which necessarily attend it, but which I conceive are greatly overbalanced by their concomitant advantages. Contemplate the vast sum of human industry to which this system so essentially contributes: seas covered with vessels, ports resounding with life, profound researches, scientific inventions, complicated mechanism, canals carried over deep valleys and through the bosoms of hills: employment and existence thus given to innumerable families, and the multiplied comforts and conveniences of life diffused over the whole community.

MR. ESCOT

You present to me a complicated picture of artificial life, and require me to admire it. Seas covered with vessels: every one of which contains two or three tyrants, and from fifty to a thousand slaves, ignorant, gross, perverted, and active only in mischief. Ports resounding with life: in other words, with noise and drunkenness, the mingled din of avarice, intemperance, and prostitution. Profound researches, scientific inventions: to what end? To contract the sum of human wants? to teach the art of living on a little? to disseminate independence, liberty, and health? No; to multiply factitious desires, to stimulate depraved appetites, to invent unnatural wants, to heap up incense on the shrine of luxury, and accumulate expedients of selfish and ruinous profusion. Complicated machinery: behold its blessings. Twenty years ago, at the door of every cottage sate the good woman with her spinning-wheel: the children, if not more profitably employed than in gathering heath and sticks, at least laid in a stock of health and strength to sustain the labours of maturer years. Where is the spinning-wheel now, and every simple and insulated occupation of the industrious cottager? Wherever this boasted machinery is established, the children of the poor are death-doomed from their cradles. Look for one moment at midnight into a cotton-mill, amidst the smell of oil, the smoke of lamps, the rattling of wheels, the dizzy and complicated motions of diabolical mechanism: contemplate the little human machines that keep play with the revolutions of the iron work, robbed at that hour of their natural rest, as of air and exercise by day: observe their pale and ghastly features, more ghastly in that baleful and malignant light, and tell me if you do not fancy yourself on the threshold of Virgil's hell, where

> continuò auditæ voces, vagitus et ingens,
> *Infantumque animæ flentes*, in limine primo,
> Quos *dulcis vitæ exsortes*, et ab ubere raptos,
> *Abstulit atra dies*, et FUNERE MERSIT ACERBO!

As Mr. Escot said this, a little rosy-cheeked girl, with a basket of heath on her head, came tripping down the side of one of the rocks on the left. The force of contrast struck even on the phlegmatic spirit of Mr. Jenkison, and he almost inclined for a moment to the doctrine of deterioration. Mr. Escot continued:

Nor is the lot of the parents more enviable. Sedentary victims of unhealthy toil, they have neither the corporeal energy of the savage, nor the mental acquisitions of the civilised man. Mind, indeed, they have none, and scarcely animal life. They are mere automata, component parts of the enormous machines which administer to the pampered appetities of the few, who consider themselves the most valuable portion of a state, because they consume in indolence the fruits of the earth, and contribute nothing to the benefit of the community.

MR. JENKISON

That these are evils cannot be denied; but they have their counterbalancing advantages. That a man should pass the day in a furnace and the night in a cellar, is bad for the individual, but good for others who enjoy the benefit of his labour.

MR. ESCOT

By what right do they so?

MR. JENKISON

By the right of all property and all possession: *le droit du plus fort.*

MR. ESCOT

Do you justify that principle?

MR. JENKISON

I neither justify nor condemn it. It is practically recognised in all societies; and, though it is certainly the source of enormous evil, I conceive it is also the source of abundant good, or it would not have so many supporters.

MR. ESCOT

That is by no means a consequence. Do we not every day see men supporting the most enormous evils, which they know to be so with respect to others, and which in reality are so with respect to themselves, though an erroneous view of their own miserable self-interest induces them to think otherwise?

MR. JENKISON

Good and evil exist only as they are perceived. I cannot therefore understand, how that which a man perceives to be good can be in reality an evil to him: indeed, the word *reality* only signifies *strong belief.*

MR. ESCOT

The views of such a man I contend are false. If he could be made to see the truth——

MR. JENKISON

He sees his own truth. Truth is that which a man *troweth.* Where there is no man there is no truth. Thus the truth of one is not the truth of another.*

* Tooke's Diversions of Purley.

MR. ESCOT

I am aware of the etymology; but I contend that there is an universal and immutable truth, deducible from the nature of things.

MR. JENKISON

By whom deducible? Philosophers have investigated the nature of things for centuries, yet no two of them will agree in *trowing* the same conclusion.

MR. FOSTER

The progress of philosophical investigation, and the rapidly increasing accuracy of human knowledge, approximate by degrees the diversities of opinion; so that, in process of time, moral science will be susceptible of mathematical demonstration; and, clear and indisputable principles being universally recognised, the coincidence of deduction will necessarily follow.

MR. ESCOT

Possibly, when the inroads of luxury and disease shall have exterminated nine hundred and ninety-nine thousand nine hundred and ninety-nine of every million of the human race, the remaining fractional units may congregate into one point, and come to something like the same conclusion.

MR. JENKISON

I doubt it much. I conceive, if only we three were survivors of the whole system of terrestrial being, we should never agree in our decisions as to the cause of the calamity.

MR. ESCOT

Be that as it may, I think you must at least assent to the following positions: that the many are sacrificed to the few; that ninety-nine in a hundred are occupied in a perpetual struggle for the preservation of a perilous and precarious existence, while the remaining one wallows in all the redundancies of luxury that can be wrung from their labours and privations; that luxury and liberty are incompatible; and that every new want you invent for civilised man is a new instrument of torture for him who cannot indulge it.

They had now regained the shores of the lake, when the conversation was suddenly interrupted by a tremendous explosion, followed by a violent splashing of water, and various sounds of tumult and confusion, which induced them to quicken their pace towards the spot whence they proceeded.

CHAPTER VIII

THE TOWER

In all the thoughts, words, and actions of Squire Headlong, there was a remarkable alacrity of progression, which almost annihilated the interval between conception and execution. He was utterly regardless of obstacles, and seemed to have expunged their very name from his vocabulary. His designs were never nipped in their infancy by the contemplation of those trivial difficulties which often turn awry the current of enterprise; and, though the rapidity of his movements was sometimes arrested by a more formidable barrier, either naturally existing in the pursuit he had undertaken, or created by his own impetuosity, he seldom failed to succeed either in knocking it down or cutting his way through it. He had little idea of gradation; he saw no interval between the first step and the last, but pounced upon his object with the impetus of a mountain cataract. This rapidity of movement, indeed, subjected him to some disasters which cooler spirits would have escaped. He was an excellent sportsman, and almost always killed his game; but now and then he killed his dog.* Rocks, streams, hedges, gates, and ditches, were

* Some readers will, perhaps, recollect the Archbishop of Prague, who also was an excellent sportsman, and who,

> Com' era scritto in certi suoi giornali,
> Ucciso avea con le sue proprie mani
> Un numero infinito d'animali:
> Cinquemila con quindici fagiani,
> Seimila lepri, ottantatrè cignali,
> E per disgrazia, ancor *tredici cani*, &c.

objects of no account in his estimation; though a dislocated shoulder, several severe bruises, and two or three narrow escapes for his neck, might have been expected to teach him a certain degree of caution in effecting his transitions. He was so singularly alert in climbing precipices and traversing torrents, that, when he went out on a shooting party, he was very soon left to continue his sport alone, for he was sure to dash up or down some nearly perpendicular path, where no one else had either ability or inclination to follow. He had a pleasure boat on the lake, which he steered with amazing dexterity; but as he always indulged himself in the utmost possible latitude of sail, he was occasionally upset by a sudden gust, and was indebted to his skill in the art of swimming for the opportunity of tempering with a copious libation of wine the unnatural frigidity introduced into his stomach by the extraordinary intrusion of water, an element which he had religiously determined should never pass his lips, but of which, on these occasions, he was sometimes compelled to swallow no inconsiderable quantity. This circumstance alone, of the various disasters that befel him, occasioned him any permanent affliction, and he accordingly noted the day in his pocket book as a *dies nefastus*, with this simple abstract, and brief chronicle of the calamity: *Mem. Swallowed two or three pints of water*: without any notice whatever of the concomitant circumstances. These days, of which there were several, were set apart in Headlong Hall for the purpose of anniversary expiation; and, as often as the day returned on which the Squire had swallowed water, he not only made a point of swallowing a treble allowance of wine himself, but imposed a heavy mulct on every one of his servants who should be detected in a state of sobriety after sunset: but their conduct on these occasions was so uniformly exemplary, that no instance of the infliction of the penalty appears on record.

The Squire and Mr. Milestone, as we have already said, had set out immediately after breakfast to examine the capabilities of the scenery. The object that most attracted Mr. Milestone's admiration was a ruined tower on a projecting point of rock, almost totally overgrown with ivy. This ivy, Mr. Milestone observed, required trimming and clearing in various parts: a little pointing and polishing was also necessary for the dilapidated walls: and the whole effect would be materially increased by a plantation of spruce fir, interspersed with cypress and juniper, the present rugged and broken ascent from the land side being first converted into a beautiful slope, which might be easily effected by blowing up a part of the rock with gunpowder, laying on a quantity of fine mould, and covering the whole with an elegant stratum of turf.

Squire Headlong caught with avidity at this suggestion; and, as he had always a store of gunpowder in the house, for the accommodation of himself and his shooting visitors, and for the supply of a small battery of cannon, which he kept for his private amusement, he insisted on commencing operations immediately. Accordingly, he bounded back to the house, and very speedily returned, accompanied by the little butler, and half a dozen servants and labourers, with pickaxes and gunpowder, a hanging stove and a poker, together with a basket of cold meat and two or three bottles of Madeira: for the Squire thought, with many others, that a copious supply of provision is a very necessary ingredient in all rural amusements.

Mr. Milestone superintended the proceedings. The rock was excavated, the powder introduced, the apertures strongly blockaded with fragments of stone: a long train was laid to a spot which Mr. Milestone fixed on as sufficiently remote from the possibility of harm: the Squire seized the poker, and, after flourishing it in the air with a degree of dexterity which induced the rest of the party to leave him in solitary possession of an extensive circumference, applied the end of it to the train; and the rapidly communicated ignition ran hissing along the surface of the soil.

At this critical moment, Mr. Cranium and Mr. Panscope appeared at the top of the tower, which, unseeing and unseen, they had ascended on the opposite side to that where the Squire and Mr. Milestone were conducting their operations. Their sudden appearance a little dismayed the Squire, who, however, comforted himself with the reflection, that the tower was perfectly safe, or at least was intended to be so, and that his friends were in no probable danger but of a knock on the head from a flying fragment of stone.

The succession of these thoughts in the mind of the Squire was commensurate in rapidity to the progress of the ignition, which having reached its extremity, the explosion took place, and the shattered rock was hurled into the air in the midst of fire and smoke.

Mr. Milestone had properly calculated the force of the explosion; for the tower remained untouched: but the Squire, in his consolatory reflections, had omitted the consideration of the influence of sudden fear, which had so violent an effect on Mr. Cranium, who was just commencing a speech concerning the very fine prospect from the top of the tower, that, cutting short the thread of his observations, he bounded, under the elastic influence of terror, several feet into the air. His ascent being unluckily a little out of the perpendicular, he descended with a proportionate curve from the apex of his projection, and alighted, not on the wall of the tower, but in an ivy-bush by its side, which, giving way beneath him, transferred him to a tuft of hazel at its base, which, after upholding him an instant, consigned him to the boughs of an ash that had rooted itself in a fissure about half way down the rock, which finally transmitted him to the waters below.

Squire Headlong anxiously watched the tower as the smoke which at first enveloped it rolled away; but when this shadowy curtain was withdrawn, and Mr. Panscope was discovered, *solus*, in a tragical attitude, his apprehensions became boundless, and he concluded that the unlucky collision of a flying fragment of rock had indeed emancipated the spirit of the craniologist from its terrestrial bondage.

Mr. Escot had considerably outstripped his companions and arrived at the scene of the disaster just as Mr. Cranium, being utterly destitute of natatorial skill, was in imminent danger of final submersion. The deteriorationist, who had cultivated this valuable art with great success, immediately plunged in to his assistance, and brought him alive and in safety to a shelving part of the shore. Their landing was hailed with a view-holla from the delighted Squire, who, shaking them both heartily by the hand, and making ten thousand lame apologies to Mr. Cranium, concluded by asking, in a pathetic tone, *How much water he had swallowed*? and without waiting for his anwer, filled a large tumbler with Madeira, and insisted on his tossing it off, which was no sooner said than done. Mr. Jenkison and Mr. Foster now made their appearance. Mr. Panscope descended the tower, which he vowed never again to approach within a quarter of a mile. The tumbler of Madeira was replenished, and handed round to recruit the spirits of the party, which now began to move towards Headlong Hall, the Squire capering for joy in the van, and the little fat butler waddling in the rear.

The Squire took care that Mr. Cranium should be seated next to him at dinner, and plied him so hard with Madeira to prevent him, as he said, from taking cold, that long before the ladies sent in their summons to coffee, every organ in his brain was in a complete state of revolution, and the Squire was under the necessity of ringing for three or four servants to carry him to bed, observing, with a smile of great satisfaction, that he was in a very excellent way for escaping any ill consequences that might have resulted from his accident.

The beautiful Cephalis, being thus freed from his *surveillance*, was enabled, during the course of the evening, to develop to his preserver the full extent of her gratitude.

CHAPTER IX

THE SEXTON

Mr. Escot passed a sleepless night, the ordinary effect of love, according to some amatory poets, who seem to have composed their whining ditties for the benevolent purpose of bestowing on others that gentle slumber of which they so pathetically lament the privation. The deteriorationist entered into a profound moral soliloquy, in which he first examined *whether a philosopher ought to be in love*? Having decided this point affirmatively against Plato and Lucretius, he next examined, *whether that passion ought to have the effect of keeping a philosopher awake*? Having decided this negatively, he resolved to go to sleep immediately: not being able to accomplish this to his satisfaction, he tossed and tumbled, like Achilles or Orlando, first on one side, then on the other; repeated to himself several hundred lines of poetry; counted a thousand; began again, and counted another thousand: in vain: the beautiful Cephalis was the predominant image in all his soliloquies, in all his repetitions: even in the numerical process from which he sought relief, he did but associate the idea of number with that of his dear tormentor, till she appeared to his mind's eye in a thousand similitudes, distinct, not different. These thousand images, indeed, were but one; and yet the one was a thousand, a sort of uni-multiplex phantasma, which will be very intelligible to some understandings.

He arose with the first peep of day, and sallied forth to enjoy the balmy breeze of morning, which any but a lover might have thought too cool; for it was an intense frost, the sun had not risen, and the wind was rather fresh from north-east and by north. But a lover, who, like Ladurlad in the curse of Kehama, always has, or at least is supposed to have, 'a fire in his heart and a fire in his brain,' feels a wintry breeze from N. E. and by N. steal over his cheek like the south over a bank of violets: therefore, on walked the philosopher, with his coat unbuttoned and his hat in his hand, careless of whither he went, till he found himself near the enclosure of a little mountain-chapel. Passing through the wicket, and stepping over two or three graves, he stood on a rustic tombstone, and peeped through the chapel window, examinging the interior with as much curiosity as if he had 'forgotten what the inside of a church was made of,' which, it is rather to be feared, was the case. Before him and beneath him were the font, the altar, and the grave; which gave rise to a train of moral reflections on the three great epochs in the course of the *featherless biped*,—birth, marriage, and death. The middle stage of the process arrested his attention; and his imagination placed before him several figures, which he thought, with the addition of his own, would make a very picturesque group; the beautiful Cephalis, 'arrayed in her bridal apparel of white;' her friend Caprioletta officiating as bridemaid; Mr. Cranium giving her away; and, last not least, the Reverend Doctor Gaster, intoning the marriage ceremony with the regular orthodox allowance of nasal recitative. Whilst he was feasting his eyes on this imaginary picture, the demon of mistrust insinuated himself into the storehouse of his conceptions, and, removing his figure from the group, substituted that of Mr. Panscope, which gave such a violent shock to his feelings, that he suddenly exclaimed, with an extraordinary elevation of voice, *Οιμοι κακοδαιμων, και τρις κακοδαιμων, και τετρακις, και πεντακις, και δωδεκακις, και μυριακις*! * to the great terror of the sexton, who was just entering the churchyard, and, not knowing from whence the voice proceeded, *pensa que fut un diableteau*. The sight of the philosopher dispelled his apprehensions, when, growing suddenly valiant, he immediately addressed him:—

'Cot pless your honour, I should n't have thought of meeting any pody here at this time of the morning, except, look you, it was the tevil—who, to pe sure, toes not often come upon consecrated cround—put for all that, I think I have seen him now and then, in former tays, when old Nanny Llwyd of Llyn-isa was living—Cot teliver us! a terriple old witch to pe sure she was—I tid n't much like tigging her crave—put I prought two cocks with me—the tevil hates cocks—and tied them py the leg on two tombstones—and I tug, and the cocks crowed, and the tevil kept at a tistance. To pe sure now, if I had n't peen very prave py nature—as I ought to pe truly—for my father was Owen Ap-Llwyd Ap-Gryffydd Ap-Shenkin Ap-Williams Ap-Thomas Ap-Morgan Ap-Parry Ap-Evan Ap-Rhys, a coot preacher and a lover of *cwrw*†—I should have thought just now pefore I saw your honour, that the foice I heard was the

* Me miserable! and thrice miserable! and four times, and five times, and twelve times, and ten thousand times miserable!

† Pronounced cooroo—the Welsh word for *ale*.

tevil's calling Nanny Llwyd—Cot pless us! to pe sure she should have been puried in the middle of the river, where the tevil can't come, as your honour fery well knows.'

'I am perfectly aware of it,' said Mr. Escot.

'True, true,' continued the sexton; 'put to pe sure, Owen Thomas of Morfa-Bach will have it that one summer evening—when he went over to Cwm Cynfael in Meirionnydd, apout some cattles he wanted to puy—he saw a strange figure—pless us!—with five horns!—Cot save us! sitting on Hugh Llwyd's pulpit, which, your honour fery well knows, is a pig rock in the middle of the river——'

'Of course he was mistaken,' said Mr. Escot.

'To pe sure he was,' said the sexton. 'For there is no toubt put the tevil, when Owen Thomas saw him, must have peen sitting on a piece of rock in a straight line from him on the other side of the river, where he used to sit, look you, for a whole summer's tay, while Hugh Llwyd was on his pulpit, and there they used to talk across the water! for Hugh Llwyd, please your honour, never raised the tevil except when he was safe in the middle of the river, which proves that Owen Thomas, in his fright, tid n't pay proper attention to the exact spot where the tevil was.'

The sexton concluded his speech with an approving smile at his own sagacity, in so luminously expounding the nature of Owen Thomas's mistake.

'I perceive,' said Mr. Escot, 'you have a very deep insight into things, and can, therefore, perhaps, facilitate the resolution of a question, concerning which, though I have little doubt on the subject, I am desirous of obtaining the most extensive and accurate information.'

The sexton scratched his head, the language of Mr. Escot not being to his apprehension quite so luminous as his own.

'You have been sexton here,' continued Mr. Escot, in the language of Hamlet, 'man and boy, forty years.'

The sexton turned pale. The period Mr. Escot named was so nearly the true one, that he began to suspect the personage before him of being rather too familiar with Hugh Llwyd's sable visitor. Recovering himself a little, he said, 'Why, thereapouts, sure enough.'

'During this period, you have of course dug up many bones of the people of ancient times.'

'Pones! Cot pless you, yes! pones as old as the 'orlt.'

'Perhaps you can show me a few.'

The sexton grinned horribly a ghastly smile. 'Will you take your Pible oath you ton't want them to raise the tevil with?'

'Willingly,' said Mr. Escot, smiling; 'I have an abstruse reason for the inquiry.'

'Why, if you have an *obtuse* reason,' said the sexton, who thought this a good opportunity to show that he could pronounce hard words as well as other people; 'if you have an *obtuse* reason, that alters the case.'

So saying he led the way to the bone-house, from which he began to throw out various bones and skulls of more than common dimensions, and amongst them a skull of very extraordinary magnitude, which he swore by St. David was the skull of Cadwallader.

'How do you know this to be his skull?' said Mr. Escot.

'He was the piggest man that ever lived, and he was puried here; and this is the piggest skull I ever found: you see now——'

'Nothing can be more logical,' said Mr. Escot. 'My good friend, will you allow me to take this skull away with me?'

'St. Winifred pless us!' exclaimed the sexton:—'would you have me haunted py his chost for taking his plessed pones out of consecrated cround? Would you have him come in the tead of the night, and fly away with the roof of my house? Would you have all the crop of my carden come to nothing? for, look you, his epitaph says,

"He that my pones shall ill pestow,
Leek in his cround shall never crow."

'You will ill bestow them,' said Mr. Escot, 'in confounding them with those of the sons of little men, the degenerate dwarfs of later generations: you will well bestow them in giving them to me; for I will have this illustrious skull bound with a silver rim, and filled with mantling wine, with this inscription, NUNC TANDEM: signifying that that pernicious liquor has at length found its proper receptacle; for, when the wine is in, the brain is out.'

Saying these words, he put a dollar into the hands of the sexton, who instantly stood spell-bound by the talismanic influence of the coin, while Mr. Escot walked off in triumph with the skull of Cadwallader.

CHAPTER X

THE SKULL

When Mr. Escot entered the breakfast-room he found the majority of the party assembled, and the little butler very active at his station. Several of the ladies shrieked at the sight of the skull; and Miss Tenorina, starting up in great haste and terror, caused the subversion of a cup of chocolate, which a servant was handing to the Reverend Doctor Gaster, into the nape of the neck of Sir Patrick O'Prism. Sir Patrick, rising impetuously, *to clap an extinguisher*, as he expressed himself, *on the farthing rushlight of the rascal's life*, pushed over the chair of Marmaduke Milestone, Esquire, who, catching for support at the first thing that came in his way, which happened unluckily to be the corner of the table-cloth, drew it instantaneously with him to the floor, involving plates, cups and saucers, in one promiscuous ruin. But, as the principal *matériel* of the breakfast apparatus was on the little butler's side-table, the confusion occasioned by this accident was happily greater than the damage. Miss Tenorina was so agitated that she was obliged to retire: Miss Graziosa accompanied her through pure sisterly affection and sympathy, not without a lingering look at Sir Patrick, who likewise retired to change his coat, but was very expeditious in returning to resume his attack on the cold partridge. The broken cups were cleared away, the cloth relaid, and the array of the table restored with wonderful celerity.

Mr. Escot was a little surprised at the scene of confusion which signalised his entrance; but, perfectly unconscious that it originated with the skull of Cadwallader, he advanced to seat himself at the table by the side of the beautiful Cephalis, first placing the skull in a corner, out of the reach of Mr. Cranium, who sate eyeing it with lively curiosity, and after several efforts to restrain his impatience, exclaimed, 'You seem to have found a rarity.'

'A rarity indeed,' said Mr. Escot, cracking an egg as he spoke; 'no less than the genuine and indubitable skull of Cadwallader.'

'The skull of Cadwallader!' vociferated Mr. Cranium: 'O treasure of treasures!'

Mr. Escot then detailed by what means he had become possessed of it, which gave birth to various remarks from the other individuals of the party: after which, rising from table, and taking the skull again in his hand,

'This skull,' said he, 'is the skull of a hero, *παλαι κατατεθνειωτος*,* and sufficiently demonstrates a point, concerning which I never myself entertained a doubt, that the human race is undergoing a gradual process of diminution in length, breadth, and thickness. Observe this skull. Even the skull of our reverend friend, which is the largest and thickest in the company, is not more than half its size. The frame this skull belonged to could scarcely have been less than nine feet high. Such is the lamentable progress of degeneracy and decay. In the course of ages, a boot of the present generation would form an ample chateau for a large family of our remote posterity. The mind, too, participates in the contraction of the body. Poets and philosophers of all ages and nations have lamented this too visible process of physical and moral deterioration. "The sons of little men," says Ossian. '*Οιοι νυν βροτοι εισιν*, says Homer: "such men as live in these degenerate days." "All things," says Virgil,† "have a retrocessive tendency, and grow worse and worse by the inevitable doom of fate." "We live in the ninth age," says Juvenal,‡ "an age worse than the age of iron; nature has no metal sufficiently pernicious to give a denomination to its wickedness." "Our fathers," says Horace,§ "worse than our grandfathers, have given birth to us, their more vicious

* Long since dead. † Georg. I. 199. ‡ Sat. XIII. 28. § Carm. III. 6. 46.

progeny, who, in our turn, shall become the parents of a still viler generation." You all know the fable of the buried Pict, who bit off the end of a pickaxe, with which sacrilegious hands were breaking open his grave, and called out with a voice like subterranean thunder, *I perceive the degeneracy of your race by the smallness of your little finger!* videlicet, the pickaxe. This, to be sure, is a fiction; but it shows the prevalent opinion, the feeling, the conviction, of absolute, universal, irremediable deterioration.'

'I should be sorry,' said Mr. Foster, 'that such an opinion should become universal, independently of my conviction of its fallacy. Its general admission would tend, in a great measure, to produce the very evils it appears to lament. What could be its effect, but to check the ardour of investigation, to extinguish the zeal of philanthropy, to freeze the current of enterprising hope, to bury in the torpor of scepticism and in the stagnation of despair, every better faculty of the human mind, which will necessarily become retrograde in ceasing to be progressive?'

'I am inclined to think, on the contrary,' said Mr. Escot, 'that the deterioration of man is accelerated by his blindness—in many respects wilful blindness—to the truth of the fact itself, and to the causes which produce it; that there is no hope whatever of ameliorating his condition but in a total and radical change of the whole scheme of human life, and that the advocates of his indefinite perfectibility are in reality the greatest enemies to the practical possibility of their own system, by so strenuously labouring to impress on his attention that he is going on in a good way, while he is really in a deplorably bad one.'

'I admit,' said Mr. Foster, 'there are many things that may, and therefore will, be changed for the better.'

'Not on the present system,' said Mr. Escot, 'in which every change is for the worse.'

'In matters of taste I am sure it is,' said Mr. Gall: 'there is, in fact, no such thing as good taste left in the world.'

'O, Mr. Gall!' said Miss Philomela Poppyseed, 'I thought my novel——'

'My paintings,' said Sir Patrick O'Prism——

'My ode,' said Mr. Mac Laurel——

'My ballad,' said Mr. Nightshade——

'My plan for Lord Littlebrain's park,' said Marmaduke Milestone, Esquire——

'My essay,' said Mr. Treacle——

'My sonata,' said Mr. Chromatic——

'My claret,' said Squire Headlong——

'My lectures,' said Mr. Cranium——

'Vanity of vanities,' said the Reverend Doctor Gaster, turning down an empty egg-shell; 'all is vanity and vexation of spirit.'

CHAPTER XI

THE ANNIVERSARY

AMONG the *dies albâ cretâ notandos*, which the beau monde of the Cambrian mountains was in the habit of remembering with the greatest pleasure, and anticipating with the most lively satisfaction, was the Christmas ball which the ancient family of the Headlongs had been accustomed to give from time immemorial. Tradition attributed the honour of its foundation to Headlong Ap-Headlong Ap-Breakneck Ap-Headlong Ap-Cataract Ap-Pistyll* Ap-Rhaidr Ap-Headlong, who lived about the time of the Trojan war. Certain it is, at least, that a grand chorus was always sung after supper in honour of this illustrious ancestor of the Squire. This ball was, indeed, an æra in the lives of all the beauty and fashion of Caenarvon, Meirionnydd, and Anglesea, and, like the Greek Olympiads and the Roman consulates, served as the main pillar of memory, round which all the events of the year were suspended and entwined. Thus, in recalling to mind any circumstance imperfectly recollected, the principal point to be ascertained was, whether it had ocurred in the year of the first, second, third, or fourth ball of Headlong Ap-Breakneck, or Headlong Ap-Torrent, or Headlong Ap-Hurricane; and, this being satisfactorily established, the

* Pistyll, in Welsh, signifies a cataract, and Rhaidr a cascade.

remainder followed of course in the natural order of its ancient association.

This eventful anniversary being arrived, every chariot, coach, barouche, and barouchette, landau and landaulet, chaise, curricle, buggy, whiskey, and tilbury, of the three counties, was in motion: not a horse was left idle within five miles of any gentleman's seat, from the high-mettled hunter to the heath-cropping galloway. The ferrymen of the Menai were at their stations before day-break, taking a double allowance of rum and *cwrw* to strengthen them for the fatigues of the day. The ivied towers of Caernarvon, the romantic woods of Tan-y-bwlch, the heathy hills of Kernioggau, the sandy shores of Tremadoc, the mountain recesses of Bedd-Gelert, and the lonely lakes of Capel Cerig, re-echoed to the voices of the delighted ostlers and postillions, who reaped on this happy day their wintry harvest. Landlords and landladies, waiters, chambermaids, and toll-gate keepers, roused themselves from the torpidity which the last solitary tourist, flying with the yellow leaves on the wings of the autumnal wind, had left them to enjoy till the returning spring: the bustle of August was renewed on all the mountain roads, and, in the meanwhile, Squire Headlong and his little fat butler carried most energetically into effect the lessons of the *savant* in the Court of Quintessence, *qui par engin mirificque jectoit les maisons par les fenestres.**

It was the custom for the guests to assemble at dinner on the day of the ball, and depart on the following morning after breakfast. Sleep during this interval was out of the question: the ancient harp of Cambria suspended the celebration of the noble race of Shenkin, and the songs of Hoel and Cyveilioc, to ring to the profaner but more lively modulation of *Voulez vous danser, Mademoiselle*? in conjunction with the symphonious scraping of fiddles, the tinkling of triangles, and the beating of tambourines. Comus and Momus were the deities of the night; and Bacchus of course was not forgotten by the male part of the assembly (with them, indeed, a ball was invariably a scene of '*tipsy dance and jollity*'): the servants flew about with wine and negus, and the

* Rabelais.

little butler was indefatigable with his corkscrew, which is reported on one occasion to have grown so hot under the influence of perpetual friction that it actually set fire to the cork.

The company assembled. The dinner, which on this occasion was a secondary object, was despatched with uncommon celerity. When the cloth was removed, and the bottle had taken its first round, Mr. Cranium stood up and addressed the company.

'Ladies and gentlemen,' said he, 'the golden key of mental phænomena, which has lain buried for ages in the deepest vein of the mine of physiological research, is now, by a happy combination of practical and speculative investigations, grasped, if I may so express myself, firmly and inexcussibly, in the hands of physiognomical empiricism.' The Cambrian visitors listened with profound attention, not comprehending a single syllable he said, but concluding he would finish his speech by proposing the health of Squire Headlong. The gentlemen accordingly tossed off their heel-taps, and Mr. Cranium proceeded: 'Ardently desirous, to the extent of my feeble capacity, of disseminating, as much as possible, the inexhaustible treasures to which this golden key admits the humblest votary of philosophical truth, I invite you, when you have sufficiently restored, replenished, refreshed, and exhilarated that osteosarchæmatosplanchnochondroneuromuelous, or to employ a more intelligible term, osseocarnisanguineoviscericartilaginonervomedullary, *compages*, or shell, the body, which at once envelopes and developes that mysterious and inestimable kernel, the desiderative, determinative, ratiocinative, imaginative, inquisitive, appetitive, comparative, reminiscent, congeries of ideas and notions, simple and compound, comprised in the comprehensive denomination of mind, to take a peep with me into the mechanical arcana of the anatomicometaphysical universe. Being not in the least dubitative of your spontaneous compliance, I proceed,' added he, suddenly changing his tone, 'to get every thing ready in the library.' Saying these words, he vanished.

The Welsh squires now imagined they had caught a

glimpse of his meaning, and set him down in their minds for a sort of gentleman conjuror, who intended to amuse them before the ball with some tricks of legerdemain. Under this impression, they became very impatient to follow him, as they had made up their minds not to be drunk before supper. The ladies, too, were extremely curious to witness an exhibition which had been announced in so singular a preamble; and the Squire, having previously insisted on every gentleman tossing off a half-pint bumper, adjourned the whole party to the library, where they were not a little surprised to discover Mr. Cranium seated, in a pensive attitude, at a large table, decorated with a copious variety of skulls.

Some of the ladies were so much shocked at this extraordinary display, that a scene of great confusion ensued. Fans were very actively exercised, and water was strenuously called for by some of the most officious of the gentlemen; on which the little butler entered with a large allowance of liquid, which bore, indeed, the name of *water*, but was in reality a very powerful spirit. This was the only species of water which the little butler had ever heard called for in Headlong Hall. The mistake was not attended with any evil effects: for the fluid was no sooner applied to the lips of the fainting fair ones, than it resuscitated them with an expedition truly miraculous.

Order was at length restored; the audience took their seats; and the craniological orator held forth in the following terms:—

CHAPTER XII

THE LECTURE

'PHYSIOLOGISTS have been much puzzled to account for the varieties of moral character in men, as well as for the remarkable similarity of habit and disposition in all the individual animals of every other respective species. A few brief sentences, perspicuously worded, and scientifically arranged, will enumerate all the characteristics of a lion, or a tiger, or a wolf, or a bear, or a squirrel, or a goat, or a horse, or an ass, or a rat, or a cat, or a hog, or a dog; and whatever is physiologically predicated of any individual lion, tiger, wolf, bear, squirrel, goat, horse, ass, hog, or dog, will be found to hold true of all lions, tigers, wolves, bears, squirrels, goats, horses, asses, hogs, and dogs, whatsoever. Now, in man, the very reverse of this appears to be the case; for he has so few distinct and characteristic marks which hold true of all his species, that philosophers in all ages have found it a task of infinite difficulty to give him a definition. Hence one has defined him to be a *featherless biped*, a definition which is equally applicable to an unfledged fowl: another, to be *an animal which forms opinions*, than which nothing can be more inaccurate, for a very small number of the species form opinions, and the remainder take them upon trust, without investigation or inquiry.

'Again, man has been defined to be *an animal that carries a stick*: an attribute which undoubtedly belongs to man only, but not to all men always; though it uniformly characterises some of the graver and more imposing varieties, such as physicians, oran-outangs, and lords in waiting.

'We cannot define man to be a reasoning animal, for we do not dispute that idiots are men; to say nothing of that very numerous description of persons who consider themselves reasoning animals, and are so denominated by the ironical courtesy of the world, who labour, nevertheless, under a very gross delusion in that essential particular.

'It appears to me, that man may be correctly defined an animal, which, without any peculiar or distinguishing faculty of its own, is, as it were, a bundle or compound of faculties of other animals, by a distinct enumeration of which any individual of the species may be satisfactorily described. This is manifest, even in the ordinary language of conversation, when, in summing up, for example, the qualities of an accomplished courtier, we say he has the vanity of a peacock, the cunning of a fox, the treachery of an hyæna, the cold-heartedness of a cat, and the servility of a jackall. That this is perfectly consentaneous to scientific truth, will appear in the further progress of these observations.

'Every particular faculty of the mind has its corresponding organ in the brain. In proportion as any particular faculty or propensity acquires paramount activity in any individual, these organs develope themselves, and their developement becomes externally obvious by corresponding lumps and bumps, exuberances and protuberances, on the osseous compages of the occiput and sinciput. In all animals but man, the same organ is equally developed in every individual of the species: for instance, that of migration in the swallow, that of destruction in the tiger, that of architecture in the beaver, and that of parental affection in the bear. The human brain, however, consists, as I have said, of a bundle or compound of all the faculties of all other animals; and from the greater developement of one or more of these, in the infinite varieties of combination, result all the peculiarities of individual character.

'Here is the skull of a beaver, and that of Sir Christopher Wren. You observe, in both these specimens, the prodigious developement of the organ of constructiveness.

'Here is the skull of a bullfinch, and that of an eminent fiddler. You may compare the organ of music.

'Here is the skull of a tiger. You observe the organ of carnage. Here is the skull of a fox. You observe the organ of plunder. Here is the skull of a peacock. You observe the organ of vanity. Here is the skull of an illustrious robber, who, after a long and triumphant process of depredation and murder, was suddenly checked in his career by means of a certain quality inherent in preparations of hemp, which, for the sake of perspicuity, I shall call *suspensiveness*. Here is the skull of a conqueror, who, after over-running several kingdoms, burning a number of cities, and causing the deaths of two or three millions of men, women, and children, was entombed with all the pageantry of public lamentation, and figured as the hero of several thousand odes and a round dozen of epics; while the poor highwayman was twice executed—

> "At the gallows first, and after in a ballad,
> Sung to a villanous tune."

You observe, in both these skulls, the combined developement of the organs of carnage, plunder, and vanity, which I have separately pointed out in the tiger, the fox, and the peacock. The greater enlargement of the organ of vanity in the hero is the only criterion by which I can distinguish them from each other. Born with the same faculties, and the same propensities, these two men were formed by nature to run the same career: the different combinations of external circumstances decided the differences of their destinies.

'Here is the skull of a Newfoundland dog. You observe the organ of benevolence, and that of attachment. Here is a human skull, in which you may observe a very striking negation of both these organs; and an equally striking development of those of destruction, cunning, avarice, and self-love. This was one of the most illustrious statesmen that ever flourished in the page of history.

'Here is the skull of a turnspit, which, after a wretched life of *dirty work*, was turned out of doors to die on a dunghill. I have been induced to preserve it, in consequence of its remarkable similarity to this, which belonged to a courtly poet, who, having grown grey in flattering the great, was cast off in the same manner to perish by the same catastrophe.'

After these, and several other illustrations, during which the skulls were handed round for the inspection of the company, Mr. Cranium proceeded thus:—

'It is obvious, from what I have said, that no man can hope for worldly honour or advancement, who is not placed in such a relation to external circumstances as may be consentaneous to his peculiar cerebral organs; and I would advise every parent, who has the welfare of his son at heart, to procure as extensive a collection as possible of the skulls of animals, and, before determining on the choice of a profession, to compare with the utmost nicety their bumps and protuberances with those of the skull of his son. If the developement of the organ of destruction point out a similarity between the youth and the tiger, let him be brought to some profession (whether that of a butcher, a soldier, or a physician, may be regulated by circumstances) in which he may be furnished with a licence to kill: as, without such licence, the indulgence of his natural propensity may lead to the untimely rescission of his vital thread, "with edge of penny cord and vile reproach." If he show an analogy with the jackall, let all possible influence be used to procure him a place at court, where he will infallibly thrive. If his skull bear a marked resemblance to that of a magpie, it cannot be doubted that he will prove an admirable lawyer; and if with this advantageous conformation be combined any similitude to that of an owl, very confident hopes may be formed of his becoming a judge.'

A furious flourish of music was now heard from the ball-room, the Squire having secretly despatched the little butler to order it to strike up, by way of a hint to Mr Cranium to finish his harangue. The company took the hint and adjourned tumultuously, having just understood as much of the lecture as furnished them with amusement for the ensuing twelvemonth, in feeling the skulls of all their acquaintance.

CHAPTER XIII

THE BALL

THE ball-room was adorned with great taste and elegance, under the direction of Miss Caprioletta and her friend Miss Cephalis, who were themselves its most beautiful ornaments, even though romantic Meirion, the pre-eminent in loveliness, sent many of its loveliest daughters to grace the festive scene. Numberless were the solicitations of the dazzled swains of Cambria for the honour of the two first dances with the one or the other of these fascinating friends; but little availed, on this occasion, the pedigree lineally traced from Caractacus or King Arthur: their two philosophical lovers, neither of whom could have given the least account of his great-great-grandfather, had engaged them many days before. Mr. Panscope chafed and fretted like Llugwy in his bed of rocks, when the object of his adoration stood up with his rival: but he consoled himself with a lively damsel from the vale of Edeirnion, having first compelled Miss Cephalis to promise him her hand for the fourth set.

The ball was accordingly opened by Miss Caprioletta and Mr. Foster, which gave rise to much speculation among the Welsh gentry, as to who this Mr. Foster could be; some of the more learned among them secretly resolving to investigate most profoundly the antiquity of the name of Foster, and ascertain what right a person so denominated could have to open the most illustrious of all possible balls with the lovely Caprioletta Headlong, the only sister of Harry Headlong, Esquire, of Headlong Hall, in the Vale of Llanberris, the only surviving male representative of the antediluvian family of Headlong Ap-Rhaiader.

When the two first dances were ended, Mr. Escot, who did not choose to dance with any one but his adorable Cephalis, looking round for a convenient seat, discovered Mr. Jenkison in a corner by the side of the Reverend Doctor Gaster, who was keeping excellent time with his nose to the lively melody of the harp and fiddle. Mr. Escot seated himself by the side of Mr. Jenkison, and inquired if he took no part in the amusement of the night?

MR. JENKISON

No. The universal cheerfulness of the company induces me to rise; the trouble of such violent exercise induces me to sit still. Did I see a young lady in want of a partner, gallantry would incite me to offer myself as her devoted knight for half an hour: but, as I perceive there are enough without me, that motive is null. I have been weighing these points *pro* and *con*, and remain *in statu quo*.

MR. ESCOT

I have danced, contrary to my system, as I have done many other things since I have been here, from a motive that you will easily guess. (*Mr. Jenkison smiled.*) I have great objections to dancing. The wild and original man is a calm and contemplative animal. The stings of natural appetite alone rouse him to action. He satisfies his hunger with roots and fruits, unvitiated by the malignant adhibition of fire, and all its diabolical processes of elixion and assation: he slakes his thirst in the mountain-stream, *συμμισγεται τη επιτυχουση*, and returns to his peaceful state of meditative repose.

MR. JENKISON

Like the metaphysical statue of Condillac.

MR. ESCOT

With all its senses and purely natural faculties developed, certainly. Imagine this tranquil and passionless being, occupied in his first meditation on the simple question of *Where am I? Whence do I come? And what is the end of my existence?* Then suddenly place before him a chandelier, a fiddler, and a magnificent beau in silk stockings and pumps, bounding, skipping, swinging, capering, and throwing himself into ten thousand attitudes, till his face glows with fever, and distils with perspiration: the first impulse excited in his mind by such an apparition will be that of violent fear, which, by the reiterated perception of its harmlessness, will subside into simple astonishment. Then let any genius, sufficiently powerful to impress on his mind all the terms of the communication, impart to him, that after a long process of ages, when his race shall have attained what some people think proper to denominate a very advanced stage of perfectibility, the most favoured and distinguished of the community shall meet by hundreds, to grin, and labour, and gesticulate, like the phantasma before him, from sunset to sunrise, while all nature is at rest, and that they shall consider this a happy and pleasurable mode of existence, and furnishing the most delightful of all possible contrasts to what they will call his vegetative state: would be not groan from his inmost soul for the lamentable condition of his posterity?

MR. JENKISON

I know not what your wild and original man might think of the matter in the abstract; but comparatively, I conceive, he would be better pleased with the vision of such a scene as this, than with that of a party of Indians (who would have all the advantage of being nearly as wild as himself), dancing their infernal war-dance round a midnight fire in a North American forest.

MR. ESCOT

Not if you should impart to him the true nature of both, by laying open to his view the springs of action in both parties.

MR. JENKISON

To do this with effect, you must make him a profound metaphysician, and thus transfer him at once from his wild and original state to a very advanced stage of intellectual progression; whether that progression be towards good or evil, I leave you and our friend Foster to settle between you.

MR. ESCOT

I wish to make no change in his habits and feelings, but to give him, hypothetically, so much mental illumination, as will enable him to take a clear view of two distinct stages of the deterioration of his posterity, that he may be enabled to compare them with each other, and with his own more happy condition. The Indian, dancing round the midnight fire, is very far deteriorated; but the magnificent beau, dancing to the light of chandeliers, is infinitely more so. The Indian is a hunter: he makes great use of fire, and subsists almost entirely on animal food. The malevolent passions that spring from these pernicious habits involve him in perpetual war. He is, therefore, necessitated, for his own preservation, to keep all the energies of his nature in constant activity: to this end his midnight war-dance is very powerfully subservient, and, though in itself a frightful spectacle, is at least justifiable on the iron plea of necessity.

MR. JENKISON

On the same iron plea, the modern system of dancing is more justifiable. The Indian dances to prepare himself for killing his enemy: but while the beaux and belles of our assemblies dance, they are in the very act of killing theirs—

TIME!— a more inveterate and formidable foe than any the Indian has to contend with; for, however completely and ingeniously killed, he is sure to rise again, 'with twenty mortal murders on his crown,' leading his army of blue devils, with ennui in the van, and vapours in the rear.

MR. ESCOT

Your observation militates on my side of the question; and it is a strong argument in favour of the Indian, that he has no such enemy to kill.

MR. JENKISON

There is certainly a great deal to be said against dancing: there is also a great deal to be said in its favour. The first side of the question I leave for the present to you: on the latter, I may venture to allege that no amusement seems more natural and more congenial to youth than this. It has the advantage of bringing young persons of both sexes together, in a manner which its publicity renders perfectly unexceptionable, enabling them to see and know each other better than, perhaps, any other mode of general association. *Tête-à-têtes* are dangerous things. Small family parties are too much under mutual observation. A ball-room appears to me almost the only scene uniting that degree of rational and innocent liberty of intercourse, which it is desirable to promote as much as possible between young persons, with that scrupulous attention to the delicacy and propriety of female conduct, which I consider the fundamental basis of all our most valuable social relations.

MR. ESCOT

There would be some plausibility in your argument, if it were not the very essence of this species of intercourse to exhibit them to each other under false colours. Here all is show, and varnish, and hypocrisy, and coquetry; they dress up their moral character for the evening at the same toilet where they manufacture their shapes and faces. Ill-temper lies buried under a studied accumulation of smiles. Envy, hatred, and malice, retreat from the countenance, to entrench themselves more deeply in the heart. Treachery lurks under the flowers of courtesy. Ignorance and folly take refuge in that unmeaning gabble which it would be profanation to call language, and which even those whom long experience in 'the dreary intercourse of daily life' has screwed up to such a pitch of stoical endurance that they can listen to it by the hour, have branded with the ignominious appellation of '*small talk.*' Small indeed!—the absolute minimum of the infinitely little.

MR. JENKISON

Go on. I have said all I intended to say on the favourable side. I shall have great pleasure in hearing you balance the argument.

MR. ESCOT

I expect you to confess that I shall have more than balanced it. A ball-room is an epitome of all that is most worthless and unamiable in the great sphere of human life. Every petty and malignant passion is called into play. Coquetry is perpetually on the alert to captivate, caprice to mortify, and vanity to take offence. One amiable female is rendered miserable for the evening by seeing another, whom she intended to outshine, in a more attractive dress than her own; while the other omits no method of giving stings to her triumph, which she enjoys with all the secret arrogance of an oriental sultana. Another is compelled to dance with a *monster* she abhors. A third has set her heart on dancing with a particular partner, perhaps for the amiable motive of annoying one of her *dear friends*: not only he does not ask her, but she sees him dancing with that identical *dear friend*, whom from that moment she hates more cordially than ever. Perhaps, what is worse than all, she has set her heart on refusing some impertinent fop, who does not give her the opportunity.—As to the men, the case is very nearly the same with them. To be sure, they have the privilege of making the first advances, and are, therefore, less liable to have an odious partner forced upon them; though this sometimes happens, as I know by woful experience: but it is seldom they can procure the very partner they prefer; and when they do, the absurd necessity of changing every two dances forces them away, and leaves them only the miserable alternative of taking up with something disagreeable perhaps in itself, and at all events rendered so by contrast, or of retreating into some solitary corner, to vent their spleen on the first idle coxcomb they can find.

MR. JENKISON

I hope that is not the motive which brings you to me.

MR. ESCOT

Clearly not. But the most afflicting consideration of all is, that these malignant and miserable feelings are masked under that uniform disguise of pretended benevolence, *that fine and delicate irony, called politeness, which gives so much ease and pliability to the mutual intercourse of civilised man, and enables him to assume the appearance of every virtue, without the reality of one.**

The second set of dances was now terminated, and Mr. Escot flew off to reclaim the hand of the beautiful Cephalis, with whom he figured away with surprising alacrity, and probably felt at least as happy among the chandeliers and silk stockings, at which he had just been railing, as he would have been in an American forest, making one in an Indian ring, by the light of a blazing fire, even though his hand had been locked in that of the most beautiful *squaw* that ever listened to the roar of Niagara.

Squire Headlong was now beset by his maiden aunt, Miss Brindle-mew Grimalkin Phœbe Tabitha Ap-Headlong, on one side, and Sir Patrick O'Prism on the other; the former insisting that he should immediately procure her a partner; the latter earnestly requesting the same interference in behalf of Miss Philomela Poppyseed. The Squire thought to emancipate himself from his two petitioners by making them dance with each other; but Sir Patrick vehemently pleading a prior engagement, the Squire threw his eyes around till they alighted on Mr. Jenkison and the Reverend Doctor Gaster; both of whom, after waking the latter, he pressed

* Rousseau, Discours sur les Sciences.

into the service. The Doctor, arising with a strange kind of guttural sound, which was half a yawn and half a groan, was handed by the officious Squire to Miss Philomela, who received him with sullen dignity: she had not yet forgotten his falling asleep during the first chapter of her novel, while she was condescending to detail to him the outlines of four superlative volumes. The Doctor, on his part, had most completely forgotten it; and though he thought there was something in her physiognomy rather more forbidding than usual, he gave himself no concern about the cause, and had not the least suspicion that it was at all connected with himself. Miss Brindle-mew was very well contented with Mr. Jenkison, and gave him two or three ogles, accompanied by a most risible distortion of the countenance which she intended for a captivating smile. As to Mr. Jenkison, it was all one to him with whom he danced, or whether he danced or not: he was therefore just as well pleased as if he had been left alone in his corner; which is probably more than could have been said of any other human being under similar circumstances.

At the end of the third set, supper was announced; and the party, pairing off like turtles, adjourned to the supper-room. The Squire was now the happiest of mortal men and the little butler the most laborious. The centre of the largest table was decorated with a model of Snowdon, surmounted with an enormous artificial leek, the leaves of angelica, and the bulb of blanc-mange. A little way from the summit was a tarn, or mountain-pool, supplied through concealed tubes with an inexhaustible flow of milk-punch, which, dashing in cascades down the miniature rocks, fell into the more capacious lake below, washing the mimic foundations of Headlong Hall. The Reverend Doctor handed Miss Philomela to the chair most conveniently situated for enjoying this interesting scene, protesting he had never before been sufficiently impressed with the magnificence of that mountain, which he now perceived to be well worthy of all the fame it had obtained.

'Now, when they had eaten and were satisfied,' Squire Headlong called on Mr. Chromatic for a song; who, with the assistance of his two accomplished daughters, regaled the ears of the company with the following

TERZETTO*

Grey Twilight, from her shadowy hill,
Discolours Nature's vernal bloom,
And sheds on grove, and field, and rill,
One placid tint of deepening gloom.

The sailor sighs 'mid shoreless seas,
Touched by the thought of friends afar,
As, fanned by ocean's flowing breeze,
He gazes on the western star.

The wanderer hears, in pensive dream,
The accents of the last farewell,
As, pausing by the mountain stream,
He listens to the evening bell.

This terzetto was of course much applauded; Mr. Milestone observing, that he thought the figure in the last verse would have been more picturesque, if it had been represented with its arms folded and its back against a tree; or leaning on its staff, with a cockle-shell in its hat, like a pilgrim of ancient times.

Mr. Chromatic professed himself astonished that a gentleman of genuine modern taste, like Mr. Milestone, should consider the words of a song of any consequence whatever, seeing that they were at the best only a species of pegs, for the more convenient suspension of crochets and quavers. This remark drew on him a very severe reprimand from Mr. Mac Laurel, who said to him, 'Dinna ye ken, sir, that soond is a thing utterly worthless in itsel, and only effectual in agreeable excitements, as far as it is an aicho to sense? Is there ony soond mair meeserable an' peetifu' than the scrape o' a feddle, when it does na touch ony chord i' the human sensorium? Is there ony mair divine than the deep note o' a bagpipe, when it breathes the auncient meelodies o' leeberty an' love? It is true, there are peculiar trains o' feeling an' sentiment, which parteecular combinations o' meelody are calculated to excite; an' sae far music can produce its effect without words: but it does na follow, that, when ye put

* Imitated from a passage in the Purgatorio of Dante.

words to it, it becomes a matter of indefference what they are; for a gude strain of impassioned poetry will greatly increase the effect, and a tessue o' nonsensical doggrel will destroy it a' thegither. Noo, as gude poetry can produce its effect without music, sae will gude music without poetry; and as gude music will be mair pooerfu' by itsel' than wi' bad poetry, sae will gude poetry than wi' bad music: but, when ye put gude music an' gude poetry thegither, ye produce the divinest compound o' sentimental harmony that can possibly find its way through the lug to the saul.'

Mr. Chromatic admitted that there was much justice in these observations, but still maintained the subserviency of poetry to music. Mr. Mac Laurel as strenuously maintained the contrary; and a furious war of words was proceeding to perilous lengths, when the Squire interposed his authority towards the reproduction of peace, which was forthwith concluded, and all animosities drowned in a libation of milk-punch, the Reverend Doctor Gaster officiating as high priest on the occasion.

Mr. Chromatic now requested Miss Caprioletta to favour the company with an air. The young lady immediately complied, and sung the following simple

BALLAD

'O Mary, my sister, thy sorrow give o'er,
I soon shall return, girl, and leave thee no more:
But with children so fair, and a husband so kind,
I shall feel less regret when I leave thee behind.

'I have made thee a bench for the door of thy cot,
And more would I give thee, but more I have not:
Sit and think of me there, in the warm summer day,
And give me three kisses, my labour to pay.'

She gave him three kisses, and forth did he fare,
And long did he wander, and no one knew where;
And long from her cottage, through sunshine and rain,
She watched his return, but he came not again.

Her children grew up, and her husband grew grey;
She sate on the bench through the long summer day:
One evening, when twilight was deep on the shore,
There came an old soldier, and stood by the door.

In English he spoke, and none knew what he said,
But her oatcake and milk on the table she spread;
Then he sate to his supper, and blithely he sung,
And she knew the dear sounds of her own native tongue:

'O rich are the feasts in the Englishman's hall,
And the wine sparkles bright in the goblets of Gaul:
But their mingled attractions I well could withstand,
For the milk and the oatcake of Meirion's dear land.'

'And art thou a Welchman, old soldier?' she cried.
'Many years have I wandered,' the stranger replied:
''Twixt Danube and Thames many rivers there be,
But the bright waves of Cynfael are fairest to me.

'I felled the grey oak, ere I hastened to roam,
And I fashioned a bench for the door of my home;
And well my dear sister my labour repaid,
Who gave me three kisses when first it was made.

'In the old English soldier thy brother appears:
Here is gold in abundance, the saving of years:
Give me oatcake and milk in return for my store,
And a seat by thy side on the bench at the door.'

Various other songs succeeded, which, as we are not composing a song book, we shall lay aside for the present.

An old squire, who had not missed one of these anniversaries, during more than half a century, now stood up, and filling a half-pint bumper, pronounced, with a stentorian voice—'To the immortal memory of Headlong Ap-Rhaiader, and to the health of his noble descendant and worthy representative!' This example was followed by all the gentlemen present. The harp struck up a triumphal strain; and, the old squire already mentioned vociferating the first stave, they sang, or rather roared, the following

CHORUS

Hail to the Headlong! the Headlong Ap-Headlong!
All hail to the Headlong, the Headlong Ap-Headlong!
  The Headlong Ap-Headlong
  Ap-Breakneck Ap-Headlong
Ap-Cataract Ap-Pistyll Ap-Rhaiader Ap-Headlong!

The bright bowl we steep in the name of the Headlong:
Let the youths pledge it deep to the Headlong Ap-Headlong,
  And the rosy-lipped lasses
  Touch the brim as it passes,
And kiss the red tide for the Headlong Ap-Headlong!

The loud harp resounds in the hall of the Headlong:
The light step rebounds in the hall of the Headlong:
  Where shall music invite us,
  Or beauty delight us,
If not in the hall of the Headlong Ap-Headlong?

Huzza! to the health of the Headlong Ap-Headlong!
Fill the bowl, fill in floods, to the health of the Headlong!
  Till the stream ruby-glowing,
  On all sides o'erflowing,
Shall fall in cascades to the health of the Headlong!
  The Headlong Ap-Headlong
  Ap-Breakneck Ap-Headlong
Ap-Cataract Ap-Pistyll Ap-Rhaiader Ap-Headlong!

Squire Headlong returned thanks with an appropriate libation, and the company re-adjourned to the ball-room, where they kept it up till sun-rise, when the little butler summoned them to breakfast.

CHAPTER XIV

THE PROPOSALS

THE chorus, which celebrated the antiquity of her lineage, had been ringing all night in the ears of Miss Brindle-mew Grimalkin Phœbe Tabitha Ap-Headlong, when, taking the Squire aside, while the visitors were sipping their tea and coffee, 'Nephew Harry,' said she, 'I have been noting your behaviour, during the several stages of the ball and supper; and, though I cannot tax you with any want of gallantry, for you are a very gallant young man, nephew Harry, very gallant—I wish I could say as much for every one' (added she, throwing a spiteful look towards a distant corner, where

Mr. Jenkison was sitting with great *nonchalance*, and at the moment dipping a rusk in a cup of chocolate); 'but I lament to perceive that you were at least as pleased with your lakes of milk-punch, and your bottles of Champagne and Burgundy, as with any of your delightful partners. Now, though I can readily excuse this degree of incombustibility in the descendant of a family so remarkable in all ages for personal beauty as ours, yet I lament it exceedingly, when I consider that, in conjunction with your present predilection for the easy life of a bachelor, it may possibly prove the means of causing our ancient genealogical tree, which has its roots, if I may so speak, in the foundations of the world, to terminate suddenly in a point: unless you feel yourself moved by my exhortations to follow the example of all your ancestors, by choosing yourself a fitting and suitable helpmate to immortalise the pedigree of Headlong Ap-Rhaiader.'

'Egad!' said Squire Headlong, 'that is very true. I'll marry directly. A good opportunity to fix on some one, now they are all here; and I'll pop the question without further ceremony.'

'What think you,' said the old lady, 'of Miss Nanny Glyn-Du, the lineal descendant of Llewelyn Ap Yorwerth?'

'She won't do,' said Squire Headlong.

'What say you, then,' said the lady, 'to Miss Williams, of Pontyglasrhydyrallt, the descendant of the ancient family of——?'

'I don't like her,' said Squire Headlong; 'and as to her ancient family, that is a matter of no consequence. I have antiquity enough for two. They are all moderns, people of yesterday, in comparison with us. What signify six or seven centuries, which are the most they can make up?'

'Why, to be sure,' said the aunt, 'on that view of the question, it is of no consequence. What think you, then, of Miss Owen, of Nidd-y-Gygfraen? She will have six thousand a year.'

'I would not have her,' said Squire Headlong, 'if she had fifty. I'll think of somebody presently. I should like to be married on the same day with Caprioletta.'

'Caprioletta!' said Miss Brindle-mew; 'without my being consulted!'

'Consulted!' said the Squire: 'I was commissioned to tell you, but somehow or other I let it slip. However, she is going to be married to my friend Mr. Foster, the philosopher.'

'Oh!' said the maiden aunt, 'that a daughter of our ancient family should marry a philosopher! It is enough to make the bones of all the Ap-Rhaiaders turn in their graves!'

'I happen to be more enlightened,' said Squire Headlong, 'than any of my ancestors were. Besides, it is Caprioletta's affair, not mine. I tell you, the matter is settled, fixed, determined; and so am I, to be married on the same day. I don't know, now I think of it, whom I can choose better than one of the daughters of my friend Chromatic.'

'A Saxon!' said the aunt, turning up her nose, and was commencing a vehement remonstrance; but the Squire, exclaiming 'Music has charms!' flew over to Mr. Chromatic, and, with a hearty slap on the shoulder, asked him 'how he should like him for a son-in-law?' Mr. Chromatic, rubbing his shoulder, and highly delighted with the proposal, answered, 'Very much indeed:' but, proceeding to ascertain which of his daughters had captivated the Squire, the Squire demurred, and was unable to satisfy his curiosity. 'I hope,' said Mr. Chromatic, 'it may be Tenorina; for I imagine Graziosa has conceived a *penchant* for Sir Patrick O'Prism.'—'Tenorina, exactly,' said Squire Headlong; and became so impatient to bring the matter to a conclusion, that Mr. Chromatic undertook to communicate with his daughter immediately. The young lady proved to be as ready as the Squire, and the preliminaries were arranged in little more than five minutes.

Mr. Chromatic's words, that he imagined his daughter Graziosa had conceived a *penchant* for Sir Patrick O'Prism, were not lost on the Squire, who at once determind to have as many companions in the scrape as possible, and who, as soon as he could tear himself from Mrs. Headlong elect, took three flying bounds across the room to the Baronet, and said, 'So, Sir Patrick, I find you and I are going to be married?'

'Are we?' said Sir Patrick: 'then sure won't I wish you joy, and myself too? for this is the first I have heard of it.'

'Well,' said Squire Headlong, 'I have made up my mind to it, and you must not disappoint me.'

'To be sure I won't, if I can help it,' said Sir Patrick; 'and I am very much obliged to you for taking so much trouble off my hands. And pray, now, who is it that I am to be metamorphosing into Lady O'Prism?'

'Miss Graziosa Chromatic,' said the Squire.

'Och violet and vermilion!' said Sir Patrick; 'though I never thought of it before, I dare say she will suit me as well as another: but then you must persuade the ould Orpheus to draw out a few *notes* of rather a more magical description than those he is so fond of scraping on his crazy violin.'

'To be sure he shall,' said the Squire; and, immediately returning to Mr. Chromatic, concluded the negotiation for Sir Patrick as expeditiously as he had done for himself.

The Squire next addressed himself to Mr. Escot: 'Here are three couple of us going to throw off together, with the Reverend Doctor Gaster for whipper-in: now, I think you cannot do better than make the fourth with Miss Cephalis; and then, as my father-in-law that is to be would say, we shall compose a very harmonious octave.'

'Indeed,' said Mr. Escot, 'nothing would be more agreeable to both of us than such an arrangement: but the old gentleman, since I first knew him, has changed, like the rest of the world, very lamentably for the worse: now, we wish to bring him to reason, if possible, though we mean to dispense with his consent, if he should prove much longer refractory.'

'I'll settle him,' said Squire Headlong; and immediately posted up to Mr. Cranium, informing him that four marriages were about to take place by way of a merry winding up of the Christmas festivities.

'Indeed!' said Mr. Cranium; 'and who are the parties?'

'In the first place,' said the Squire, 'my sister and Mr. Foster: in the second, Miss Graziosa Chromatic and Sir Patrick O'Prism: in the third, Miss Tenorina Chromatic and your humble servant: and in the fourth—to which, by the by, your consent is wanted——'

'Oho!' said Mr. Cranium.

'Your daughter,' said Squire Headlong.

'And Mr. Panscope?' said Mr. Cranium.

'And Mr. Escot,' said Squire Headlong. 'What would you have better? He has ten thousand virtues.'

'So has Mr. Panscope,' said Mr. Cranium; 'he has ten thousand a year.'

'Virtues?' said Squire Headlong.

'Pounds,' said Mr. Cranium.

'I have set my mind on Mr. Escot,' said the Squire.

'I am much obliged to you,' said Mr. Cranium, 'for dethroning me from my paternal authority.'

'Who fished you out of the water?' said Squire Headlong.

'What is that to the purpose?' said Mr. Cranium. 'The whole process of the action was mechanical and necessary. The application of the poker necessitated the ignition of the powder: the ignition necessitated the explosion: the explosion necessitated my sudden fright, which necessitated my sudden jump, which, from a necessity equally powerful, was in a curvilinear ascent: the descent, being in a corresponding curve, and commencing at a point perpendicular to the extreme line of the edge of the tower, I was, by the necessity of gravitation, attracted, first, through the ivy, and secondly through the hazel, and thirdly through the ash, into the water beneath. The motive or impulse thus adhibited in the person of a drowning man, was as powerful on his material compages as the force of gravitation on mine; and he could no more help jumping into the water than I could help falling into it.'

'All perfectly true,' said Squire Headlong; 'and, on the same principle, you make no distinction between the man who knocks you down and him who picks you up.'

'I make this distinction,' said Mr. Cranium, 'that I avoid the former as a machine containing a peculiar *cataballitive* quality, which I have found to be not consentaneous to my mode of pleasurable existencc; but I attach no moral merit or demerit to either of them, as these terms are usually employed, seeing that they are equally creatures of necessity, and must act as they do from the nature of their organisation. I no more blame or praise a man for what is called vice or virtue, than I tax a tuft of hemlock with malevolence, or discover great philanthropy in a field of potatoes, seeing that the men and the plants are equally incapacitated, by their original internal organisation, and the combinations and modifications of external circumstances, from being any thing but what they are. *Quod victus fateare necesse est.*'

'Yet you destroy the hemlock,' said Squire Headlong, 'and cultivate the potatoe: that is my way, at least.'

'I do,' said Mr. Cranium; 'because I know that the farinaceous qualities of the potatoe will tend to preserve the great requisites of unity and coalescence in the various constituent portions of my animal republic; and that the hemlock, if gathered by mistake for parsley, chopped up small with butter, and eaten with a boiled chicken, would necessitate a great derangement, and perhaps a total decomposition, of my corporeal mechanism.'

'Very well,' said the Squire; 'then you are necessitated to like Mr. Escot better than Mr. Panscope?'

'That is a *non sequitur*,' said Mr. Cranium.

'Then this is a *sequitur*,' said the Squire: 'your daughter and Mr. Escot are necessitated to love one another; and, unless you feel necessitated to adhibit your consent, they will feel necessitated to dispense with it; since it does not appear to moral and political economists to be essentially inherent in the eternal fitness of things.'

Mr. Cranium fell into a profound reverie: emerging from which, he said, looking Squire Headlong full in the face, 'Do you think Mr. Escot would give me that skull?'

'Skull!' said Squire Headlong.

'Yes,' said Mr. Cranium, 'the skull of Cadwallader.'

'To be sure he will,' said the Squire.

'Ascertain the point,' said Mr. Cranium.

'How can you doubt it?' said the Squire.

'I simply know,' said Mr. Cranium, 'that if it were once in my possession, I would not part with it for any acquisition on earth, much less for a wife. I have had one: and, as marriage has been compared to a pill, I can very safely assert that *one is a dose*; and my reason for thinking that he will not part with it is, that its extraordinary magnitude tends to support his system, as much as its very marked protuberances tend to support mine; and you know his own system is of all things the dearest to every man of liberal thinking and a philosophical tendency.'

The Squire flew over to Mr. Escot. 'I told you,' said he, 'I would settle him: but there is a very hard condition attached to his compliance.'

'I submit to it,' said Mr. Escot, 'be it what it may.'

'Nothing less,' said Squire Headlong, 'than the absolute and unconditional surrender of the skull of Cadwallader.'

'I resign it,' said Mr. Escot.

'The skull is yours,' said the Squire, skipping over to Mr. Cranium.

'I am perfectly satisfied,' said Mr. Cranium.

'The lady is yours,' said the Squire, skipping back to Mr. Escot.

'I am the happiest man alive,' said Mr. Escot.

'Come,' said the Squire, 'then there is an amelioration in the state of the sensitive man.'

'A slight oscillation of good in the instance of a solitary individual,' answered Mr. Escot, 'by no means affects the solidity of my opinions concerning the general deterioration of the civilised world; which when I can be induced to contemplate with feelings of satisfaction, I doubt not but that I may be persuaded *to be in love with tortures, and to think charitably of the rack*.'*

Saying these words, he flew off as nimbly as Squire Headlong himself, to impart the happy intelligence to his beautiful Cephalis.

Mr. Cranium now walked up to Mr. Panscope, to condole with him on the disappointment of their mutual hopes. Mr. Panscope begged him not to distress himself on the subject, observing, that the monotonous system of female education brought every individual of the sex to so remarkable an approximation of similarity, that no wise man would suffer himself to be annoyed by a loss so easily repaired; and that there was much truth, though not much elegance, in a

* Jeremy Taylor.

remark which he had heard made on a similar occasion by a post-captain of his acquaintance, 'that there never was a fish taken out of the sea, but left another as good behind.'

Mr. Cranium replied, that no two individuals having all the organs of the skull similarly developed, the universal resemblance of which Mr. Panscope had spoken could not possibly exist. Mr. Panscope rejoined; and a long discussion ensued, concerning the comparative influence of natural organisation and artificial education, in which the beautiful Cephalis was totally lost sight of, and which ended, as most controversies do, by each party continuing firm in his own opinion, and professing his profound astonishment at the blindness and prejudices of the other.

In the meanwhile, a great confusion had arisen at the outer doors, the departure of the ball-visitors being impeded by a circumstance which the experience of ages had discovered no means to obviate. The grooms, coachmen, and postillions, were all drunk. It was proposed that the gentlemen should officiate in their places: but the gentlemen were almost all in the same condition. This was a fearful dilemma: but a very diligent investigation brought to light a few servants and a few gentlemen not above *half-seas-over*; and by an equitable distribution of these rarities, the greater part of the guests were enabled to set forward, with very nearly an even chance of not having their necks broken before they reached home.

CHAPTER XV

THE CONCLUSION

THE Squire and his select party of philosophers and dilettanti were again left in peaceful possession of Headlong Hall: and, as the former made a point of never losing a moment in the accomplishment of a favourite object, he did not suffer many days to elapse, before the spiritual metamorphosis of eight into four was effected by the clerical dexterity of the Reverend Doctor Gaster.

Immediately after the ceremony, the whole party dispersed, the Squire having first extracted from every one of his chosen guests a positive promise to re-assemble in August, when they would be better enabled, in its most appropriate season, to form a correct judgment of Cambrian hospitality.

Mr. Jenkison shook hands at parting with his two brother philosophers. 'According to your respective systems,' said he, 'I ought to congratulate *you* on a change for the better, which I do most cordially: and to condole with *you* on a change for the worse, though, when I consider whom you have chosen, I should violate every principle of probability in doing so.'

'You will do well,' said Mr. Foster, 'to follow our example. The extensive circle of general philanthropy, which, in the present advanced stage of human nature, comprehends in its circumference the destinies of the whole species, originated, and still proceeds, from that narrower circle of domestic affection, which first set limits to the empire of selfishness, and, by purifying the passions and enlarging the affections of mankind, has given to the views of benevolence an increasing and illimitable expansion, which will finally diffuse happiness and peace over the whole surface of the world.'

'The affection,' said Mr. Escot, 'of two congenial spirits, united not by legal bondage and superstitious imposture, but by mutual confidence and reciprocal virtues, is the only counterbalancing consolation in this scene of mischief and misery. But how rarely is this the case according to the present system of marriage! So far from being a central point of expansion to the great circle of universal benevolence, it serves only to concentrate the feelings of natural sympathy in the reflected selfishness of family interest, and to substitute for the *humani nihil alienum puto* of youthful philanthropy, the *charity begins at home* of maturer years. And what accession of individual happiness is acquired by this oblivion of the general good? Luxury, despotism, and avarice have so seized and entangled nine hundred and ninety-nine out of every thousand of the human race, that the matrimonial compact, which ought to be the most easy, the most free, and the most simple of all engagements, is become the most slavish and complicated,—a mere question of finance,—a system of bargain, and barter, and commerce, and trick, and chicanery, and dissimulation, and fraud. Is there one instance in ten thousand, in which the buds of first affection are not most cruelly and hopelessly blasted, by avarice, or ambition, or arbitrary power? Females, condemned during the whole flower of their youth to a worse than monastic celibacy, irrevocably debarred from the hope to which their first affections pointed, will, at a certain period of life, as the natural delicacy of taste and feeling is gradually worn away by the attrition of society, become willing to take up with any coxcomb or scoundrel, whom that merciless and mercenary gang of cold-blooded slaves and assasins, called, in the ordinary prostitution of language, *friends*, may agree in designating as a *prudent choice*. Young men, on the other hand, are driven by the same vile superstitions from the company of the most amiable and modest of the opposite sex, to that of those miserable victims and outcasts of a world which dares to call itself virtuous, whom that very society whose pernicious institutions first caused their aberrations,—consigning them, without one tear of pity or one struggle of remorse, to penury, infamy, and disease,—condemns to bear the burden of its own atrocious absurdities! Thus, the youth of one sex is consumed in slavery, disappointment, and spleen; that of the other, in frantic folly and selfish intemperance: till at length, on the necks of a couple so enfeebled, so perverted, so distempered both in body and soul, society throws the yoke of marriage: that yoke which, once rivetted on the necks of its victims, clings to them like the poisoned garments of Nessus or Medea. What can be expected from these ill-assorted yoke-fellows, but that, like two ill-tempered hounds, coupled by a tyrannical sportsman, they should drag on their indissoluble fetter, snarling and growling, and pulling in different directions? What can be expected for their wretched offspring, but sickness and suffering, premature decrepitude, and untimely death? In this, as in every other institution of civilised society, avarice, luxury, and disease constitute the TRIANGULAR HARMONY of the life of man. Avarice conducts him to the abyss of toil and crime; luxury seizes on his ill-gotten spoil; and, while he revels in her enchantments, or groans beneath her tyranny, disease bursts upon him, and sweeps him from the earth.'

'Your theory,' said Mr. Jenkison, 'forms an admirable counterpoise to your example. As far as I am attracted by the one, I am repelled by the other. Thus, the scales of my philosophical balance remain eternally equiponderant, and I see no reason to say of either of them, *ΟΙΧΕΤΑΙ ΕΙΣ ΑΪΔΑΟ*.'*

* *It descends to the shades*: or, in other words, *it goes to the devil.*

GRYLL GRANGE.

BY THE

AUTHOR OF 'HEADLONG HALL'

Opinion governs all mankind,
Like the blind leading of the blind:—
And like the world, men's jobbernoles
Turn round upon their ears the poles,
And what they're confidently told
By no sense else can be control'd.

BUTLER.

LONDON:
PARKER, SON, AND BOURN, WEST STRAND.
1861.

In the following pages, the New Forest is always mentioned as if it were still unenclosed. This is the only state in which the Author has been acquainted with it. Since its enclosure, he has never seen it, and purposes never to do so.

The mottoes are sometimes specially apposite to the chapters to which they are prefixed; but more frequently to the general scope, or to borrow a musical term, the *motivo* of the *operetta*.

CONTENTS

GRYLL GRANGE

CHAPTER I

Ego sic semper et ubique vixi, ut ultimam quamque lucem, tanquam non redituram, consumerem.—PETRONIUS ARBITER.

Always and everywhere I have so lived, that I might consume the passing light, as if it were not to return.

'PALESTINE soup!' said the Reverend Doctor Opimian, dining with his friend Squire Gryll; 'a curiously complicated misnomer. We have an excellent old vegetable, the artichoke, of which we eat the head; we have another of subsequent introduction, of which we eat the root, and which we also call artichoke, because it resembles the first in flavour, although, *me judice*, a very inferior affair. This last is a species of the helianthus, or sunflower genus of the *Syngenesia frustranea* class of plants. It is therefore a girasol, or turn-to-the-sun. From this girasol we have made Jerusalem, and from the Jerusalem artichoke we make Palestine soup.'

MR. GRYLL

A very good thing, Doctor.

THE REVEREND DOCTOR OPIMIAN

A very good thing; but a palpable misnomer.

MR. GRYLL

I am afraid we live in a world of misnomers, and of a worse kind than this. In my little experience I have found that a gang of swindling bankers is a respectable old firm; that men who sell their votes to the highest bidder, and want only 'the protection of the ballot' to sell the promise of them to both parties, are a free and independent constituency; that a man who successively betrays everybody that trusts him, and abandons every principle he ever professed, is a great statesman, and a Conservative, forsooth, *à nil conservando;* that schemes for breeding pestilence are sanitary improvements; that the test of intellectual capacity is in swallow, and not in digestion; that the art of teaching everything, except what will be of use to the recipient, is national education; and that a change for the worse is reform. Look across the Atlantic. A Sympathizer would seem to imply a certain degree of benevolent feeling. Nothing of the kind. It signifies a ready-made accomplice in any species of political villany. A Know-Nothing would seem to imply a liberal self-diffidence—on the scriptural principle that the beginning of knowledge is to know that thou art ignorant. No such thing. It implies furious political dogmatism, enforced by bludgeons and revolvers. A Locofoco is the only intelligible term: a fellow that would set any place on fire to roast his own eggs. A Filibuster is a pirate under national colours; but I suppose the word in its origin implies something virtuous: perhaps a friend of humanity.

THE REVEREND DOCTOR OPIMIAN

More likely a friend of roaring—*Φιλοβωστρὴς*—in the sense in which roaring is used by our old dramatists; for which see Middleton's *Roaring Girl*, and the commentators thereon.*

MR. GRYLL

While we are on the subject of misnomers, what say you to the wisdom of Parliament?

THE REVEREND DOCTOR OPIMIAN

Why, sir, I do not call that a misnomer. The term wisdom is used in a parliamentary sense. The wisdom of Parliament is a wisdom *sui generis*. It is not like any other wisdom. It is not the wisdom of Socrates, nor the wisdom of Solomon. It is the wisdom of Parliament. It is not easily analysed or defined; but it is very easily understood. It has achieved wonderful things by itself, and still more when Science has come to its aid. Between them, they have poisoned the Thames, and killed the fish in the river. A little further development of the same wisdom and science will complete the poisoning of the air, and kill the dwellers on the banks. It is pleasant that the precious effluvium has been brought so efficiently under the Wisdom's own wise nose. Thereat the nose, like Trinculo's, has been in great indignation. The Wisdom has ordered the Science to do

* '*Roaring boys* was a cant term for the riotous, quarrelsome blades of the time, who abounded in London, and took pleasure in annoying its quieter inhabitants. Of *Roaring Girls*, the heroine of the present play was the choicest specimen. Her real name was *Mary Frith*, but she was most commonly known by that of *Moll Cutpurse*.'—Dyce. She wore male apparel, smoked, fought, robbed on the highway, kept all minor thieves in subjection, and compelled the restitution of stolen goods, when duly paid for her services.

something. The Wisdom does not know what, nor the Science either. But the Wisdom has empowered the Science to spend some millions of money; and this, no doubt, the Science will do. When the money has been spent, it will be found that the something has been worse than nothing. The Science will want more money to do some other something, and the Wisdom will grant it. *Redit labor actus in orbem.** But you have got on moral and political ground. My remark was merely on a perversion of words, of which we have an inexhaustible catalogue.

MR. GRYLL

Whatever ground we take, Doctor, there is one point common to most of these cases: the word presents an idea, which does not belong to the subject, critically considered. Palestine Soup is not more remote from the true Jerusalem, than many an honourable friend from public honesty and honour. However, Doctor, what say you to a glass of old Madeira, which I really believe is what it is called?

THE REVEREND DOCTOR OPIMIAN

In vino veritas. I accept with pleasure.

MISS GRYLL

You and my uncle, Doctor, get up a discussion on everything that presents itself; dealing with your theme like a series of variations in music. You have run half round the world *àpropos* of the soup. What say you to the fish?

THE REVEREND DOCTOR OPIMIAN

Premising that this is a remarkably fine slice of salmon, there is much to be said about fish: but not in the way of misnomers. Their names are single and simple. Perch, sole, cod, eel, carp, char, skate, tench, trout, brill, bream, pike, and many others, plain monosyllables: salmon, dory, turbot, gudgeon, lobster, whitebait, grayling, haddock, mullet, herring, oyster, sturgeon, flounder, turtle, plain dissyllables: only two trisyllables worth naming, anchovy and mackerel; unless any one should be disposed to stand up for halibut, which for my part I have excommunicated.

* The labour returns, compelled into a circle.

MR. GRYLL

I agree with you on that point; but I think you have named one or two, that might as well keep it company.

THE REVEREND DOCTOR OPIMIAN

I do not think I have named a single unpresentable fish.

MR. GRYLL

Bream, Doctor: there is not much to be said for bream.

THE REVEREND DOCTOR OPIMIAN

On the contrary, sir, I think there is much to be said for him. In the first place, there is the authority of the monastic brotherhoods, who are universally admitted to have been connoisseurs in fish, and in the mode of preparing it; and you will find bream pie set down as a prominent item of luxurious living in the indictments prepared against them at the dissolution of the monasteries. The work of destruction was rather too rapid, and I fear the receipt is lost. But he can still be served up as an excellent stew, provided always that he is full-grown, and has swum all his life in clear running water. I call everything fish that seas, lakes, and rivers furnish to cookery; though, scientifically a turtle is a reptile, and a lobster an insect. Fish, Miss Gryll—I could discourse to you on fish by the hour: but for the present I will forbear: as Lord Curryfin is coming down to Thornback Bay, to lecture the fishermen on fish and fisheries, and to astonish them all with the science of their art. You will, no doubt, be curious to hear him. There will be some reserved seats.

MISS GRYLL

I shall be very curious to hear him, indeed. I have never heard a lecturing lord. The fancy of lords and gentlemen to lecture everybody on everything, everywhere, seems to me something very comical; but perhaps it is something very serious, gracious in the lecturer, and instructive to the audience. I shall be glad to be cured of my unbecoming propensity to laugh, whenever I hear of a lecturing lord.

THE REVEREND DOCTOR OPIMIAN

I hope, Miss Gryll, you will not laugh at Lord Curryfin: for you may be assured, nothing will be farther from his lordship's intention than to say anything in the slightest degree droll.

MR. GRYLL

Doctor Johnson was astonished at the mania for lectures, even in his day, when there were no lecturing lords. He thought little was to be learned from lectures, unless where, as in chemistry, the subject required illustration by experiment. Now, if your lord is going to exhibit experiments, in the art of cooking fish, with specimens in sufficient number for all his audience to taste, I have no doubt his lecture will be well attended, and a repetition earnestly desired.

THE REVEREND DOCTOR OPIMIAN

I am afraid the lecture will not have the aid of such pleasant adventitious attractions. It will be a pure scientific exposition, carefully classified, under the several divisions and subdivisions of Ichthyology, Entomology, Herpetology, and Conchology. But I agree with Doctor Johnson, that little is to be learned from lectures. For the most part, those who do not already understand the subject will not understand the lecture, and those who do will learn nothing from it. The latter will hear many things they would like to contradict, which the *bienséance* of the lecture-room does not allow. I do not comprehend how people can find amusement in lectures. I should much prefer a *tenson* of the twelfth century, when two or three masters of the *Gai Saber* discussed questions of love and chivalry.

MISS GRYLL

I am afraid, Doctor, our age is too prosy for that sort of thing. We have neither wit enough, nor poetry enough, to furnish the disputants. I can conceive a state of society in which such *tensons* would form a pleasant winter evening amusement: but that state of society is not ours.

THE REVEREND DOCTOR OPIMIAN

Well, Miss Gryll, I should like, some winter evening, to challenge you to a *tenson*, and your uncle should be umpire. I think you have wit enough by nature, and I have poetry enough by memory, to supply a fair portion of the requisite materials, without assuming an absolute mastery of the *Gai Saber*.

MISS GRYLL

I shall accept the challenge, Doctor. The wit on one side will, I am afraid, be very shortcoming; but the poetry on the other will no doubt be abundant.

MR. GRYLL

Suppose, Doctor, you were to get up a *tenson* a little more relative to our own wise days. Spirit-rapping, for example, is a fine field. *Nec pueri credunt. . . . Sed tu vera puta.** You might go beyond the limits of a *tenson*. There is ample scope for an Aristophanic comedy. In the contest between the Just and the Unjust in the *Clouds*, and in other scenes of Aristophanes, you have ancient specimens of something very like *tensons*, except that Love has not much share in them. Let us for a moment suppose this same spirit-rapping to be true—dramatically so, at least. Let us fit up a stage for the purpose: make the invoked spirits visible as well as audible: and calling before us some of the illustrious of former days, ask them what they think of us and our doings? Of our astounding progress of intellect? Our march of mind? Our higher tone of morality? Our vast diffusion of education? Our art of choosing the most unfit man by competitive examination?

THE REVEREND DOCTOR OPIMIAN

You had better not bring on many of them at once, nor ask many similar questions, or the chorus of ghostly laughter will be overwhelming. I imagine the answer would be something like Hamlet's: 'You yourselves, sirs, shall be as wise as we were, if, like crabs, you could go backward.' It is thought something wonderful that uneducated persons should believe in witchcraft in the nineteenth century: as if educated persons did not believe in grosser follies: such as this same spirit-rapping, unknown tongues, clairvoyance, table-turning, and all sorts of fanatical impositions, having for the present their climax in Mormonism. Herein all times are alike. There is nothing too monstrous for human credulity. I like the notion of the Aristophanic comedy. But it would require a numerous company, especially as the chorus is indispensable. The *tenson* may be carried on by two.

MR. GRYLL

I do not see why we should not have both.

MISS GRYLL

Oh pray, Doctor! let us have the comedy. We hope to have a houseful at Christmas, and I think we may get it up well, chorus and all. I should so like to hear what my great ancestor, Gryllus,

* Not even boys believe it: but suppose it to be true.

thinks of us: and Homer, and Dante, and Shakspeare, and Richard the First, and Oliver Cromwell.

THE REVEREND DOCTOR OPIMIAN

A very good *dramatis personæ*. With these, and the help of one or two Athenians and Romans, we may arrive at a tolerable judgment on our own immeasurable superiority to everything that has gone before us.

Before we proceed further, we will give some account of our interlocutors.

CHAPTER II

FORTUNA . SPONDET . MULTA . MULTIS . PRAESTAT . NEMINI . VIVE . IN . DIES . ET . HORAS . NAM . PROPRIUM . EST . NIHIL.*

Marmor vetus apud Feam, ad Hor. Epist. i. 11, 23.

Fortune makes many promises to many,
Keeps them to none. Live to the days and hours,
For nothing is your own.

GREGORY GRYLL, Esq., of Gryll Grange in Hampshire, on the borders of the New Forest, in the midst of a park which was a little forest in itself, reaching nearly to the sea, and well stocked with deer, having a large outer tract, where a numerous light-rented and well-conditioned tenantry fattened innumerable pigs, considered himself well located for what he professed to be, *Epicuri de grege porcus*,† and held, though he found it difficult to trace the pedigree, that he was lineally descended from the ancient and illustrious Gryllus, who maintained against Ulysses the superior happiness of the life of other animals to that of the life of man.‡

It might seem, that to a man who traced his ancestry from the Palace of Circe, the first care would be the continuance of his ancient race; but a wife presented to him the forethought of a perturbation of his equanimity, which he never could bring himself to encounter. He liked to dine well, and withal to dine quietly, and to have quiet friends at his table,

* This inscription appears to consist of comic senarii, slightly dislocated for the inscriptional purpose.

Spondet ¯ ˘ ˘
Fortuna multa multis, praestat nemini.
Vive in dies et horas: nam propriụm est nihil.

† *A pig from the herd of Epicurus.* The old philosophers accepted good-humouredly the disparaging terms attached to them by their enemies or rivals. The Epicureans acquiesced in the pig, the Cynics in the dog, and Cleanthes was content to be called the Ass of Zeno, as being alone capable of bearing the burthen of the Stoic philosophy.

‡ PLUTARCH. *Bruta animalia ratione uti.* Gryllus, in this dialogue, seems to have the best of the argument. Spenser, however, did not think so,

with whom he could discuss qucstions which might afford ample room for pleasant conversation and none for acrimonious dispute. He feared that a wife would interfere with his dinner, his company, and his after-dinner bottle of port. For the perpetuation of his name, he relied on an orphan niece, whom he had brought up from a child, who superintended his household, and sate at the head of his table. She was to be his heiress, and her husband was to take his name. He left the choice to her, but reserved to himself a veto if he should think the aspirant unworthy of the honourable appellation.

The young lady had too much taste, feeling, and sense to be likely to make a choice which her uncle would not approve; but time, as it rolled on, foreshadowed a result which the Squire had not anticipated. Miss Gryll did not seem likely to make any choice at all. The atmosphere of quiet enjoyment in which she had grown up seemed to have

when he introduced his Gryll, in the Paradise of Acrasia, reviling Sir Guyon's Palmer for having restored him to the human form.

> Streightway he with his virtuous staff them strooke,
> And streight of beasts they comely men became:
> Yet being men they did unmanly looke,
> And stared ghastly, some for inward shame,
> And some for wrath to see their captive dame:
> But one above the rest in speciall,
> That had an hog been late, hight Grylle by name,
> Repyned greatly, and did him miscall,
> That had from hoggish forme him brought to naturall.
> Said Guyon: 'See the mind of beastly man,
> That hath so soon forgot the excellence
> Of his creation, when he life began,
> That now he chooseth, with vile difference,
> To be a beast, and lacke intelligence.'

Fairy Queen, Book ii. Canto 12.

In Plutarch's dialogue, Ulysses, after his own companions have been restored to the human form, solicits Circe to restore in the same manner any other Greeks who may be under her enchantments. Circe consents, provided they desire it. Gryllus, endowed with speech for the purpose, answers for all, that they had rather remain as they are; and supports the decision by showing the greater comfort of their condition as it is, to what it would probably be if they were again sent forth to share the common lot of mankind. We have unfortunately only the beginning of the dialogue, of which the greater portion has perished.

steeped her feelings in its own tranquillity; and still more, the affection which she felt for her uncle, and the conviction that, though he had always premeditated her marriage, her departure from his house would be the severest blow that fate could inflict on him, led her to postpone what she knew must be an evil day to him, and might peradventure not be a good one to her.

'Oh, the ancient name of Gryll!' sighed the Squire to himself. 'What if it should pass away in the nineteenth century, after having lived from the time of Circe!'

Often indeed, when he looked at her at the head of his table, the star of his little circle, joyous herself and the source of joy in others, he thought the actual state of things admitted no change for the better, and the perpetuity of the old name became a secondary consideration; but though the purpose was dimmed in the evening it usually brightened in the morning. In the meantime the young lady had many suitors, who were permitted to plead their cause, though they made little apparent progress.

Several young gentlemen of fair promise, seemingly on the point of being accepted, had been, each in his turn, suddenly and summarily dismissed. Why, was the young lady's secret. If it were known, it would be easy, she said, in these days of artificial manners, to counterfeit the presence of the qualities she liked, and, still more easy, the absence of the qualities she disliked. There was sufficient diversity in the characters of the rejected to place conjecture at fault, and Mr. Gryll began to despair.

The uncle and niece had come to a clear understanding on this subject. He might present to her attention any one whom he might deem worthy to be her suitor, and she might reject the suitor without assigning a reason for so doing. In this way several had appeared, and passed away like bubbles on a stream.

Was the young lady over fastidious, or were none among the presented worthy, or had that which was to touch her heart not yet appeared?

Mr. Gryll was the godfather of his niece, and to please him, she had been called Morgana. He had had some thoughts of calling her Circe, but acquiesced in the name of a sister enchantress, who had worked out her own idea of a beautiful garden, and exercised similar power over the minds and forms of men.

CHAPTER III

Τέγγε πνεύμονας οἴνῳ· τὸ γὰρ ἄστρον περιτέλλεται·
Ἁ δ' ὥρα χαλεπά, πάντα δὲ διψᾷ ὑπὸ καύματος.—ALCAEUS.

Moisten your lungs with wine. The dog star's sway
Returns, and all things thirst beneath his ray.

FALERNUM . OPIMIANUM . ANNORUM . CENTUM.

Heu! heu! inquit Trimalchio, ergo diutius vivit vinum quam homuncio! Quare *τέγγε πνεύμονας* faciamus. Vita vinum est.—PETRONIUS ARBITER.

FALERNIAN OPIMIAN WINE AN HUNDRED YEARS OLD.

Alas! Alas! exclaimed Trimalchio. Thus wine lives longer than man! Wherefore, let us sing 'moisten your lungs.' Wine is life.

WORDSWORTH'S question, in his *Poet's Epitaph*,

Art thou a man of purple cheer,
A rosy man, right plump to see?

might have been answered in the affirmative by the Reverend Doctor Opimian. The worthy divine dwelt in an agreeably situated vicarage, on the outskirts of the New Forest. A good living, a comfortable patrimony, a moderate dowry with his wife, placed him sufficiently above the cares of the world to enable him to gratify all his tastes without minute calculations of cost. His tastes in fact were four: a good library, a good dinner, a pleasant garden, and rural walks. He was an athlete in pedestrianism. He took no pleasure in riding, either on horseback or in a carriage; but he kept a brougham for the service of Mrs. Opimian, and for his own occasional use in dining out.

Mrs. Opimian was domestic. The care of the Doctor had supplied her with the best books on cookery, to which his own inventive genius and the kindness of friends had added a large and always increasing manuscript volume. The lady studied them carefully, and by diligent superintendence left the Doctor nothing to desire in the service of his table. His cellar was well stocked with a selection of the best vintages, under his own especial charge. In all its arrangements his house was a model of order and comfort; and the whole establishment partook of the genial physiognomy of the master. From the master and mistress to the cook, and from the cook to the tom cat, there was about the inhabitants of the vicarage a sleek and purring rotundity of face and figure that denoted community of feelings, habits, and diet; each in its kind, of course, for the Doctor had his port, the cook her ale, and the cat his milk, in sufficiently liberal allowance. In the morning, while Mrs. Opimian found ample occupation in the details of her household duties and the care of her little family, the Doctor, unless he had predestined the whole day to an excursion, studied in his library. In the afternoon he walked; in the evening he dined; and after dinner read to his wife and family, or heard his children read to him. This was his home life. Now and then he dined out; more frequently than at any other place with his friend and neighbour Mr. Gryll, who entirely sympathized with him in his taste for a good dinner.

Beyond the limits of his ordinary but within those of his occasional range was a solitary round tower on an eminence backed with wood, which had probably in old days been a landmark for hunters; but having in modern days no very obvious use, was designated, as many such buildings are, by the name of the Folly. The country people called it 'the Duke's Folly,' though who the Duke in question was nobody could tell. Tradition had dropped his name.

One fine Midsummer day, with a southerly breeze and a cloudless sky, the Doctor, having taken an early breakfast, in the process of which he had considerably reduced the altitude of a round of beef, set out with a good stick in his hand and a Newfoundland dog at his heels for one of his longest walks, such as he could only take in the longest days.

Arriving at the Folly, which he had not visited for a long time, he was surprised to find it enclosed, and having at the back the novelty of a covered passage, built of the same grey stone as the tower itself. This passage passed away into the wood at the back, whence was ascending a wreath of smoke which immediately recalled to him the dwelling of Circe.* Indeed, the change before him had much the air of enchantment; and the Circean similitude was not a little enhanced by the antique masonry,† and the expanse of sea which was visible from the eminence. He leaned over the gate, repeated aloud the lines of the *Odyssey*, and fell into a brown study, from which he was aroused by the approach of a young gentleman from within the enclosure.

'I beg your pardon, sir,' said the Doctor, 'but my curiosity is excited by what I see here; and if you do not think it impertinent, and would inform me how these changes have come about, I should be greatly obliged.'

'Most willingly, sir,' said the other; 'but if you will walk in, and see what has been done, the obligation will be mine.'

The Doctor readily accepted the proposal. The stranger led the way, across an open space in the wood, to a circular hall, from each side of which a wide passage led, on the left hand to the tower, and on the right to the new building,

* *Καὶ τότ' ἐγὼν ἐμὸν ἔγχος ἑλὼν καὶ φάσγανον ὀξὺ*
Καρπαλίμως παρὰ νηὸς ἀνήϊον ἐς περιωπὴν,
Εἴπως ἔργα ἴδοιμι βροτῶν ἐνοπήν τε πυθοίμην.
Ἔστην δὲ, σκοπιὴν ἐς παιπαλόεσσαν ἀνελθὼν,
Καὶ μοι ἐείσατο καπνὸς ἀπὸ χθονὸς εὐρυοδείης,
Κίρκης ἐν μεγάροισι, διὰ δρυμὰ πυκνὰ καὶ ὕλην.
Μερμήριξα δ' ἔπειτα κατὰ φρένα καὶ κατὰ θυμὸν
Ἐλθεῖν, ἠδὲ πυθέσθαι, ἐπεὶ ἴδον αἴθοπα καπνόν.
Od. K. 145–162.

I climbed a cliff with spear and sword in hand,
Whose ridge o'erlooked a shady length of land:
To learn if aught of mortal works appear,
Or cheerful voice of mortal strike the ear.
From the high point I marked, in distant view,
A stream of curling smoke ascending blue,
And spiry tops, the tufted trees above,
Of Circe's palace bosomed in the grove.
Thither to haste, the region to explore,
Was first my thought . . .

† *Εὗρον δ' ἐν βήσσῃσι τετυγμένα δώματα Κίρκης*
Ξεστοῖσιν λάεσσι, περισκέπτῳ ἐνὶ χώρῳ.—Ib. 210, 211.

The palace in a woody vale they found,
High-raised of stone, a shaded space around.—POPE.

which was so masked by the wood, as not to be visible except from within the glade. It was a square structure of plain stone, much in the same style as that of the tower.

The young gentleman took the left-hand passage, and introduced the Doctor to the lower floor of the tower.

'I have divided the tower,' he observed, 'into three rooms: one on each floor. This is the dining-room; above it is my bedroom; above it again is my library. The prospect is good from all the floors, but from the library it is most extensive, as you look over the woods far away into the open sea.'

'A noble dining-room,' said the Doctor. 'The height is well proportioned to the diameter. That circular table well becomes the form of the room, and gives promise of a fine prospect in its way.'

'I hope you will favour me by forming a practical judgment on the point,' said his new acquaintance, as he led the way to the upper floor, the Doctor marvelling at the extreme courtesy with which he was treated. 'This building,' thought he, 'might belong to the age of chivalry, and my young host might be Sir Calidore himself.' But the library brought him back to other days.

The walls were covered with books, the upper portion accessible by a gallery, running entirely round the apartment. The books of the lower circle were all classical; those of the upper, English, Italian, and French, with a few volumes in Spanish.

The young gentleman took down a Homer, and pointed out to the Doctor the passage which, as he leaned over the gate, he had repeated from the *Odyssey*. This accounted to the Doctor for the deference shown to him. He saw at once into the Greek sympathy.

'You have a great collection of books,' said the Doctor.

'I believe,' said the young gentleman, 'I have all the best books in the languages I cultivate. Horne Tooke says: "Greek, Latin, Italian, and French, are unfortunately the usual bounds of an English scholar's acquisition." I think any scholar fortunate whose acquisition extends so far. These languages and our own comprise, I believe, with a few rare exceptions, all the best books in the world. I may add

Spanish, for the sake of Cervantes, Lope de Vega, and Calderon. It was a *dictum* of Porson, that "Life is too short to learn German:" meaning, I apprehend, not that it is too difficult to be acquired within the ordinary space of life, but that there is nothing in it to compensate for the portion of life bestowed on its acquirement, however little that may be.'*

The Doctor was somewhat puzzled what to say. He had some French and more Italian, being fond of romances of chivalry; and in Greek and Latin he thought himself a match for any man; but he was more occupied with speculations on the position and character of his new acquaintance, than on the literary opinions he was enunciating. He marvelled to find a young man, rich enough to do what he here saw done, doing anything of the kind, and fitting up a library in a solitary tower, instead of passing his time in clubs and *réunions*, and other pursuits and pleasures of general society. But he thought it necessary to say something to the point, and rejoined:

'Porson was a great man, and his *dictum* would have weighed with me if I had had a velleity towards German; but I never had any. But I rather wonder you should have placed your library on the upper instead of the middle floor. The prospect, as you have observed, is fine from all the floors; but here you have the sea and the sky to the greatest advantage; and I would assign my best look-out to the hours of dressing and undressing; the first thing in the morning, the last at night, and the half-hour before dinner. You can give

* Mr. Hayward's French hotel-keeper in Germany had a different, but not less cogent reason for not learning German. 'Whenever a dish attracts attention by the art displayed in its conception or preparation, apart from the material, the artist will commonly be discovered to be French. Many years ago we had the curiosity to inquire at the Hotel de France, at Dresden, to whom our party were indebted for the enjoyment they had derived from a *suprême de volaille*, and were informed the cook and the master of the hotel were one and the same person: a Frenchman, *ci-devant chef* of a Russian minister. He had been eighteen years in Germany, but knew not a word of any language but his own. "*A quoi bon, messieurs*," was his reply to our expression of astonishment; "*à quoi bon, apprendre la langue d'un peuple qui ne possède pas une cuisine?*"'—*Art of Dining*, pp. 69, 70.

greater attention to the views before you, when you are following operations, important certainly, but mechanical from repetition, and uninteresting in themselves, than when you are engaged in some absorbing study, which probably shuts out all perception of the external world.'

'What you say is very true, sir,' said the other; 'but you know the lines of Milton——

> Or let my lamp, at midnight hour,
> Be seen in some high lonely tower,
> Where I may oft outwatch the Bear,
> With thrice great Hermes.

'These lines have haunted me from very early days, and principally influenced me in purchasing this tower, and placing my library on the top of it. And I have another association with such a mode of life.'

A French clock in the library struck two, and the young gentleman proposed to his visitor to walk into the house. They accordingly descended the stairs, and crossed the entrance-hall to a large drawing-room, simply but handsomely furnished; having some good pictures on the walls, an organ at one end of the room, a piano and harp at the other, and an elegantly disposed luncheon in the middle.

'At this time of the year,' said the young gentleman, 'I lunch at two, and dine at eight. This gives me two long divisions of the morning, for any in-door and out-door purposes. I hope you will partake with me. You will not find a precedent in Homer for declining the invitation.'

'Really,' said the Doctor, 'that argument is cogent and conclusive. I accept with pleasure: and indeed my long walk has given me an appetite.'

'Now you must know,' said the young gentleman, 'I have none but female domestics. You will see my two waiting-maids.'

He rang the bell, and the specified attendants appeared: two young girls about sixteen and seventeen; both pretty, and simply, but very becomingly, dressed.

Of the provision set before him the Doctor preferred some cold chicken and tongue. Madeira and sherry were on the table, and the young attendants offered him hock and claret. The Doctor took a capacious glass from each of the fair cup-bearers, and pronounced both wines excellent, and deliciously cool. He declined more, not to over-heat himself in walking, and not to infringe on his anticipations of dinner. The dog, who had behaved throughout with exemplary propriety, was not forgotten. The Doctor rose to depart.

'I think,' said his host, 'I may now ask you the Homeric question—*Τίς πόθεν εἶς ἀνδρῶν;*'*

'Most justly,' said the Doctor. 'My name is Theophilus Opimian. I am a Doctor of Divinity, and the incumbent of Ashbrook-cum-Ferndale.'

'I am simply,' said the other, 'Algernon Falconer. I have inherited some money, but no land. Therefore having the opportunity, I made this purchase, to fit it up in my own fashion, and live in it in my own way.'

The Doctor preparing to depart, Mr. Falconer proposed to accompany him part of the way, and calling out another Newfoundland dog, who immediately struck up a friendship with his companion, he walked away with the Doctor, the two dogs gambolling before them.

* Who, and whence, are you?

CHAPTER IV

> Mille hominum species, et rerum discolor usus:
> Velle suum cuique est, nec voto vivitur uno.—PERSIUS.
>
> In mind and taste men differ as in frame:
> Each has his special will, and few the same.

THE REVEREND DOCTOR OPIMIAN

IT strikes me as singular that, with such a house, you should have only female domestics.

MR. FALCONER

It is not less singular perhaps that they are seven sisters, all the children of two old servants of my father and mother. The eldest is about my own age, twenty-six, so that they have all grown up with me in time and place. They live in great harmony together, and divide among them the charge of all the household duties. Those whom you saw are the two youngest.

THE REVEREND DOCTOR OPIMIAN

If the others acquit themselves as well, you have a very efficient staff; but seven young women as the establishment of one young bachelor, for such I presume you to be (*Mr. Falconer assented*), is something new and strange. The world is not over charitable.

MR. FALCONER

The world will never suppose a good motive, where it can suppose a bad one. I would not willingly offend any of its prejudices. I would not affect eccentricity. At the same time, I do not feel disposed to be put out of my way because it is not the way of the world—*Le Chemin du Monde*, as a Frenchman entitled Congreve's comedy*—but I assure you these seven young women live here as they might do in the Temple of Vesta. It was a singular combination of circumstances that induced and enabled me to form such an establishment; but I would not give it up, nor alter it, nor diminish it, nor increase it, for any earthly consideration.

* Congreve, le meilleur auteur comique d'Angleterre: ses pièces les plus estimées sont *Le Fourbe, Le Vieux Garçon, Amour pour Amour, L'Épouse du Matin, Le Chemin du Monde.—Manuel Bibliographique.* Par G. Peigniot. Paris. 1800.

THE REVEREND DOCTOR OPIMIAN

You hinted that, besides Milton's verses, you had another association of ideas with living in the top of a tower.

MR. FALCONER

I have read of somebody who lived so, and admitted to his *sanctum* only one young person, a niece or a daughter, I forget which, but on very rare occasions would descend to speak to some visitor who had previously propitiated the young lady to obtain him an interview. At last the young lady introduced one who proposed for her, and gained the consent of the recluse (I am not sure of his name, but I always call him Lord Noirmont) to carry her off. I think this was associated with some affliction that was cured or some mystery that was solved, and that the hermit returned into the every-day world. I do not know where I read it, but I have always liked the idea of living like Lord Noirmont, when I shall have become a sufficiently disappointed man.

THE REVEREND DOCTOR OPIMIAN

You look as little like a disappointed man as any I have seen; but as you have neither daughter nor niece, you would have seven links instead of one between the top of your tower and the external world.

MR. FALCONER

We are all born to disappointment. It is as well to be prospective. Our happiness is not in what is, but in what is to be. We may be disappointed in our every-day realities, and if not, we may make an ideality of the unattainable, and quarrel with nature for not giving what she has not to give. It is unreasonable to be so disappointed, but it is disappointment not the less.

THE REVEREND DOCTOR OPIMIAN

It is something like the disappointment of the men of Gotham when they could not fish up the moon from the sea.

MR. FALCONER

It is very like it, and there are more of us in the predicament of the men of Gotham than are ready to acknowledge the similitude.

THE REVEREND DOCTOR OPIMIAN

I am afraid I am too matter-of-fact to sympathize very clearly with this form of æstheticism; but here is a charming bit of forest scenery. Look at that old oak with the deer under it; the long and deep range of fern running up from it to that beech-grove on the upland, the lights and shadows on the projections and recesses of the wood, and the blaze of foxglove in its foreground. It is a place in which a poet might look for a glimpse of a Hamadryad.

MR. FALCONER

Very beautiful for the actual present—too beautiful for the probable future. Some day or other the forest will be disforested; the deer will be either banished or destroyed; the wood will be either shut up or cut down. Here is another basis for disappointment. The more we admire it now, the more we shall regret it then. The admiration of sylvan and pastoral scenery is at the mercy of an inclosure act, and instead of the glimpse of a Hamadryad you will sometime see a large board warning you off the premises under penalty of rigour of law.

THE REVEREND DOCTOR OPIMIAN

But, my dear young friend, you have yourself enclosed a favourite old resort of mine and of many others. I did not see such a board as you speak of; but there is an effective fence which answers the purpose.

MR. FALCONER

True; but when the lot of crown land was put up for sale, it was sure to be purchased and shut up by somebody. At any rate, I have not interfered with the external picturesque; and I have been much more influenced by an intense desire of shutting up myself than of shutting up the place, merely because it is my property.

About half way from their respective homes the two new friends separated, the Doctor having promised to walk over again soon to dine and pass the night.

The Doctor soliloquized as he walked.

Strange metamorphosis of the old tower. A good dining-room. A good library. A bed-room between them: he did not show it me. Good wine: excellent. Pretty waiting-maids: exceedingly pretty. Two of seven Vestals, who maintain the domestic fire on the hearth of this young Numa. By the way, they had something of the Vestal costume: white dresses with purple borders. But they had nothing on their heads but their own hair, very gracefully arranged. The Vestals had head-dresses, which hid their hair, if they had any. They were shaved on admission. Perhaps the hair was allowed to grow again. Perhaps not. I must look into the point. If not, it was a wise precaution. 'Hair, the only grace of form,'* says the *Arbiter Elegantiarum*, who compares a bald head to a fungus.† A head without hair, says Ovid, is as a field without grass, and a shrub without leaves.‡ Venus herself, if she had appeared with a bald head, would not have tempted Apuleius:§ and I am of his mind. A husband, in Menander,|| in a fit of jealous madness, shaves his wife's head; and when he sees what he has made of her, rolls at her feet in a paroxysm of remorse. He was at any rate safe from jealousy till it grew again. And here is a subtlety of Euripides, which none of his commentators have seen into. Ægisthus has married Electra to a young farmer, who cultivates his own land. He respects the Princess from magnanimity, and restores her a pure virgin to her brother Orestes. 'Not probable,' say some

* Quod solum formæ decus est, cecidere capilli.—PETRONIUS, c. 109.

† . . . . lævior . . . . rotundo
Horti tubere, quod creavit unda.—*Ibid.*

'A head, to speak in the gardener's style, is a bulbous excrescence, growing up between the shoulders.'—G. A. STEEVENS: *Lecture on Heads.*

‡ Turpe pecus mutilum; turpe est sine gramine campus;
Et sine fronde frutex; et sine crine caput.
OVID. *Artis Amatoriæ,* iii, 249.

§ At vero, quod nefas dicere, neque sit ullum hujus rei tam dirum exemplum: si cujuslibet eximiæ pulcherrimæque fœminæ caput capillo exspoliaveris, et faciem nativâ specie nudaveris, licet illa cœlo dejecta, mari edita, fluctibus educata, licet, inquam, Venus ipsa fuerit, licet omni Gratiarum choro stipata, et toto Cupidinum populo comitata, et balteo suo cincta, cinnama fragrans, et balsama rorans, calva processerit, placere non poterit nec Vulcano suo.—APULEIUS, *Metamorph.* ii. 25.

But, indeed, what it is profanation to speak, nor let there be hereof any so dire example, if you despoil of its hair the head of any most transcendent and perfectly beautiful woman, and present her face thus denuded of its native loveliness, though it were even she, the descended from heaven, the born of the sea, the educated in the waves, though, I say, it were Venus herself, attended by the Graces, surrounded by the Loves, cinctured with her girdle, fragrant with spices, and dewy with balsams, yet, if she appeared with a bald head, she could not please even her own Vulcan.

|| *Περικειρομένη.*

critics. But I say, highly probable: for she comes on with her head shaved. There is the talisman, and the consummate artifice of the great poet. It is ostensibly a symbol of grief; but not the less a most efficient ally of the aforesaid magnanimity. 'In mourning,' says Aristotle, 'sympathizing with the dead, we deform ourselves by cutting off our hair.' And truly, it is sympathy in approximation. A woman's head shaved is a step towards a death's head. As a symbol of grief, it was not necessary to the case of Electra; for in the sister tragedies of Æschylus and Sophocles, her grief is equally great, and she appears with flowing hair; but in them she is an unmarried maid, and there is no dramatic necessity for so conspicuous an antidote to her other charms. Neither is it according to custom; for in recent grief the whole hair was sacrificed, but in the memory of an old sorrow only one or two curls were cut off.* Therefore it was the dramatic necessity of a counter-charm that influenced Euripides. Helen knew better than to shave her head in a case where custom required it. Euripides makes Electra reproach Helen for thus preserving her beauty;† which further illustrates his purpose in shaving the head of Electra where custom did not require it. And Terence showed his taste in not shaving the head of his heroine in the *Phormio*, though the severity of Athenian custom would have required it. Her beauty shone through her dishevelled hair, but with no hair at all she would not have touched the heart of Antipho. *Ἀλλὰ τίη μοι ταῦτα φίλος διελέξατο θυμός;* But wherefore does my mind discourse these things to me? suspending dismal images on lovely realities? for the luxuriant hair of these young girls is of no ordinary beauty. Their tresses have not been deposited under the shadow of the sacred lotus, as Pliny tells us those of the Vestals were. Well, this young gentleman's establishment may be perfectly moral, strictly correct, but in one sense it is morality thrown away: the world will give him no credit for it. I am sure Mrs. Opimian will not. If he were married it would be different. But I think, if he were to marry now,

* SOPHOCLES: *Electra*, v. 449.
† EURIPIDES: *Orestes*, v. 128.

there would be a fiercer fire than Vesta's among his Lares. The temple would be too hot for the seven virgins. I suppose, as he is so resolute against change, he does not mean to marry. Then he talks about anticipated disappointment in some unrealizable ideality, leading him to live like Lord Noirmont, whom I never heard of before. He is far enough off from that while he lunches and walks as he does, and no doubt dines in accordance. He will not break his heart for any moon in the water, if his cooks are as good as his waiting-maids, and the wine which he gave me is a fair specimen of his cellar. He is learned too. Greek seems to be the strongest chord in his sympathies. If it had not been for the singular accident of his overhearing me repeat half a dozen lines of Homer, I should not have been asked to walk in. I might have leaned over the gate till sunset, and have had no more notice taken of me than if I had been a crow.

At dinner the Doctor narrated his morning adventure to Mrs. Opimian, and found her, as he had anticipated, most virtuously uncharitable with respect to the seven sisters. She did not depart from her usual serenity, but said, with equal calmness and decision, that she had no belief in the virtue of young men.

'My dear,' said the Doctor, 'it has been observed, though I forget by whom, that there is in every man's life a page which is usually doubled down. Perhaps there is such a page in the life of our young friend; but if there be, the volume which contains it is not in the same house with the seven sisters.'

The Doctor could not retire to rest without verifying his question touching the hair of the Vestals; and stepping into his study was taking out an old folio to consult *Lipsius de Vestalibus*, when a passage flashed across his memory, which seemed decisive on the point. 'How could I overlook it?' he thought.

> 'Ignibus Iliacis aderam: cum lapsa capillis
> Decidit ante sacros lanea vitta focos:*

says Rhea Sylvia in the *Fasti*.'

* The wollen wreath, by Vesta's inmost shrine,
Fell from my hair before the fire divine.

He took down the *Fasti*, and turning over the leaves lighted on another line:—

Attonitæ flebant demisso crine ministræ.*

With the note of an old commentator: 'This will enlighten those who doubt if the Vestals wore their hair.' 'I infer,' said the Doctor, 'that I have doubted in good company; but it is clear that the Vestals did wear their hair of second growth. But if it was wrapped up in wool, it might as well not have been there. The *vitta* was at once the symbol and the talisman of chastity. Shall I recommend my young friend to wrap up the heads of his Vestals in a *vitta*? It would be safer for all parties. But I cannot imagine a piece of advice for which the giver would receive less thanks. And I had rather see them as they are. So I shall let well alone.'

* With hair dishevelled wept the Vestal train.

CHAPTER V

Εὔφραινε σαυτὸν· πίνε· τὸν καθ' ἡμέραν
Βίον λογίζου σὸν, τὰ δ' ἄλλα τῆς Τύχης.
EURIPIDES: Alcestis.

Rejoice thy spirit: drink: the passing day
Esteem thine own, and all beyond as Fortune's.

THE Doctor was not long without remembering his promise to revisit his new acquaintance, and purposing to remain till the next morning, he set out later in the day. The weather was intensely hot; he walked slowly, and paused more frequently than usual, to rest under the shade of trees. He was shown into the drawing-room, where he was shortly joined by Mr. Falconer, and very cordially welcomed.

The two friends dined together in the lower room of the Tower. The dinner and wine were greatly to the Doctor's mind. In due time they adjourned to the drawing-room, and the two young handmaids who had waited at dinner attended with coffee and tea. The Doctor then said—'You are well provided with musical instruments. Do you play?'

MR. FALCONER

No. I have profited by the observation of Doctor Johnson: 'Sir, once on a time I took to fiddling; but I found that to fiddle well I must fiddle all my life, and I thought I could do something better.'

THE REVEREND DOCTOR OPIMIAN

Then, I presume, these are pieces of ornamental furniture, for the use of occasional visitors?

MR. FALCONER

Not exactly. My maids play on them, and sing to them.

THE REVEREND DOCTOR OPIMIAN

Your maids!

MR. FALCONER

Even so. They have been thoroughly well educated, and are all accomplished musicians.

THE REVEREND DOCTOR OPIMIAN

And at what time do they usually play on them?

MR. FALCONER

Every evening about this time, when I am alone.

THE REVEREND DOCTOR OPIMIAN

And why not when you have company?

MR. FALCONER

La Morgue Aristocratique, which pervades all society, would not tolerate such a proceeding on the part of young women, of whom some had superintended the preparation of the dinner, and others attended on it. It would not have been incongruous in the Homeric age.

THE REVEREND DOCTOR OPIMIAN

Then I hope you will allow it to be not incongruous this evening, Homer being the original *vinculum* between you and me.

MR. FALCONER

Would you like to hear them?

THE REVEREND DOCTOR OPIMIAN

Indeed I should.

The two younger sisters having answered the summons, and the Doctor's wish having been communicated, the seven appeared together, all in the same dress of white and purple.

'The Seven Pleiads!' thought the Doctor. 'What a constellation of beauty!' He stood up and bowed to them, which they gracefully acknowledged.

They then played on, and sang to, the harp and piano. The Doctor was enchanted.

After a while, they passed over to the organ, and performed some sacred music of Mozart and Beethoven. They then paused and looked round, as if for instructions.

'We usually end,' said Mr. Falconer, 'with a hymn to St. Catharine, but perhaps it may not be to your taste; although Saint Catharine is a saint of the English Church Calendar.'

'I like all sacred music,' said the Doctor. 'And I am not disposed to object to a saint of the English Church Calendar.'

'She is also,' said Mr. Falconer, 'a most perfect emblem of purity, and in that sense alone there can be no fitter image to be presented to the minds of young women.'

'Very true,' said the Doctor. 'And very strange withal,' he thought to himself.

The sisters sang their hymn, made their obeisance, and departed.

THE REVEREND DOCTOR OPIMIAN

The hands of those young women do not show signs of menial work.

MR. FALCONER

They are the regulating spirits of the household. They have a staff of their own for the coarser and harder work.

THE REVEREND DOCTOR OPIMIAN

Their household duties, then, are such as Homeric damsels discharged in the homes of their fathers, with *δμωαὶ* for the lower drudgery.

MR. FALCONER

Something like it.

THE REVEREND DOCTOR OPIMIAN

Young ladies, in short, in manners and accomplishments, though not in social position; only more useful in a house than young ladies generally are.

MR. FALCONER

Something like that, too. If you know the tree by its fruit, the manner in which this house is kept may reconcile you to the singularity of the experiment.

THE REVEREND DOCTOR OPIMIAN

I am perfectly reconciled to it. The experiment is eminently successful.

The Doctor always finished his day with a tumbler of brandy and water: soda water in summer, and hot water in winter. After his usual draught he retired to his chamber, where he slept like a top, and dreamed of Electra and

Nausicaa, Vestals, Pleiads, and Saint Catharine, and woke with the last words he had heard sung on the preceding night still ringing in his ears:—

Dei virgo Catharina,
Lege constans in divinâ,
Cœli gemma preciosa,
Margarita fulgida,
Sponsa Christi gloriosa,
Paradisi viola!*

* Virgin bride, supremely bright,
Gem and flower of heavenly light,
Pearl of the empyreal skies,
Violet of Paradise!

CHAPTER VI

> Despairing beside a clear stream
> A shepherd forsaken was laid.

THE next morning, after a comfortable breakfast, the Doctor set out on his walk home. His young friend accompanied him part of the way, and did not part with him till he had obtained a promise of another and longer visit.

The Doctor, as usual, soliloquized as he walked. 'No doubt these are Vestals. The purity of the establishment is past question. This young gentleman has every requisite which her dearest friends would desire in a husband for Miss Gryll. And she is in every way suited to him. But these seven damsels interpose themselves, like the sevenfold shield of Ajax. There is something very attractive in these damsels:

> facies non omnibus una,
> Nec diversa tamen: qualem decet esse sororum.*

If I had such an establishment, I should be loth to break it up. It is original, in these days of monotony. It is satisfactory, in these days of uncongenial relations between master and servant. It is effective, in the admirable arrangements of the household. It is graceful, in the personal beauty and tasteful apparel of the maidens. It is agreeable, in their manners, in their accomplishments, in their musical skill. It is like an enchanted palace. Mr. Gryll, who talks so much of Circe, would find himself at home; he might fancy himself waited on by her handmaids, the daughters of fountains, groves, and rivers. Miss Gryll might fancy herself in the dwelling of her namesake, Morgana. But I fear she would be for dealing with it as Orlando did with Morgana, breaking the talisman and dissolving the enchantment. This would be a pity; but it would also be a pity that these two young

* Though various features did the sisters grace,
A sister's likeness was in every face.
ADDISON: *Ovid. Met.* l. ii.

persons should not come together. But why should I trouble myself with match-making? It is always a thankless office. If it turns out well, your good service is forgotten. If it turns out ill, you are abused by both parties.'

The Doctor's soliloquy was cut short by a sound of lamentation, which, as he went on, came to him in louder and louder bursts. He was attracted to the spot whence the sounds proceeded, and had some difficulty in discovering a doleful swain, who was ensconced in a mass of fern, taller than himself if he had been upright; and but that, by rolling over and over in the turbulence of his grief, he had flattened a large space down to the edge of the forest brook near which he reclined, he would have remained invisible in his lair. The tears in his eyes, and the passionate utterances of his voice, contrasted strangely with a round russetin face, which seemed fortified by beef and ale against all possible furrows of care; but against love, even beef and ale, mighty talismans as they are, are feeble barriers. Cupid's arrows had pierced through the *æs triplex* of treble X, and the stricken deer lay mourning by the stream.

The Doctor approaching, kindly inquired, 'What is the matter?' but was answered only by a redoubled burst of sorrow and an emphatic rejection of all sympathy.

'You can't do me any good.'

'You do not know that,' said the Doctor. 'No man knows what good another can do him till he communicates his trouble.'

For some time the Doctor could obtain no other answer than the repetition of 'You can't do me any good.' But at length the patience and kind face of the inquirer had their effect on the sad shepherd, and he brought out with a desperate effort and a more clamorous explosion of grief,

'She wont have me!'

'Who wont have you?' said the Doctor.

'Well, if you must know,' said the swain, 'you must. It's one of the young ladies up at the Folly.'

'Young ladies?' said the Doctor.

'Servants they call themselves,' said the other; 'but they are more like ladies, and hold their heads high enough when one of them wont have me. Father's is one of the best farms for miles round, and it's all his own. He's a true old yeoman, father is. And there's nobody but him and me. And if I had a nice wife, that would be a good housekeeper for him, and play and sing to him of an evening—for she can do anything, she can—read, and write, and keep accounts, and play and sing—I've heard her—and make a plum pudding—I've seen her—we should be as happy as three crickets—four, perhaps, at the year's end: and she wont have me.'

'You have put the question?' said the Doctor.

'Plump,' said the other. 'And she looked at first as if she was going to laugh. She didn't, though. Then she looked serious, and said she was sorry for me. She said she saw I was in earnest. She knew I was a good son, and deserved a good wife; but she couldn't have me. Miss, said I, do you like anybody better? No, she said, very heartily.'

'That is one comfort,' said the Doctor.

'What comfort,' said the other, 'when she wont have me?

'She may alter her mind,' said the Doctor, 'if she does not prefer any one else. Besides, she only says she can't.'

'Can't,' said the other, 'is civil for wont. That's all.'

'Does she say why she can't?' said the Doctor.

'Yes,' said the other. 'She says she and her sisters wont part with each other and their young master.

'Now,' said the Doctor, 'you have not told me which of the seven sisters is the one in question.'

'It's the third,' said the other. 'What they call the second cook. There's a housekeeper and two cooks, and two housemaids and two waiting-maids. But they only manage for the young master. There are others that wait on them.'

'And what is her name?' said the Doctor.

'Dorothy,' said the other; 'her name is Dorothy. Their names follow like A B C, only that A comes last. Betsey, Catharine, Dorothy, Eleanor, Fanny, Grace, Anna. But they told me it was not the alphabet they were christened from; it was the key of A minor, if you know what that means.'

'I think I do,' said the Doctor, laughing. 'They were christened from the Greek diatonic scale, and make up two conjunct tetrachords, if you know what that means.'

'I can't say I do,' said the other, looking bewildered.

'And so,' said the Doctor, 'the young gentleman, whose name is Algernon, is the Proslambanomenos, or key-note, and makes up the octave. His parents must have designed it as a foretelling, that he and his seven foster-sisters were to live in harmony all their lives. But how did you become acquainted?'

'Why,' said the other, 'I take a great many things to the house from our farm, and it's generally she that takes them in.'

'I know the house well,' said the Doctor, 'and the master, and the maids. Perhaps he may marry, and they may follow the example. Live in hope. Tell me your name.'

'Hedgerow,' said the other; 'Harry Hedgerow. And if you know her, ain't she a beauty?'

'Why, yes,' said the Doctor, 'they are all good looking.'

'And she wont have me,' cried the other, but with a more subdued expression. The Doctor had consoled him, and given him a ray of hope. And they went on their several ways.

The Doctor resumed his soliloquy.

'Here is the semblance of something towards a solution of the difficulty. If one of the damsels should marry, it would break the combination. One will not by herself. But what if seven apple-faced Hedgerows should propose simultaneously, seven notes in the key of A minor, an octave below? Stranger things have happened. I have read of six brothers who had the civility to break their necks in succession, that the seventh, who was the hero of the story, might inherit an estate. But, again and again, why should I trouble myself with match-making? I had better leave things to take their own course.'

Still in his interior *speculum*, the Doctor could not help seeing a dim reflection of himself pronouncing the nuptial benediction on his two young friends.

CHAPTER VII

Indulge Genio: carpamus dulcia: nostrum est
Quod vivis: cinis, et manes, et fabula fies.
Vive memor lethi: fugit hora: hoc quod loquor, inde est.

PERSIUS.

Indulge thy Genius, while the hour's thine own:
Even while we speak, some part of it has flown.
Snatch the swift-passing good: 'twill end ere long
In dust, and shadow, and an old wife's song.

'AGAPÊTUS and Agapêtae,'* said the Reverend Doctor Opimian, the next morning at breakfast, 'in the best sense of the words: that, I am satisfied, is the relation between this young gentleman and his handmaids.'

MRS. OPIMIAN

Perhaps, Doctor, you will have the goodness to make your view of this relation a little more intelligible to me.

THE REVEREND DOCTOR OPIMIAN

Assuredly, my dear. The word signifies 'beloved,' in its purest sense. And in this sense it was used by Saint Paul in reference to some of his female co-religionists and fellow-labourers in the vineyard, in whose houses he occasionally dwelt. And in this sense it was applied to virgins and holy men, who dwelt under the same roof in spiritual love.

MRS. OPIMIAN

Very likely, indeed. You are a holy man, Doctor, but I think, if you were a bachelor, and I were a maid, I should not trust myself to be your aga—aga—

THE REVEREND DOCTOR OPIMIAN

Agapêtê. But I never pretended to this sort of spiritualism. I followed the advice of Saint Paul, who says it is better to marry—

MRS. OPIMIAN

You need not finish the quotation.

* Ἀγαπητὸς καὶ ἀγαπηταί.

THE REVEREND DOCTOR OPIMIAN

Agapêtê is often translated 'adoptive sister.' A very possible relation, I think, where there are vows of celibacy, and inward spiritual grace.

MRS. OPIMIAN

Very possible, indeed: and equally possible where there are none.

THE REVEREND DOCTOR OPIMIAN

But more possible where there are seven adoptive sisters, than where there is only one.

MRS. OPIMIAN

Perhaps.

THE REVEREND DOCTOR OPIMIAN

The manners, my dear, of these damsels towards their young master, are infallible indications of the relations between them. Their respectful deference to him is a symptom in which I cannot be mistaken.

MRS. OPIMIAN

I hope you are not.

THE REVEREND DOCTOR OPIMIAN

I am sure I am not. I would stake all my credit for observation and experience on the purity of the seven Vestals. I am not strictly accurate in calling them so: for in Rome the number of Vestals was only six. But there were seven Pleiads, till one disappeared. We may fancy she became a seventh Vestal. Or as the planets used to be seven, and are now more than fifty, we may pass a seventh Vestal in the name of modern progress.

MRS. OPIMIAN

There used to be seven deadly sins. How many has modern progress added to them?

THE REVEREND DOCTOR OPIMIAN

None, I hope, my dear. But this will be due, not to its own tendencies, but to the comprehensiveness of the old definitions.

MRS. OPIMIAN

I think I have heard something like your Greek word before.

THE REVEREND DOCTOR OPIMIAN

Agapêmonê, my dear. You may have heard the word Agapêmonê.

MRS. OPIMIAN

That is it. And what may it signify?

THE REVEREND DOCTOR OPIMIAN

It signifies Abode of Love: spiritual love, of course.

MRS. OPIMIAN

Spiritual love, which rides in carriages and four, fares sumptuously, like Dives, and protects itself with a high wall from profane observation.

THE REVEREND DOCTOR OPIMIAN

Well, my dear, and there may be no harm in all that.

MRS. OPIMIAN

Doctor, you are determined not to see harm in anything.

THE REVEREND DOCTOR OPIMIAN

I am afraid I see more harm in many things than I like to see. But one reason for not seeing harm in this Agapêmonê matter is, that I hear so little about it. The world is ready enough to promulgate scandal; but that which is quietly right may rest in peace.

MRS. OPIMIAN

Surely, Doctor, you do not think this Agapêmonê right?

THE REVEREND DOCTOR OPIMIAN

I only say I do not know whether it is right or wrong. It is nothing new. Three centuries ago there was a Family of Love, on which Middleton wrote a comedy. Queen Elizabeth persecuted this family; Middleton made it ridiculous; but it outlived them both, and there may have been no harm in it after all.

MRS. OPIMIAN

Perhaps, Doctor, the world is too good to see any novelty except in something wrong.

THE REVEREND DOCTOR OPIMIAN

Perhaps it is only wrong that arrests attention, because right is common, and wrong is rare. Of the many thousand persons who walk daily through a street you only hear of one who has been robbed or knocked down. If ever Hamlet's news—'that the world has grown honest'—should prove true, there would be an end of our newspaper. For, let us see, what is the epitome of a newspaper? In the first place, specimens of all the deadly sins, and infinite varieties of violence and fraud; a great quantity of talk, called by courtesy legislative wisdom, of which the result is 'an incoherent and undigested mass of law, shot down, as from a rubbish-cart, on the heads of the people;* lawyers barking at each other in that peculiar style of hylactic delivery which is called forensic eloquence, and of which the first and most distinguished practitioner was Cerberus;† bear-garden meetings of mismanaged companies, in which directors and shareholders abuse each other in choice terms, not all to be found even in Rabelais; burstings of bank bubbles, which, like a touch of harlequin's wand, strip off their masks and dominoes from 'highly respectable' gentlemen, and leave them in their true figures of cheats and pickpockets; societies of all sorts, for teaching everybody everything, meddling with everybody's business, and mending everybody's morals; mountebank advertisements promising the beauty of Helen in a bottle of cosmetic, and the age of Old Parr in a box of pills; folly all alive in things called reunions; announcements that some exceedingly stupid fellow has been 'entertaining' a select company; matters, however multiform, multifarious, and multitudinous, all brought into family likeness by the varnish of false pretension with which they are all overlaid.

MRS. OPIMIAN

I did not like to interrupt you, Doctor; but it struck me, while you were speaking, that in reading the newspaper you do not hear the bark of the lawyers.

THE REVEREND DOCTOR OPIMIAN

True; but no one who has once heard the wow-wow can fail to reproduce it in imagination.

MRS. OPIMIAN

You have omitted accidents, which occupy a large space in the newspaper. If the world grew ever so honest, there would still be accidents.

* Jeremy Bentham.

† Cerberus forensis erat causidicus.—PETRONIUS ARBITER.

THE REVEREND DOCTOR OPIMIAN

But honesty would materially diminish the number. High-pressure steam boilers would not scatter death and destruction around them, if the dishonesty of avarice did not tempt their employment, where the more costly low pressure would ensure absolute safety. Honestly built houses would not come suddenly down and crush their occupants. Ships, faithfully built and efficiently manned, would not so readily strike on a lee shore, nor go instantly to pieces on the first touch of the ground. Honestly made sweetmeats would not poison children; honestly compounded drugs would not poison patients. In short, the larger portion of what we call accidents are crimes.

MRS. OPIMIAN

I have often heard you say, of railways and steam vessels, that the primary cause of their disasters is the insane passion of the public for speed. That is not crime, but folly.

THE REVEREND DOCTOR OPIMIAN

It is crime in those who ought to know better than to act in furtherance of the folly. But when the world has grown honest, it will no doubt grow wise. When we have got rid of crime, we may consider how to get rid of folly. So that question is adjourned to the Greek kalends.

MRS. OPIMIAN

There are always in a newspaper some things of a creditable character.

THE REVEREND DOCTOR OPIMIAN

When we are at war, naval and military heroism abundantly; but in time of peace, these virtues sleep. They are laid up like ships in ordinary. No doubt, of the recorded facts of civil life some are good, and more are indifferent, neither good nor bad; but good and indifferent together are scarcely more than a twelfth part of the whole. Still, the matters thus presented are all exceptional cases. A hermit reading nothing but a newspaper might find little else than food for misanthropy; but living among friends, and in the bosom of our family, we see that the dark side of life is the occasional picture, the bright is its everyday aspect. The occasional is the matter of curiosity, of incident, of adventure, of things that really happen to few, and may possibly happen to any. The interest attendant on any action or event is in just proportion to its rarity; and, happily, quiet virtues are all around us, and obtrusive vices seldom cross our path. On the whole, I agree in opinion with Theseus,* that there is more good than evil in the world.

MRS. OPIMIAN

I think, Doctor, you would not maintain any opinion if you had not an authority two thousand years old for it.

THE REVEREND DOCTOR OPIMIAN

Well, my dear, I think most opinions worth maintaining have an authority of about that age.

* Eurip. *Suppl.* 207: Herm.

CHAPTER VIII

Ψῦξον τὸν οἶνον, Δῶρι.——
——Ἔγχεον σὺ δὴ πιεῖν·
Εὐζωρότερόν γε νὴ Δί', ὦ παῖ, δός· τὸ γάρ
Ὕδαρες ἅπαν τοῦτ' ἐστὶ τῇ ψυχῇ κακόν.

DIPHILUS.

Cool the wine, Doris. Pour it in the cup
Simple, unmixed with water. Such dilution
Serves only to wash out the spirit of man.

THE Doctor, under the attraction of his new acquaintance, had allowed more time than usual to elapse between his visits to Gryll Grange, and when he resumed them, he was not long without communicating the metamorphosis of the Old Tower, and the singularities of its inhabitants. They dined well as usual, and drank their wine cool.

MISS GRYLL

There are many things in what you have told us that excite my curiosity; but first, what do you suppose is the young gentleman's religion?

THE REVEREND DOCTOR OPIMIAN

From the great liking he seems to have taken to me, I should think he was of the Church of England, if I did not rather explain it by our Greek sympathy. At the same time, he kept very carefully in view that Saint Catharine is a Saint of the English Church Calendar. I imagine there is less of true piety than of an abstract notion of ideal beauty, even in his devotion to her. But it is so far satisfactory that he wished to prove his religion, such as it is, to be within the pale of the Church of England.

MISS GRYLL

I like the idea of his closing the day with a hymn, sung in concert by his seven Vestals.

THE REVEREND DOCTOR OPIMIAN

I am glad that you think charitably of the damsels. It is not every lady that would. But I am satisfied they deserve it.

MR. GRYLL

I should like to know the young gentleman. I wish you could manage to bring him here. Should not you like to see him, Morgana?

MISS GRYLL

Yes, uncle.

MR. GRYLL

Try what you can do, Doctor. We shall have before long some poetical and philosophical visitors. That may tempt him to join us.

THE REVEREND DOCTOR OPIMIAN

It may; but I am not confident. He seems to me to be indisposed to general society, and to care for nothing but woods, rivers, and the sea; Greek poetry, Saint Catharine, and the seven Vestals. However, I will try what can be done.

MR. GRYLL

But, Doctor, I think he would scarcely have provided such a spacious dining-room, and so much domestic accommodation, if he had intended to shut himself up from society altogether. I expect that some day when you go there you will find a large party. Try if he will co-operate in the Aristophanic comedy.

THE REVEREND DOCTOR OPIMIAN

A good idea. That may be something to his mind.

MISS GRYLL

Talking of comedy, Doctor, what has become of Lord Curryfin, and his lecture on fish?

THE REVEREND DOCTOR OPIMIAN

Why, Lord Michin Malicho,* Lord Facing-bothways, and two or three other arch-quacks, have taken to merry-andrewising in a new arena, which they call the Science of Pantopragmatics, and they have bitten Lord Curryfin into tumbling with them; but the mania will subside when the weather grows cool; and no doubt we shall still have him at Thornback Bay, teaching the fishermen how to know a herring from a halibut.

* 'Marry, this is *miching mallecho*: it means mischief.'—*Hamlet.*

MISS GRYLL

But pray, Doctor, what is this new science?

THE REVEREND DOCTOR OPIMIAN

Why that, Miss Gryll, I cannot well make out. I have asked several professors of the science, and have got nothing in return but some fine varieties of rigmarole, of which I can make neither head nor tail. It seems to be a real art of talking about an imaginary art of teaching every man his own business. Nothing practical comes of it, and indeed so much the better. It will be at least harmless, as long as it is like Hamlet's reading, 'words, words, words.' Like most other science, it resolves itself into lecturing, lecturing, lecturing, about all sorts of matters, relevant and irrelevant: one enormous bore prating about jurisprudence, another about statistics, another about education, and so forth; the *crambe repetita* of the same rubbish, which has already been served up 'twiës hot and twiës cold',* at as many other associations nick-named scientific.

MISS GRYLL

Then, Doctor, I should think Lord Curryfin's lecture would be a great relief to the unfortunate audience.

THE REVEREND DOCTOR OPIMIAN

No doubt more amusing, and equally profitable. Not a fish more would be caught for it, and this will typify the result of all such scientific talk. I had rather hear a practical cook lecture on bubble and squeak: no bad emblem of the whole affair.

MR. GRYLL

It has been said a man of genius can discourse on anything. Bubble and squeak seems a limited subject; but in the days of the French revolution there was an amusing poem with that title;† and there might be an amusing lecture; especially if it were like the poem, discursive and emblematical. But men so dismally far gone in the affection of earnestness would scarcely relish it.

* And many a Jacke of Dover hast thou sold,
That hath been twiës hot and twiës cold.
CHAUCER: *The Coke's Prologue.*

† 'Bubble and squeak: a Gallimaufry of British Beef with the Chopped Cabbage of Gallic Philosophy.' By HUDDLESTON.

CHAPTER IX

——gli occhi su levai,
E vidi lei che si facea corona,
Riflettendo da sè gli eterni rai.
DANTE: *Paradiso,* xxxi. 70–72.

I lifted up my gaze,
And looked on her who made herself a crown,
Reflecting from herself the eternal rays.

IT was not long before the Doctor again walked over to the Tower, to propose to his young friend to co-operate in the Aristophanic comedy.

He found him well disposed to do so, and they passed a portion of the afternoon in arranging their programme.

They dined, and passed the evening much as before. The next morning, as they were ascending to the library to resume their pleasant labour, the Doctor said to himself, 'I have passed along galleries wherein were many chambers, and the doors in the day were more commonly open than shut, yet this chamber door of my young friend is always shut. There must be a mystery in it.' And the Doctor, not generally given to morbid curiosity, found himself very curious about this very simple matter.

At last he mustered up courage to say, 'I have seen your library, dining-room, and drawing-room; but you have so much taste in internal arrangements, I should like to see the rest of the house.'

MR. FALCONER

There is not much more to see. You have occupied one of the best bedrooms. The rest do not materially differ.

THE REVEREND DOCTOR OPIMIAN

To say the truth, I should like to see your own.

MR. FALCONER

I am quite willing. But I have thought, perhaps erroneously, it is decorated in a manner you might not altogether approve.

THE REVEREND DOCTOR OPIMIAN

Nothing indecorous, I hope.

MR. FALCONER

Quite the contrary. You may, perhaps, think it too much devoted to my peculiar views of the purity of ideal beauty, as developed in Saint Catharine.

THE REVEREND DOCTOR OPIMIAN

You have not much to apprehend on that score.

MR. FALCONER

You see, there is an altar, with an image of Saint Catharine, and the panels of the room are painted with subjects from her life, mostly copied from Italian masters. The pictures of St. Catharine and her legend very early impressed her on my mind as the type of ideal beauty—of all that can charm, irradiate, refine, exalt, in the best of the better sex.

THE REVEREND DOCTOR OPIMIAN

You are enthusiastic; but indeed, though she is retained as a saint in the Reformed Church, I am not very familiar with her history. And to me some of these pictures require explanation.

MR. FALCONER

I will tell you her legend as briefly as I may. And we will pass from picture to picture as the subjects arise.

THE LEGEND OF SAINT CATHARINE

Catharine was a Princess of Alexandria in the third century. She embraced the Christian religion by divine inspiration. She was pre-eminent in beauty, learning, and discourse. She converted her father and mother, and all with whom she came into communication. The Emperor Maxentius brought together the fifty wisest men of the empire to convert her from the error of her way, and she converted them all to the new faith. Maxentius burned her proselytes, and threatened her with a similar death. She remained firm. He had her publicly scourged, and cast her into prison to perish by famine. Going on an expedition, he left the execution of his orders to the empress and his chief general, Porphyrius. Angels healed her wounds and supplied her with food; and in a beatific vision the Saviour of the world placed a ring on her finger, and called her his bride.* The presence of the ring showed to her the truth of the visitation. The empress and Porphyrius visited the prison, and she converted them also. The emperor, returning, put the empress and Porphyrius to death; and after many ineffectual expostulations with Catharine, determined on putting her to death by the wheel which bears her name. Four of these wheels, armed with iron teeth, and revolving towards each other, were to cut her to pieces. Angels broke the wheels. He then brought her to the stake, and the angels extinguished the flames. He then ordered her to be beheaded by the sword. This was permitted, and in the meantime the day had closed. The body, reserved for exposure to wild beasts, was left under guard at the place of execution. Intense darkness fell on the night, and in the morning the body had disappeared. The angels had borne it to the summit of the loftiest mountain of the Horeb range, where still a rock, bearing the form of a natural sarcophagus, meets the eye of the traveller. Here it was watched by angel-guards, and preserved in unchanging beauty, till, in the fulness of time, it was revealed to a holy man, who removed it to the shrine, under which it lies to this day, with the ring still on its hand, in the convent which was then founded, and which bears her name—the convent of Saint Catharine of Mount Sinai.

THE REVEREND DOCTOR OPIMIAN

Most of this is new to me. Yet I am not unfamiliar with pictures of the Marriage of Saint Catharine, which was a favourite subject with the great Italian masters. But here is a picture which the legend, as you have related it, does not illustrate. What is this tomb, with flames bursting from it, and monks and others recoiling in dismay?

* Maria, Vergine delle Vergini, e Misericordia delle Misericordie, vestita de i lampi del Sole, e coronata de i raggi delle Stelle, prese il sottile, il delicato, ed il sacro dito di Catarina, humile di core e mansueta di vita, ed il largo, il clemente, ed il pietoso figliuol suo lo cinse con lo anello.—*Vita di Santa Catarina*, L. II. Vinezia, 1541.

MR. FALCONER

It represents a remarkable incident at the tomb of the saint. The Empress Catharine II. was a great benefactress to the Convent of Mount Sinai, and desired to possess Saint Catharine's ring. She sent a mitred Abbot as an envoy to request it from the brotherhood. The monks, unwilling to displease the Empress, replied that they did not dare to remove it themselves, but that they would open the tomb, and the envoy might take it. They opened the tomb accordingly, and the envoy looked on the hand and the ring. He approached to draw it off, but flames burst forth: he recoiled, and the tomb closed. Under such a manifestation of the saint's displeasure, the fathers could not again attempt to open it.*

THE REVEREND DOCTOR OPIMIAN

I should like to have seen the Empress receiving the envoy's report.

MR. FALCONER

Her reception of it would depend on the degree of faith which she either actually felt, or might have thought it politic to assume. At any rate, the fathers had shown their devotion, and afforded her a good opportunity for exhibiting hers. She did not again seek to obtain the ring.

THE REVEREND DOCTOR OPIMIAN

Now, what are these three pictures in one frame, of chapels on hills?

MR. FALCONER

These chapels are here represented as they may be supposed to have been in the Catholic days of England. Three sisters, named Catharine, Martha, and Anne, built them to their namesake saints, on the summits of three hills, which took from these dedications the names they still bear. From the summit of each of these chapels the other two were visible. The sisters thought the chapels would long remain memorials of Catholic piety and sisterly love. The Reformation laid them in ruins. Nothing remains of the chapel of Saint Anne but a few grey stones, built into an earthen wall, which, some half century ago, enclosed a plantation. The hill is now better known by the memory of Charles Fox, than by that of its ancient

* *Illustrations of Jerusalem and Mount Sinai* (1837), p. 27.

saint. The chapel of Saint Martha has been restored and applied to Protestant worship. The chapel of Saint Catharine remains a picturesque ruin, on the banks of the Wey, near Guildford.

THE REVEREND DOCTOR OPIMIAN

And that old church?

MR. FALCONER

That was the church of St. Catharine, which was pulled down to make way for the dock by which her name is now profaned; an act of desecration which has been followed by others, and will be followed by many more, whenever it may suit the interests of commerce to commit sacrilege on consecrated ground and dissipate the ashes of the dead; an act which, even when that of a barbarian invader, Horace thought it would be profanation even to look on.* Whatever may be in other respects the superiority of modern piety, we are far inferior to the ancients in reverence for temples and tombs.

THE REVEREND DOCTOR OPIMIAN

I am afraid I cannot gainsay that observation. But what is that stained glass window?

MR. FALCONER

It is copied on a smaller scale, and with more of Italian artistic beauty in the principal figure, from the window in West Wickham church. She is trampling on the Emperor Maxentius. You see all her emblems: the palm, which belongs to all sainted martyrs; the crown, the wheel, the fire, the sword, which belong especially to her; and the book, with which she is always represented, as herself a miracle of learning, and its chosen universal patroness in the schools of the Middle Ages.

THE REVEREND DOCTOR OPIMIAN

Unquestionably the legend is interesting. At present, your faith is simply poetical. But take care, my young friend, that you do not finish by becoming the dupe of your own mystification.

MR. FALCONER

I have no fear of that. I think I can clearly distinguish devotion

* *Epod.*, 16, 13.

to ideal beauty from superstitious belief. I feel the necessity of some such devotion, to fill up the void which the world, as it is, leaves in my mind. I wish to believe in the presence of some local spiritual influence; genius or nymph; linking us by a medium of something like human feeling, but more pure and more exalted, to the all-pervading, creative, and preservative spirit of the universe; but I cannot realize it from things as they are. Everything is too deeply tinged with sordid vulgarity. There can be no intellectual power resident in a wood, where the only inscription is not '*Genio loci*,' but 'Trespassers will be prosecuted;' no Naiad in a stream that turns a cotton-mill; no Oread in a mountain dell, where a railway train deposits a cargo of Vandals; no Nereids or Oceanitides along the sea-shore, where a coast-guard is watching for smugglers. No; the intellectual life of the material world is dead. Imagination cannot replace it. But the intercession of saints still forms a link between the visible and invisible. In their symbols I can imagine their presence. Each in the recess of our own thought we may preserve their symbols from the intrusion of the world. And the saint, whom I have chosen, presents to my mind the most perfect ideality of physical, moral, and intellectual beauty.

THE REVEREND DOCTOR OPIMIAN

I cannot object to your taste. But I hope you will not be led into investing the ideality with too much of the semblance of reality. I should be sorry to find you far gone in hagiolatry. I hope you will acquiesce in Martin, keeping equally clear of Peter and Jack.

MR. FALCONER

Nothing will more effectually induce me so to acquiesce, than your company, dear Doctor. A tolerant liberality like yours has a very persuasive influence.

From this digression, the two friends proceeded to the arrangement of their Aristophanic comedy, and divided their respective shares after the manner of Beaumont and Fletcher.

CHAPTER X

Si bene calculum ponas, ubique naufragium est.
PETRONIUS ARBITER.

If you consider well the events of life, shipwreck is everywhere.

AFTER luncheon the Doctor thought of returning home, when a rumbling of distant thunder made him pause. They reascended the tower, to reconnoitre the elements from the library. The windows were so arranged as to afford a panoramic view.

The thunder muttered far off, but there was neither rain nor visible lightning.

'The storm is at a great distance,' said the Doctor, 'and it seems to be passing away on the verge of the sky.'

But on the opposite horizon appeared a mass of dark-blue cloud, which rose rapidly, and advanced in the direct line of the Tower. Before it rolled a lighter but still lurid volume of vapour, which curled and wreathed like eddying smoke before the denser blackness of the unbroken cloud.

Simultaneously followed the flashing of lightning, the rolling of thunder, and a deluge of rain like the bursting of a waterspout.

They sate some time in silence, watching the storm as it swept along, with wind, and driving rain, and whirling hail, bringing for a time almost the darkness of night, through which the forked lightning poured a scarcely interrupted blaze.

Suddenly came a long dazzling flash, that seemed to irradiate the entire circumference of the sky, followed instantaneously by one of those crashing peals of thunder, which always indicate that something very near has been struck by the lightning.

The Doctor turned round to make a remark on the awful grandeur of the effect, when he observed that his young friend had disappeared. On his return, he said he had been looking for what had been struck.

'And what was?' said the Doctor.

'Nothing in the house,' said his host.

'The Vestals,' thought the Doctor; 'these were all his solicitude.'

But though Mr. Falconer had looked no further than to the safety of the seven sisters, his attention was soon drawn to a tumult below, which seemed to indicate that some serious mischief had resulted from the lightning; and the youngest of the sisters, appearing in great trepidation, informed him that one of two horses in a gentleman's carriage had been struck dead, and that a young lady in the carriage had been stunned by the passing flash, though how far she was injured by it could not be immediately known. The other horse, it appeared, had been prancing in terror, and had nearly overthrown the carriage; but he had been restrained by the vigorous arm of a young farmer, who had subsequently carried the young lady into the house, where she was now resting on a couch in the female apartments, and carefully attended by the sisters.

Mr. Falconer and the Doctor descended into the hall, and were assured that the young lady was doing well, but that she would be much the better for being left some time longer undisturbed. An elderly gentleman issued from the female apartments, and the Doctor with some amazement recognised his friend Mr. Gryll, to whom and his niece this disaster had occurred.

The beauty of the morning had tempted them to a long drive; and they thought it would be a good opportunity to gratify at least a portion of the curiosity which the Doctor's description of the Folly and its inhabitants had excited in them. They had therefore determined on taking a circuit, in which they would pass under the walls of the Tower. They were almost at the extremity of their longest radius when the storm burst over them, and were just under the Tower when the lightning struck one of their horses. Harry Hedgerow was on his way with some farm produce when the accident occurred, and was the young farmer who had subdued the surviving horse and carried the young lady into the house. Mr. Gryll was very panegyrical of this young man's behaviour, and the Doctor when he recognised him shook him heartily by the hand, and told him he felt sure that he was a lad who would make his way: a remark which Harry received as a good omen: for Dorothy heard it, and looked at him with a concurrent, though silent, approbation.

The drawing-room and the chambers for visitors were between the tower and the *gynæceum*, or female apartments, which were as completely separated from the rest of the house as they could have been in Athens.

After some anxious inquiries, it was reported that the young lady was sleeping, and that one or other of the sisters would keep constant watch by her. It was therefore arranged that Mr. Gryll should dine and pass the night where he was. Before dinner he had the satisfaction of hearing from medical authority that all would be well after a little time.

Harry Hedgerow had bethought him of a retired physician, who lived with a maiden sister in a cottage at no great distance from the Tower, and who often gave gratuitous advice to his poorer neighbours. If he prescribed anything beyond their means, himself or his sister was always ready to supply it. Though their own means were limited, they were the good angels of a small circumference

The old physician confirmed the opinion already given by the sisters, that the young lady for the present only required repose: but he accepted the invitation to remain till the morning, in the event of his advice being needed.

So Miss Gryll remained with the elder sisters. Mr. Gryll and the two Doctors, spiritual and temporal, sat down to dinner with Mr. Falconer, and were waited on, as usual, by the younger handmaids.

CHAPTER XI

Οἴνου μὴ παρεόντος, ἀτερπέα δεῖπνα τραπέζης·
Οἴνου μὴ παρεόντος, ἀθελγέες εἰσὶ χορεῖαι.
Ἀνὴρ πένθος ἔχων, ὅτε γεύσεται ἡδέος οἴνου,
Στυγνὸν ἀεξομένης ἀποσείσεται ὄγκον ἀνίης.

Where wine is not, no mirth the banquet knows:
Where wine is not, the dance all joyless goes.
The man, oppressed with cares, who tastes the bowl,
Shall shake the weight of sorrow from his soul.

BACCHUS, on the birth of the vine, predicting its benefits: in the twelfth book of the *Dionysiaca* of NONNUS

THE conversation at dinner turned on the occurrences of the morning and the phenomena of electricity. The physician, who had been a traveller, related many anecdotes from his own observation; especially such as tended to show by similarity that the injury to Miss Gryll would not be of long duration. He had known, in similar cases, instances of apparent total paralysis; but he had always found it temporary. Perhaps in a day or two, but at most in a very few days, it would certainly pass away. In the meantime, he recommended absolute repose. Mr. Falconer entreated Mr. Gryll to consider the house as his own. Matters were arranged accordingly; and it was determined that the next morning a messenger should be despatched to Gryll Grange for a supply of apparel. The Reverend Dr. Opimian, who was as fond as the Squire himself of the young lady, had been grievously discomposed by the accident of the morning, and felt that he should not thoroughly recover his serenity till he could again see her in her proper character, the light and life of her society. He quoted Homer, Æschylus, Aristotle, Plutarch, Athenæus, Horace, Persius, and Pliny, to show that all which is practically worth knowing on the subject of electricity had been known to the ancients. The electric telegraph he held to be a nuisance, as disarranging chronology, and giving only the heads of a chapter, of which the details lost their interest before they arrived, the heads of another chapter having intervened to destroy it. Then, what an amount of misery it inflicted, when, merely saying that there had been a great battle, and that thousands had been wounded or killed, it maintained an agony of suspense in all who had friends on the field, till the ordinary channels of intelligence brought the names of the sufferers. No Sicilian tyrant had invented such an engine of cruelty. This declamation against a supposed triumph of modern science, which was listened to with some surprise by the physician, and with great respect by his other auditors, having somewhat soothed his troubled spirit, in conjunction with the physician's assurance, he propitiated his Genius by copious libations of claret, pronouncing high panegyrics on the specimen before him, and interspersing quotations in praise of wine, as the one great panacea for the cares of this world.

A week passed away and the convalescent had made good progress. Mr. Falconer had not yet seen his fair guest. Six of the sisters, one remaining with Miss Gryll, performed every evening, at the earnest request of Mr. Gryll, a great variety of music, but always ending with the hymn to their master's saint. The old physician came once or twice and stayed the night. The Reverend Doctor Opimian went home for his Sunday duties, but took too much interest in the fair Morgana not to return as soon as he could to the Tower. Arriving one morning in the first division of the day, and ascending to the library, he found his young friend writing. He asked him if he were working on the Aristophanic comedy? Mr. Falconer said, he got on best with that in the Doctor's company. 'But I have been writing,' he said, 'on something connected with the Athenian drama. I have been writing a ballad on the death of Philemon, as told by Suidas and Apuleius.' The Doctor expressed a wish to hear it, and Mr. Falconer read it to him.

THE DEATH OF PHILEMON.*

I

Closed was Philemon's hundredth year:
The theatre was thronged to hear
His last completed play:
In the mid scene, a sudden rain
Dispersed the crowd—to meet again
On the succeeding day.

* SUIDAS: *sub voce Φιλήμων*. APULEIUS: *Florid.* 16.

He sought his home, and slept, and dreamed,
Nine maidens through his door, it seemed,
Passed to the public street.
He asked them, 'Why they left his home?'
They said, 'A guest will hither come,
We must not stay to meet.'

He called his boy with morning light,
Told him the vision of the night,
And bade his play be brought.
His finished page again he scanned,
Resting his head upon his hand,
Absorbed in studious thought.

He knew not what the dream foreshowed:
That nought divine may hold abode
Where death's dark shade is felt:
And therefore were the Muses nine
Leaving the old poetic shrine,
Where they so long had dwelt.

II

The theatre was thronged once more,
More thickly than the day before,
To hear the half-heard song.
The day wore on. Impatience came.
They called upon Philemon's name,
With murmurs loud and long.

Some sought at length his studious cell,
And to the stage returned, to tell
What thousands strove to ask.
'The poet we have been to seek
Sate with his hand upon his cheek,
As pondering o'er his task.

'We spoke. He made us no reply.
We reverentially drew nigh,
And twice our errand told.
He answered not. We drew more near:
The awful mystery then was clear:
We found him stiff and cold.

'Struck by so fair a death, we stood
Awhile in sad admiring mood;
Then hastened back, to say
That he, the praised and loved of all,
Is deaf for ever to your call:
That on this self-same day,

'When here presented should have been
The close of his fictitious scene,
His life's true scene was o'er:
We seemed, in solemn silence awed,
To hear the "Farewell and applaud,"
Which he may speak no more.

'Of tears the rain gave prophecy:
The nuptial dance of comedy
Yields to the funeral train.
Assemble where his pyre must burn:
Honour his ashes in their urn:
And on another day return
To hear his songs again.'

THE REVEREND DOCTOR OPIMIAN

A beautiful fiction.

MR. FALCONER

If it be a fiction. The supernatural is confined to the dream. All the rest is probable; and I am willing to think it true, dream and all.

THE REVEREND DOCTOR OPIMIAN

You are determined to connect the immaterial with the material world, as far as you can.

MR. FALCONER

I like the immaterial world. I like to live among thoughts and images of the past and the possible, and even of the impossible, now and then.

THE REVEREND DOCTOR OPIMIAN

Certainly, there is much in the material world to displease sensitive and imaginative minds; but I do not know any one who has less cause to complain of it than you have. You are surrounded with all possible comforts, and with all the elements of beauty, and of intellectual enjoyment.

MR. FALCONER

It is not my own world that I complain of. It is the world on which I look 'from the loop-holes of retreat.' I cannot sit here, like one of the Gods of Epicurus, who, as Cicero says, was satisfied with thinking, through all eternity, 'how comfortable he was.'* I look with feelings of intense pain on the mass of poverty and crime; of unhealthy, unavailing, unremunerated toil, blighting childhood in its blossom, and womanhood in its prime; of 'all the oppressions that are done under the sun.'

THE REVEREND DOCTOR OPIMIAN

I feel with you on all these points; but there is much good in the world; more good than evil, I have always maintained.

They would have gone off in a discussion on this point, but the French clock warned them to luncheon.

In the evening the young lady was sufficiently recovered to join the little party in the drawing-room, which consisted, as before, of Mr. Falconer, Mr. Gryll, Doctor Anodyne, and the Reverend Doctor Opimian. Miss Gryll was introduced to Mr. Falconer. She was full of grateful encomium for the kind attention of the sisters, and expressed an earnest desire to hear their music. The wish was readily complied with. She heard them with great pleasure, and, though not yet equal to much exertion, she could not refrain from joining in with them in their hymn to Saint Catharine.

She accompanied them when they retired.

THE REVEREND DOCTOR OPIMIAN

I presume those Latin words are genuine old monastic verses: they have all the air of it.

MR. FALCONER

They are so, and they are adapted to old music.

DOCTOR ANODYNE

There is something in this hymn very solemn and impressive. In an age like ours, in which music and pictures are the predominant tastes, I do not wonder that the forms of the old Catholic worship are received with increasing favour. There is a sort of adhesion to the old religion, which results less from faith than from a certain

* Comprehende igitur animo, et propone ante oculos, deum nihil aliud in omni æternitate, nisi, Mihi pulchre est, et, Ego beatus sum, cogitantem.—Cicero: *De Naturâ Deorum;* l. i. c. 41.

feeling of poetry; it finds its disciples; but it is of modern growth; and has very essential differences from what it outwardly resembles.

THE REVEREND DOCTOR OPIMIAN

It is, as I have frequently had occasion to remark, and as my young friend here will readily admit, one of the many forms of the love of ideal beauty, which, without being in itself religion, exerts on vivid imaginations an influence that is very often like it.

MR. FALCONER

An orthodox English Churchman was the poet who sang to the Virgin:

> Thy image falls to earth. Yet some, I ween,
> Not unforgiven the suppliant knee might bend,
> As to a visible Power, in which did blend
> All that was mixed and reconciled in thee,
> Of mother's love with maiden purity,
> Of high with low, celestial with terrene.*

THE REVEREND DOCTOR OPIMIAN

Well, my young friend, the love of ideal beauty has exercised none but a benignant influence on you, whatever degree of orthodoxy there may be in your view of it.

The little party separated for the night.

* WORDSWORTH: *Ecclesiastical Sonnets*, i. 21.

CHAPTER XII

Τί δεῖ γὰρ ὄντα θνητόν, ἱκετεύω, ποιεῖν,
Πλὴν ἡδέως ζῆν τὸν βίον καθ' ἡμέραν,
Ἐὰν ἔχῃ τις ὁπόθεν. . . . . .
Εἰς αὔριον δὲ μηδὲ φροντίζειν ὅ τι
Ἔσται . .

PHILETAERUS: *Cynagis.*

I pray you, what can mortal man do better,
Than live his daily life as pleasantly
As daily means avail him? Life's frail tenure
Warns not to trust to-morrow.

THE next day Mr. Falconer was perfectly certain that Miss Gryll was not yet well enough to be removed. No one was anxious to refute the proposition; they were all so well satisfied with the place and the company they were in, that they felt, the young lady included, a decided unwillingness to go. That day Miss Gryll came to dinner, and the next day she came to breakfast, and in the evening she joined in the music, and in short she was once more altogether herself; but Mr. Falconer continued to insist that the journey home would be too much for her. When this excuse failed, he still entreated his new friends to remain; and so passed several days. At length Mr. Gryll found he must resolve on departing, especially as the time had arrived when he expected some visitors. He urgently invited Mr. Falconer to visit him in return. The invitation was cordially accepted, and in the meantime considerable progress had been made in the Aristophanic comedy.

Mr. Falconer, after the departure of his visitors, went up into his library. He took down one book after another, but they did not fix his attention as they had used to do; he turned over the leaves of Homer, and read some passages about Circe; then took down Bojardo, and read of Morgana and Falerina and Dragontina; then took down Tasso and read of Armida. He would not look at Ariosto's Alcina, because her change into an old woman destroyed all the charm of the previous picture. He dwelt on the enchantresses, who remained in unaltered beauty. But even this he did only by fits and starts, and found himself continually wandering away towards a more enchanting reality.

He descended to his bedroom, and meditated on ideal beauty in the portraits of Saint Catharine. But he could not help thinking that the ideal might be real, at least in one instance, and he wandered down into his drawing-room. There he sat absorbed in thought, till his two young handmaids appeared with his luncheon. He smiled when he saw them, and sat down to the table as if nothing had disturbed him. Then, taking his stick and his dog, he walked out into the forest.

There was within moderate distance a deep dell, in the bottom of which ran a rivulet, very small in dry weather, but in heavy rains becoming a torrent, which had worn itself a high-banked channel, winding in fantastic curves from side to side of its narrow boundaries. Above this channel old forest trees rose to a great height on both sides of the dell. The slope every here and there was broken by promontories which during centuries the fall of the softer portions of the soil had formed; and on these promontories were natural platforms, covered, as they were more or less accessible to the sun, with grass and moss and fern and foxglove, and every variety of forest vegetation. These platforms were favourite resorts of deer, which imparted to the wild scene its own peculiar life.

This was a scene in which, but for the deeper and deeper wear of the floods and the bolder falls of the promontories, time had made little change. The eyes of the twelfth century had seen it much as it appeared to those of the nineteenth. The ghosts of departed ages might seem to pass through it in succession, with all their changes of faith and purpose and manners and costume. To a man who loved to dwell in the past, there could not be a more congenial scene. One old oak stood in the centre of one of the green platforms, and a portion of its gnarled roots presented a convenient seat. Mr. Falconer had frequently passed a day here when alone. The deer had become too accustomed to him to fly at his approach, and the dog had been too well disciplined to molest them. There he had sat for hours at a time, reading his favourite poets. There was no great poet with some of whose scenes this scenery did not harmonize. The deep woods that surrounded the dwelling of Circe, the obscure sylvan valley in which Dante met Virgil, the forest depths through which Angelica fled, the enchanted wood in which Rinaldo met the semblance of Armida, the forest-brook by which Jaques moralized over the wounded deer, were all reproduced in this single spot, and fancy peopled it at pleasure with nymphs and genii, fauns and satyrs, knights and ladies, friars, foresters, hunters, and huntress maids, till the whole diurnal world seemed to pass away like a vision. There, for him, Matilda had gathered flowers on the opposite bank;* Laura had risen from one of the little pools—resting-places of the stream—to seat herself in the shade;† Rosalind and Maid Marian had peeped forth from their alleys green; all different in form, in feature, and in apparel; but now they were all one; each, as she rose in imagination, presented herself under the aspect of the newly-known Morgana.

Finding his old imaginations thus disturbed, he arose and walked home. He dined alone, drank a bottle of Madeira as if it had been so much water, summoned the seven sisters to the drawing-room earlier, and detained them later than usual, till their music and its old associations had restored him to something like tranquillity. He had always placed the *summum bonum* of life in tranquillity and not in excitement. He felt that his path was now crossed by a disturbing force, and determined to use his utmost exertions to avoid exposing himself again to its influence.

In this mood the Reverend Doctor Opimian found him one morning in the library, reading. He sprang up to meet the divine, exclaiming, 'Ah, dear Doctor, I am very glad to

* DANTE: *Purgatorio*, c. 28.

† Or in forma di Ninfa o d'altra Diva,
Che del più chiaro fondo di Sorga esca,
E pongasi a seder in sulla riva.
PETRARCA: *Sonetto* 240.

see you. Have you any especial favourite among the Odes of Pindar?'

The Doctor thought this an odd question for the first salutation. He had expected that the first inquiry would have been for the fair convalescent. He divined that the evasion of this subject was the result of an inward struggle. He thought it would be best to fall in with the mood of the questioner, and said, 'Charles Fox's favourite is said to have been the second Olympic; I am not sure that there is, or can be, anything better. What say you?'

MR. FALCONER

It may be that something in it touches a peculiar tone of feeling; but to me there is nothing like the ninth Pythian.

THE REVEREND DOCTOR OPIMIAN

I can understand your fancy for that ode. You see an image of ideal beauty in the nymph Cyrene.

MR. FALCONER

'Hidden are the keys of wise persuasion of sacred endearments,'* seems a strange phrase in English; but in Greek the words invest a charming sentiment with singular grace. Fit words to words as closely as we may, the difference of the mind which utters them fails to reproduce the true semblance of the thought. The difference of the effect, produced, as in this instance, by exactly corresponding words, can only be traced to the essential difference of the Greek and the English mind.

THE REVEREND DOCTOR OPIMIAN

And indeed, as with the words so with the image. We are charmed by Cyrene wrestling with the lion; but we should scarcely choose an English girl so doing as the type of ideal beauty.

MR. FALCONER

We must draw the image of Cyrene, not from an English girl, but from a Greek statue.

THE REVEREND DOCTOR OPIMIAN

Unless a man is in love, and then to him all images of beauty take something of the form and features of his mistress.

* *Κρυπταὶ κλαΐδες ἐντι σοφᾶς Πειθοῦς ἱερᾶν φιλοτάτων.*

MR. FALCONER

That is to say, a man in love sees everything through a false medium. It must be a dreadful calamity to be in love.

THE REVEREND DOCTOR OPIMIAN

Surely not, when all goes well with it.

MR. FALCONER

To me it would be the worst of all mischances.

THE REVEREND DOCTOR OPIMIAN

Every man must be subject to Love once in his life. It is useless to contend with him. 'Love', says Sophocles, 'is unconquered in battle, and keeps his watch in the soft cheeks of beauty.'*

MR. FALCONER

I am afraid, Doctor, the Morgana to whom you have introduced me is a veritable enchantress. You find me here, determined to avoid the spell.

THE REVEREND DOCTOR OPIMIAN

Pardon me. You were introduced, as Jupiter was to Semele, by thunder and lightning, which was, happily, not quite as fatal.

MR. FALCONER

I must guard against its being as fatal in a different sense; otherwise I may be myself the *triste bidental*.† I have aimed at living, like an ancient Epicurean, a life of tranquillity. I had thought myself armed with triple brass against the folds of a three-formed Chimæra. What with classical studies, and rural walks, and a domestic society peculiarly my own, I led what I considered the perfection of life: 'days so like each other they could not be remembered.'‡

THE REVEREND DOCTOR OPIMIAN

It is vain to make schemes of life. The world will have its slaves, and so will Love.

* *Ἔρως ἀνίκατε μάχαν, κ.τ.λ.—Antigone.*

† *Bidental* is usually a place struck by lightning: thence enclosed, and the soil forbidden to be moved. Persius uses it for a person so killed.

‡ WORDSWORTH: *The Brothers.*

Say, if you can, in what you cannot change.
For such the mind of man, as is the day
The Sire of Gods and men brings over him.*

MR. FALCONER

I presume, Doctor, from the complacency with which you speak of Love, you have had no cause to complain of him.

THE REVEREND DOCTOR OPIMIAN

Quite the contrary. I have been an exception to the rule, that 'The course of true love never did run smooth.' Nothing could run more smooth than mine. I was in love. I proposed. I was accepted. No crossings before. No bickerings after. I drew a prize in the lottery of marriage.

MR. FALCONER

It strikes me, Doctor, that the lady may say as much.

THE REVEREND DOCTOR OPIMIAN

I have made it my study to give her cause to say so. And I have found my reward.

MR. FALCONER

Still, yours is an exceptional case. For, as far as my reading and limited observation have shown me, there are few happy marriages. It has been said by an old comic poet, that 'a man, who brings a wife into his house, brings into it with her either a good or an evil genius.'† And I may add from Juvenal: 'The Gods only know which it will be.'‡

---

* Quid placet aut odio est, quod non mutabile credas?
Τοῖος γαρ νόος ἐστὶν ἐπιχθονίων ἀνθρώπων,
Οἷον ἐπ' ἦμαρ ἄγῃσι πατὴρ ἀνδρῶν τε θεῶν τε.

These two quotations form the motto of KNIGHT's *Principles of Taste*.

† *Ὅταν γὰρ ἄλοχον εἰς δόμους ἄγῃ πόσις,*
Οὐχ ὡς δοκεῖ γυναῖκα λαμβάνει μόνον,
Ὁμοῦ δὲ τῇδ' ἐπεισκομίζεται λαβὼν
Καὶ δαίμον' ἤτοι χρηστὸν ἢ τοὐναντίον.
THEODECTES: *apud Stobaeum*.

‡ Conjugium petimus partumque uxoris, at illis
Notum, qui pueri, qualisque futura sit uxor.
JUV. *Sat*. x. 352–3.

THE REVEREND DOCTOR OPIMIAN

Well, the time advances for the rehearsals of our Aristophanic comedy, and independently of your promise to visit the Grange, and their earnest desire to see you, you ought to be there to assist in the preliminary arrangements.

MR. FALCONER

Before you came, I had determined not to go; for to tell you the truth, I am afraid of falling in love.

THE REVEREND DOCTOR OPIMIAN

It is not such a fearful matter. Many have been the better for it. Many have been cured of it. It is one of those disorders which every one must have once.

MR. FALCONER

The later the better.

THE REVEREND DOCTOR OPIMIAN

No; the later the worse, if it falls into a season when it cannot be reciprocated.

MR. FALCONER

That is just the season for it. If I were sure it would not be reciprocated, I think I should be content to have gone through it.

THE REVEREND DOCTOR OPIMIAN

Do you think it would be reciprocated?

MR. FALCONER

Oh! no. I only think it possible that it might be.

THE REVEREND DOCTOR OPIMIAN

Well, there is a gentleman doing his best to bring about your wish.

MR. FALCONER

Indeed! who?

THE REVEREND DOCTOR OPIMIAN

A visitor at the Grange, who seems in great favour with both uncle and niece—Lord Curryfin.

MR. FALCONER

Lord Curryfin! I never heard you speak of him, but as a person to be laughed at.

THE REVEREND DOCTOR OPIMIAN

That was my impression of him, before I knew him. Barring his absurdities, in the way of lecturing on fish, and of shining in absurd company in the science of pantopragmatics, he has very much to recommend him: and I discover in him one quality which is invaluable. He does all he can to make himself agreeable to all about him, and he has great tact in seeing how to do it. In any intimate relation of life—with a reasonable wife, for instance, he would be the pink of a good husband.

The Doctor was playing, not altogether unconsciously, the part of an innocent Iago. He said only what was true, and he said it with a good purpose; for with all his repeated resolutions against match-making, he could not dismiss from his mind the wish to see his young friends come together; and he would not have liked to see Lord Curryfin carry off the prize through Mr. Falconer's neglect of his opportunity. Jealousy being the test of love, he thought a spice of it might be not unseasonably thrown in.

MR. FALCONER

Notwithstanding your example, Doctor, love is to be avoided, because marriage is at best a dangerous experiment. The experience of all time demonstrates that it is seldom a happy condition. Jupiter and Juno, to begin with; Venus and Vulcan. Fictions to be sure, but they show Homer's view of the conjugal state. Agamemnon in the shades, though he congratulates Ulysses on his good fortune in having an excellent wife, advises him not to trust even her too far. Come down to realities, even to the masters of the wise: Socrates with Xantippe; Euripides with his two wives, who made him a woman-hater; Cicero, who was divorced; Marcus Aurelius.—Travel downwards: Dante, who, when he left Florence, left his wife behind him; Milton, whose first wife ran away from him; Shakspeare, who scarcely shines in the light of a happy husband. And if such be the lot of the lights of the world, what can humbler men expect?

THE REVEREND DOCTOR OPIMIAN

You have given two or three heads of a catalogue which, I admit, might be largely extended. You can never read a history, you can never open a newspaper, without seeing some example of unhappy marriage. But the conspicuous are not the frequent. In the quiet path of everyday life—the *secretum iter et fallentis semita vitæ*—I could show you many couples who are really comforts and helpmates to each other. Then, above all things, children. The great blessing of old age, the one that never fails, if all else fail, is a daughter.

MR. FALCONER

All daughters are not good.

THE REVEREND DOCTOR OPIMIAN

Most are. Of all relations in life, it is the least disappointing: where parents do not so treat their daughters as to alienate their affections, which unhappily many do.

MR. FALCONER

You do not say so much for sons.

THE REVEREND DOCTOR OPIMIAN

Young men are ambitious, self-willed, self-indulgent, easily corrupted by bad example, of which there is always too much. I cannot say much for those of the present day, though it is not absolutely destitute of good specimens.

MR. FALCONER

You know what Paterculus says of those of his own day.

THE REVEREND DOCTOR OPIMIAN

'The faith of wives towards the proscribed was great; of freedmen, middling; of slaves, some; of sons, none.'* So he says; but there was some: for example, of the sons of Marcus Oppius and Quintus Cicero.† You may observe, by the way, he gives the first place to the wives.

* Id tamen notandum est, fuisse in proscriptos uxorum fidem summam, libertorum mediam, servorum aliquam, filiorum nullam. PATERCULUS: l. ii. c. 67.

† A compendious and comprehensive account of these and other instances of filial piety, in the proscription of the second triumvirate, will be found in *Freinshemius*; *Supplementa Liviana*; cxx: 77–80.

MR. FALCONER

Well, that is a lottery in which every man must take his chance. But my scheme of life was perfect.

THE REVEREND DOCTOR OPIMIAN

Perhaps there is something to be said against condemning seven young women to celibacy.

MR. FALCONER

But if such were their choice—

THE REVEREND DOCTOR OPIMIAN

No doubt there are many reasons why they should prefer the condition they are placed in to the ordinary chances of marriage: but after all, to be married is the natural aspiration of a young woman, and if favourable conditions presented themselves—

MR. FALCONER

Conditions suitable to their education are scarcely compatible with their social position.

THE REVEREND DOCTOR OPIMIAN

They have been educated to be both useful and ornamental. The ornamental need not, and in their case certainly does not, damage the useful, which in itself would procure them suitable matches.

Mr. Falconer shook his head, and after a brief pause poured out a volume of quotations, demonstrating the general unhappiness of marriage. The Doctor responded by as many demonstrating the contrary. He paused to take breath. Both laughed heartily. But the result of the discussion and the laughter was, that Mr. Falconer was curious to see Lord Curryfin, and would therefore go to Gryll Grange.

CHAPTER XIII

Ille potens sui
Laetusque deget, cui licet in diem
Dixisse, Vixi: cras vel atrâ
Nube polum pater occupato,
Vel sole puro: non tamen irritum
Quodcumque retro est efficiet; neque
Diffinget infectumque reddet,
Quod fugiens semel hora vexit.
HOR. *Carm.* iii. 29.

Happy the man, and happy he alone,
He who can call to-day his own:
He who, secure within, can say,
To-morrow do thy worst, for I have lived to-day.
Be storm, or calm, or rain, or shine,
The joys I have possessed in spite of fate are mine.
Not heaven itself upon the past has power,
But what has been has been, and I have had my hour.
DRYDEN.

A LARGE party was assembled at the Grange. Among them were some of the young ladies who were to form the chorus; one elderly spinster, Miss Ilex, who passed more than half her life in visits, and was everywhere welcome, being always good-humoured, agreeable in conversation, having much knowledge of society, good sense in matters of conduct, good taste and knowledge in music; sound judgment in dress, which alone sufficed to make her valuable to young ladies; a fair amount of reading, old and new; and on most subjects an opinion of her own, for which she had always something to say; Mr. MacBorrowdale, an old friend of Mr. Gryll, a gentleman who comprised in himself all that Scotland had ever been supposed to possess of mental, moral, and political philosophy; 'And yet he bore it not about;' not 'as being loth to wear it out,'* but because he held that there was a time for all things, and that dinner was the time for joviality, and not for argument; Mr. Minim, the amateur composer of the music for the comedy; Mr. Pallet, the amateur painter of the scenery; and last, not least, the newly-made acquaintance, Lord Curryfin.

Lord Curryfin was a man on the younger side of thirty, with a good person, handsome features, a powerful voice, and an agreeable delivery. He had a strong memory, much power of application, and a facility of learning rapidly whatever he turned his mind to. But with all this, he valued what he learned less for the pleasure which he derived from the acquisition, than from the effect which it enabled him to produce on others. He liked to shine in conversation, and there was scarcely a subject which could be mooted in any society, on which his multifarious attainments did not qualify him to say something. He was readily taken by novelty in doctrine, and followed a new lead with great pertinacity; and in this way he had been caught by the science of pantopragmatics, and firmly believed for a time, that a scientific organization for teaching everybody everything, would cure all the evils of society. But being one of those 'over sharp wits whose edges are very soon turned', he did not adhere to any opinion with sufficient earnestness to be on any occasion betrayed into intemperance in maintaining it. So far from this, if he found any unfortunate opinion in a hopeless minority of the company he happened to be in, he was often chivalrous enough to come to its aid, and see what could be said for it. When lecturing became a mania, he had taken to lecturing; and looking about for an unoccupied subject, he had lighted on the natural history of fish, in

* 'Tis true, although he had much wit,
He was very shy of using it,
As being loth to wear it out;
And therefore bore it not about,
Except on holidays or so,
As men their best apparel do.
HUDIBRAS.

which he soon became sufficiently proficient to amuse the ladies, and astonish the fishermen in any seaside place of fashionable resort. Here he always arranged his lecture-room, so that the gentility of his audience could sit on a platform, and the natives in a gallery above, and that thus the fishy and tarry odours which the latter were most likely to bring with them, might ascend into the upper air, and not mingle with the more delicate fragrances that surrounded the select company below. He took a summer tour to several watering-places, and was thoroughly satisfied with his success. The fishermen at first did not take cordially to him; but their wives attended from curiosity, and brought their husbands with them on nights not favourable to fishing; and by degrees he won on their attention, and they took pleasure in hearing him, though they learned nothing from him that was of any use in their trade. But he seemed to exalt their art in the eyes of themselves and others, and he told them some pleasant anecdotes of strange fish, and of perilous adventures of some of their own craft, which led in due time to the crowding of his gallery. The ladies went, as they always will go, to lectures, where they fancy they learn something, whether they learn anything or not; and on these occasions, not merely to hear the lecturer, but to be seen by him. To them, however attractive the lecture might have been, the lecturer was more so. He was an irresistible temptation to matrons with marriageable daughters, and wherever he sojourned he was overwhelmed with invitations. It was a contest who should have him to dinner, and in the simplicity of his heart, he ascribed to admiration of his science and eloquence, all the courtesies and compliments with which he was everywhere received. He did not like to receive unreturned favours, and never left a place in which he had accepted many invitations, without giving in return a ball and supper on a scale of great munificence; which filled up the measure of his popularity, and left on all his guests a very enduring impression of a desire to see him again.

So his time passed pleasantly, with a heart untouched by either love or care, till he fell in at a dinner party with the Reverend Doctor Opimian. The Doctor spoke of Gryll

Grange and the Aristophanic comedy which was to be produced at Christmas, and Lord Curryfin, with his usual desire to have a finger in every pie, expressed an earnest wish to be introduced to the Squire. This was no difficult matter. The Doctor had quickly brought it about, and Lord Curryfin had gone over in the Doctor's company to pass a few days at the Grange. Here, in a very short time, he had made himself completely at home; and had taken on himself the office of architect, to superintend the construction of the theatre, receiving with due deference instructions on the subject from the Reverend Doctor Opimian.

Sufficient progress had been made in the comedy for the painter and musician to begin work on their respective portions; and Lord Curryfin, whose heart was in his work, passed whole mornings in indefatigable attention to the progress of the building. It was near the house, and was to be approached by a covered way. It was a miniature of the Athenian theatre, from which it differed in having a roof, but it resembled it in the arrangements of the stage and orchestra, and in the graduated series of semicircular seats for the audience.

When dinner was announced, Mr. Gryll took in Miss Ilex. Miss Gryll, of course, took the arm of Lord Curryfin. Mr. Falconer took in one of the young ladies and placed her on the left hand of the host. The Reverend Doctor Opimian took in another, and was consequently seated between her and Miss Ilex. Mr. Falconer was thus as far removed as possible from the young lady of the house, and was consequently, though he struggled as much as possible against it, frequently *distrait*, unconsciously and unwillingly observing Miss Gryll and Lord Curryfin, and making occasional observations very wide of the mark to the fair damsels on his right and left, who set him down in their minds for a very odd young man. The soup and fish were discussed in comparative silence; the entrées not much otherwise; but suddenly a jubilant expression from Mr. MacBorrowdale hailed the disclosure of a large sirloin of beef which figured before Mr. Gryll.

MR. MACBORROWDALE

You are a man of taste, Mr. Gryll. That is a handsomer ornament of a dinner-table than clusters of nosegays, and all sorts of uneatable decorations. I detest and abominate the idea of a Siberian dinner, where you just look on fiddle-faddles, while your dinner is behind a screen, and you are served with rations like a pauper.

THE REVEREND DOCTOR OPIMIAN

I quite agree with Mr. MacBorrowdale. I like to see my dinner. And herein I rejoice to have Addison on my side; for I remember a paper, in which he objects to having roast beef placed on a sideboard. Even in his day it had been displaced to make way for some incomprehensible French dishes, among which he could find nothing to eat.* I do not know what he would have said to its being placed altogether out of sight. Still there is something to be said on the other side. There is hardly one gentleman in twenty who knows how to carve; and as to ladies, though they did know once on a time, they do not now. What can be more pitiable than the right-hand man of the lady of the house, awkward enough in himself, with the dish twisted round to him in the most awkward possible position, digging in unutterable mortification for a joint which he cannot find, and wishing the unanatomisable *volaille* behind a Russian screen with the footmen?

MR. MACBORROWDALE

I still like to see the *volaille*. It might be put on table with its joints divided.

MR. GRYLL

As that turkey-poult is, Mr. MacBorrowdale; which gives my niece no trouble; but the precaution is not necessary with such a right-hand man as Lord Curryfin, who carves to perfection.

* I was now in great hunger and confusion, when I thought I smelled the agreeable savour of roast beef; but could not tell from which dish it arose, though I did not question but it lay disguised in one of them. Upon turning my head I saw a noble sirloin on the side-table, smoking in the most delicious manner. I had recourse to it more than once, and could not see without some indignation that substantial English dish banished in so ignominious a manner, to make way for French kickshaws.—*Tatler*, No. 148.

MR. MACBORROWDALE

Your arrangements are perfect. At the last of these Siberian dinners at which I had the misfortune to be present, I had offered me, for two of my rations, the tail of a mullet and the drum-stick of a fowl. Men who carve behind screens ought to pass a competitive examination before a jury of gastronomers. Men who carve at a table are drilled by degrees into something like tolerable operators by the mere shame of the public process.

MR. GRYLL

I will guarantee you against a Siberian dinner, whenever you dine with me.

THE REVEREND DOCTOR OPIMIAN

Mr. Gryll is a true conservative in dining.

MR. GRYLL

A true conservative, I hope. Not what a *soi-disant* conservative is practically: a man who sails under national colours, hauls them down, and hoists the enemy's. I like old customs. I like a glass of wine with a friend. What say you, Doctor? Mr. MacBorrowdale will join us?

MR. MACBORROWDALE

Most willingly.

MISS GRYLL

My uncle and the Doctor have got as usual into a discussion, to the great amusement of the old lady, who sits between them and says nothing.

LORD CURRYFIN

Perhaps their discussion is too recondite for her.

MISS GRYLL

No; they never talk before ladies of any subject in which ladies cannot join. And she has plenty to say for herself when she pleases. But when conversation pleases her, she likes to listen and be silent. It strikes me, by a few words that float this way, that they are discussing the Art of Dining. She ought to be a proficient in it, for she lives much in the world, and has met as many persons whom she is equally willing either to meet to-morrow, or never to meet again, as any regular *dineur en ville*. And indeed that is the price that must be paid for society. Whatever difference of character may lie under the surface, the persons you meet in its circles are externally others yet the same: the same dress, the same manners, the same tastes and opinions, real or assumed. Strongly defined characteristic differences are so few, and artificial general resemblances so many, that in every party you may always make out the same theatrical company. It is like the flowing of a river: it is always different water, but you do not see the difference.

LORD CURRYFIN

For my part I do not like these monotonous exteriors. I like visible character. Now, in your party here, there is a good deal of character. Your uncle and Mr. MacBorrowdale are characters. Then the Reverend Doctor Opimian. He is not a man made to pattern. He is simple-minded, learned, tolerant, and the quintessence of *bonhomie*. The young gentleman who arrived to-day, the Hermit of the Folly, is evidently a character. I flatter myself, I am a character (*laughing*).

MISS GRYLL (*laughing*)

Indeed you are, or rather many characters in one. I never knew a man of such infinite variety. You seem always to present yourself in the aspect in which those you are with would best wish to see you.

There was some ambiguity in the compliment; but Lord Curryfin took it as implying that his aspect in all its variety was agreeable to the young lady. He did not then dream of a rival in the Hermit of the Folly.

CHAPTER XIV

Οὐ φίλος, ὅς κρατῆρι παρὰ πλέῳ οἰνοποτάζων
Νείκεα καὶ πόλεμον δακρυόεντα λέγει,
’Αλλ’ ὅστις, Μουσέων τε καὶ ἀγλαὰ δωρ’ ’Αφροδίτης
Συμμίσγων, ἐρατῆς μνήσκεται εὐφροσύνης.

ANACREON.

I love not him, who o'er the wine-cup's flow
Talks but of war, and strife, and scenes of woe:
But him, who can the Muses' gifts employ,
To mingle love and song with festal joy.

THE dinner and dessert passed away. The ladies retired to the drawing-room: the gentlemen discoursed over their wine. Mr. MacBorrowdale pronounced an eulogium on the port, which was cordially echoed by the divine in regard to the claret.

MR. FALCONER

Doctor, your tastes and sympathies are very much with the Greeks; but I doubt if you would have liked their wine. Condiments of sea-water and turpentine must have given it an odd flavour; and mixing water with it, in the proportion of three to one, must have reduced the strength of merely fermented liquor to something like the smallest ale of Christophero Sly.

THE REVEREND DOCTOR OPIMIAN

I must say I should not like to put either salt-water or turpentine into this claret: they would not improve its bouquet; nor to dilute it with any portion of water: it has to my mind, as it is, just the strength it ought to have, and no more. But the Greek taste was so exquisite in all matters in which we can bring it to the test, as to justify a strong presumption that in matters in which we cannot test it, it was equally correct. Salt-water and turpentine do not suit our wine: it does not follow that theirs had not in it some basis of contrast, which may have made them pleasant in combination. And it was only a few of their wines that were so treated.

LORD CURRYFIN

Then it could not have been much like their drink of the present day. 'My master cannot be right in his mind,' said Lord Byron's man Fletcher, 'or he would not have left Italy, where we had everything, to go to a country of savages; there is nothing to eat in Greece but tough billy-goats, or to drink but spirits of turpentine.'*

THE REVEREND DOCTOR OPIMIAN

There is an ambiguous present, which somewhat perplexes me, in an epigram of Rhianus, 'Here is a vessel of half-wine, half-turpentine, and a singularly lean specimen of kid: the sender, Hippocrates, is worthy of all praise.'† Perhaps this was a Doctor's present to a patient. Alcæus, Anacreon, and Nonnus could not have sung as they did under the inspiration of spirit of turpentine. We learn from Athenæus, and Pliny, and the old comedians, that the Greeks had a vast variety of wine, enough to suit every variety of taste. I infer the unknown from the known. We know little of their music. I have no doubt it was as excellent in its kind as their sculpture.

MR. MINIM

I can scarcely think that, sir. They seem to have had only the minor key, and to have known no more of counterpoint than they did of perspective.

THE REVEREND DOCTOR OPIMIAN

Their system of painting did not require perspective. Their main subject was on one foreground. Buildings, rocks, trees, served simply to indicate, not to delineate, the scene.

MR. FALCONER

I must demur to their having only the minor key. The natural ascent of the voice is in the major key, and with their exquisite sensibility to sound they could not have missed the obvious expression of cheerfulness. With their three scales, diatonic, chromatic, and enharmonic, they must have exhausted every possible expression of feeling. Their scales were in true intervals; they had really major and minor tones; we have neither, but a confusion of both.

* Trelawny's *Recollections*.

† *Ἥμισυ μὲν πίσσης κωνίτιδος, ἥμισυ δ' οἴνου,*
Ἀρχῖν', ἀτρεκέως ἥδε λάγυνος ἔχει·
Λεπτοτέρης δ' οὐκ οἶδ' ἐρίφου κρέας· πλὴν ὅγε πέμψας
Αἰνεῖσθαι πάντων ἄξιος Ἱπποκράτης.

Anthologia Palatina: Appendix. 72.

They had both sharps and flats: we have neither, but a mere set of semitones, which serve for both. In their enharmonic scale the fineness of their ear perceived distinctions, which are lost on the coarseness of ours.

MR. MINIM

With all that they never got beyond melody. They had no harmony, in our sense. They sang only in unisons and octaves.

MR. FALCONER

It is not clear that they did not sing in fifths. As to harmony in our sense, I will not go so far as to say with Ritson, that the only use of the harmony is to spoil the melody; but I will say, that to my taste a simple accompaniment, in strict subordination to the melody, is far more agreeable than that Niagara of sound under which it is now the fashion to bury it.

MR. MINIM

In that case, you would prefer a song with a simple pianoforte accompaniment to the same song on the Italian stage.

MR. FALCONER

A song sung with feeling and expression is good, however accompanied. Otherwise, the pianoforte is not much to my mind. All its intervals are false, and temperament is a poor substitute for natural intonation. Then its incapability of sustaining a note has led, as the only means of producing effect, to those infinitesimal subdivisions of sound, in which all sentiment and expression are twittered and frittered into nothingness.

THE REVEREND DOCTOR OPIMIAN

I quite agree with you. The other day a band passed my gate playing 'The Campbells are coming;' but instead of the fine old Scotch lilt, and the emphasis on 'Ohó! ohó!' what they actually played was, 'The Ca-a-a-ampbells are co-o-o-o-ming, Oh-o-ho-o-o! Oh-o-ho-o-o:' I thought to myself: There is the essence and quintessence of modern music. I like the old organ-music such as it was, when there were no keys but C and F, and every note responded to a syllable. The effect of the prolonged and sustained sound must have been truly magnificent:

> Where, through the long-drawn aisle and fretted vault,
> The pealing anthem swelled the note of praise.

Who cares to hear sacred music on a piano?

MR. MINIM

Yet I must say that there is a great charm in that brilliancy of execution, which is an exclusively modern and very modern accomplishment.

MR. FALCONER

To those who perceive it. All things are as they are perceived. To me music has no charm without expression.

LORD CURRYFIN (*who, having observed* MR. MACBORROWDALE'S *determination not to be drawn into an argument, amused himself with asking his opinion on all subjects*)

What is your opinion, Mr. MacBorrowdale?

MR. MACBORROWDALE

I hold to the opinion I have already expressed, that this is as good a glass of port as ever I tasted.

LORD CURRYFIN

I mean your opinion of modern music and musical instruments.

MR. MACBORROWDALE

The organ is very good for psalms, which I never sing, and the pianoforte for jigs, which I never dance. And if I were not to hear either of them from January to December, I should not complain of the privation.

LORD CURRYFIN

You are an utilitarian, Mr. MacBorrowdale. You are all for utility—public utility—and you see none in music.

MR. MACBORROWDALE

Nay, not exactly so. If devotion is good, if cheerfulness is good, and if music promotes each of them in proper time and place, music is useful. If I am as devout without the organ, and as cheerful without the piano, as I ever should be with them, that may be the defect of my head or my ear. I am not for forcing my tastes or no-tastes on other people. Let every man enjoy himself in his own way, while he does not annoy others. I would not deprive you of your enjoyment of a brilliant symphony, and I hope you would not deprive me of my enjoyment of a glass of old wine.

THE REVEREND DOCTOR OPIMIAN

'Tres mihi convivæ prope dissentire videntur,
Poscentes vario multum diversa palato.'*

MR. FALCONER

Nor our reverend friend of the pleasure of a classical quotation.

THE REVEREND DOCTOR OPIMIAN

And the utility, too, sir: for I think I am indebted to one for the pleasure of your acquaintance.

MR. FALCONER

When you did me the honour to compare my house to the Palace of Circe. The gain was mine.

MR. PALLET

You admit, sir, that the Greeks had no knowledge of perspective.

THE REVEREND DOCTOR OPIMIAN

Observing, that they had no need of it. Their subject was a foreground like a relievo. Their background was a symbol, not a representation. 'No knowledge' is perhaps too strong. They had it where it was essential. They drew a peristyle, as it appeared to the eye, as accurately as we can do. In short, they gave to each distinct object its own proper perspective, but to separate objects they did not give their relative perspective, for the reason I have given, that they did not need it.

MR. FALCONER

There is to me one great charm in their painting, as we may judge from the specimens in Pompeii, which, though not their greatest works, indicate their school. They never crowded their canvas with figures. They presented one, two, three, four, or, at most, five persons, preferring one, and rarely exceeding three. These persons were never lost in the profusion of scenery, dress, and decoration. They had clearly-defined outlines, and were agreeable objects from any part of the room in which they were placed.

* Three guest dissent most widely in their wishes:
With different taste they call for different dishes.

MR. PALLET

They must have lost much in beauty of detail.

THE REVEREND DOCTOR OPIMIAN

Therein is the essential difference of ancient and modern taste. Simple beauty—of idea in poetry, of sound in music, of figure in painting—was their great characteristic. Ours is detail in all these matters, overwhelming detail. We have not grand outlines for the imagination of the spectator or hearer to fill up: his imagination has no play of its own: it is overloaded with *minutiæ* and kaleidoscopical colours.

LORD CURRYFIN

Detail has its own beauty. I have admired a Dutch picture of a butcher's shop, where all the charm was in detail.

THE REVEREND DOCTOR OPIMIAN

I cannot admire anything of the kind. I must take pleasure in the thing represented before I can derive any from the representation.

MR. PALLET

I am afraid, sir, as our favourite studies all lead us to extreme opinions, you think the Greek painting was the better for not having perspective, and the Greek music for not having harmony.

THE REVEREND DOCTOR OPIMIAN

I think they had as much perspective and as much harmony as was consistent with that simplicity, which characterized their painting and music as much as their sculpture and poetry.

LORD CURRYFIN

What is your opinion, Mr. MacBorrowdale?

MR. MACBORROWDALE

I think you may just buz the bottle before you.

LORD CURRYFIN

I mean your opinion of Greek perspective?

MR. MACBORROWDALE

Troth, I am of opinion that a bottle looks smaller at a distance than when it is close by, and I prefer it as a full-sized object in the foreground.

LORD CURRYFIN

I have often wondered that a gentleman so well qualified as you are to discuss all subjects should so carefully avoid discussing any.

MR. MACBORROWDALE

After dinner, my lord, after dinner. I work hard all the morning at serious things, sometimes till I get a headache, which, however, does not often trouble me. After dinner I like to crack my bottle and chirp and talk nonsense, and fit myself for the company of Jack of Dover.

LORD CURRYFIN

Jack of Dover! Who was he?

MR. MACBORROWDALE

He was a man who travelled in search of a greater fool than himself, and did not find him.*

THE REVEREND DOCTOR OPIMIAN

He must have lived in odd times. In our days he would not have gone far without falling in with a teetotaller, or a decimal coinage man, or a school-for-all man, or a competitive examination man, who would not allow a drayman to lower a barrel into a cellar unless he could expound the mathematical principles by which he performed the operation.

MR. MACBORROWDALE

Nay, that is all pragmatical fooling. The fooling Jack looked for was jovial fooling, fooling to the top of his bent, excellent fooling, which, under the semblance of folly, was both merry and wise. He did not look for mere unmixed folly, of which there never was a deficiency. The fool he looked for was one which it takes a wise man to make—a Shakspearian fool.†

* *Jacke of Dover, His Quest of Inquirie, or His Privy Search for the Veriest Foole in England.* London. 1604. Reprinted for the Percy Society. 1842.

† Oeuvre, ma foi, où n'est facile atteindre:
Pourtant qu'il faut parfaitement sage être,
Pour le vrai fol bien naïvement feindre.
EUTRAPEL, p. 28

THE REVEREND DOCTOR OPIMIAN

In that sense he might travel far, and return, as he did in his own day, without having found the fool he looked for.

MR. MACBORROWDALE

A teetotaller! Well! He is the true Heautontimorumenos, the self-punisher, with a jug of toast-and-water for his Christmas wassail. So far his folly is merely pitiable, but his intolerance makes it offensive. He cannot enjoy his own tipple unless he can deprive me of mine. A fox that has lost his tail. There is no tyrant like a thorough-paced reformer. I drink to his own reformation.

MR. GRYLL

He is like Bababec's faquir, who sat in a chair full of nails, *pour avoir de la considération.* But the faquir did not want others to do the same. He wanted all the consideration for himself, and kept all the nails for himself. If these meddlers would do the like by their toast-and-water, nobody would begrudge it them.

THE REVEREND DOCTOR OPIMIAN

Now, sir, if the man who has fooled the greatest number of persons to the top of their bent were to be adjudged the fittest companion for Jack of Dover, you would find him a distinguished meddler with everything, who has been for half a century the merry-andrew of a vast arena, which he calls moral and political science, but which has in it a dash of everything that has ever occupied human thought.

LORD CURRYFIN

I know whom you mean; but he is a great man in his way, and has done much good.

THE REVEREND DOCTOR OPIMIAN

He has helped to introduce much change; whether for good or for ill remains to be seen. I forgot he was your lordship's friend. I apologize, and drink to his health.

LORD CURRYFIN

Oh! pray, do not apologize to me. I would not have my friendships, tastes, pursuits, and predilections interfere in the slightest degree with the fullest liberty of speech on all persons and things. There are many who think with you that he is a moral and political Jack of Dover. So be it. Time will bring him to his level.

MR. MACBORROWDALE

I will only say of the distinguished personage, that Jack of Dover would not pair off with him. This is the true universal science, the oracle of *La Dive Bouteille*.

MR. GRYLL

It is not exactly Greek music, Mr. Minim, that you are giving us for our Aristophanic choruses.

MR. MINIM

No, sir; I have endeavoured to give you a good selection, as appropriate as I can make it.

MR. PALLET

Neither am I giving you Greek painting for the scenery. I have taken the liberty to introduce perspective.

THE REVEREND DOCTOR OPIMIAN

Very rightly both, for Aristophanes in London.

MR. MINIM

Besides, sir, we must have such music as your young ladies can sing.

THE REVEREND DOCTOR OPIMIAN

Assuredly; and so far as we have yet heard them rehearse, they sing it delightfully.

After a little more desultory conversation, they adjourned to the drawing-rooms.

CHAPTER XV

Τοῦτο βίος, τοῦτ' αὐτό· τρυφὴ βίος· ἔρρετ' ἀνίαι·
Ζωῆς ἀνθρώποις ὀλίγος χρόνος· ἄρτι Λύαιος,
Ἄρτι χοροί, στέφανοί τε φιλανθέες, ἄρτι γυναῖκες.
Σήμερον ἐσθλὰ πάθω, τὸ γὰρ αὔριον οὐδενὶ δῆλον.
Anthologia Palatina: V. 72.

This, this is life, when pleasure drives out care.
Short is the span of time we each may share.
To-day, while love, wine, song, the hours adorn,
To-day we live: none know the coming mourn.

LORD CURRYFIN's assiduities to Miss Gryll had discomposed Mr. Falconer more than he chose to confess to himself. Lord Curryfin, on entering the drawing-rooms, went up immediately to the young lady of the house; and Mr. Falconer, to the amazement of the reverend Doctor, sat down, in the outer drawing-room, on a sofa by the side of Miss Ilex, with whom he entered into conversation.

In the inner drawing-room some of the young ladies were engaged with music, and were entreated to continue their performance. Some of them were conversing, or looking over new publications.

After a brilliant symphony, performed by one of the young visitors, in which runs and crossings of demisemiquavers in *tempo prestissimo* occupied the principal share, Mr. Falconer asked Miss Ilex how she liked it.

MISS ILEX

I admire it as a splendid piece of legerdemain; but it expresses nothing.

MR. FALCONER

It is well to know that such things can be done; and when we have reached the extreme complications of art, we may hope to return to nature and simplicity.

MISS ILEX

Not that it is impossible to reconcile execution and expression.

Rubini identified the redundancies of ornament with the overflowings of feeling, and the music of Donizetti furnished him most happily with the means of developing this power. I never felt so transported out of myself as when I heard him sing *Tu che al ciel spiegasti l'ali*.

MR. FALCONER

Do you place Donizetti above Mozart?

MISS ILEX

Oh, surely not. But for supplying expressive music to a singer like Rubini, I think Donizetti has no equal; at any rate no superior. For music that does not require, and does not even suit, such a singer, but which requires only to be correctly interpreted to be universally recognized as the absolute perfection of melody, harmony, and expression, I think Mozart has none. Beethoven perhaps: he composed only one opera, *Fidelio*: but what an opera that is. What an effect in the sudden change of the key, when Leonora throws herself between her husband and Pizarro: and again, in the change of the key with the change of the scene, when we pass from the prison to the hall of the palace. What pathos in the songs of affection, what grandeur in the songs of triumph, what wonderful combinations in the accompaniments, where a perpetual stream of counter-melody creeps along in the bass, yet in perfect harmony with the melody above.

MR. FALCONER

What say you to Haydn?

MISS ILEX

Haydn has not written operas, and my principal experience is derived from the Italian theatre. But his music is essentially dramatic. It is a full stream of perfect harmony in subjection to exquisite melody; and in simple ballad-strains, that go direct to the heart, he is almost supreme and alone. Think of that air with which every one is familiar, 'My mother bids me bind my hair:' the graceful flow of the first part, the touching effect of the semitones in the second: with true intonation and true expression, the less such an air is accompanied the better.

MR. FALCONER

There is a beauty and an appeal to the heart in ballads which will never lose its effect except on those with whom the pretence of fashion overpowers the feelings of nature.*

MISS ILEX

It is strange, however, what influence that pretence has, in overpowering all natural feelings, not in music alone.

'Is it not curious,' thought the Doctor, 'that there is only one old woman in the room, and that my young friend should have selected her for the object of his especial attention?'

But a few simple notes struck on the ear of his young friend, who rose from the sofa and approached the singer. The Doctor took his place to cut off his retreat.

Miss Gryll, who, though a proficient in all music, was particularly partial to ballads, had just begun to sing one.

THE DAPPLED PALFREY.†

'My traitorous uncle has wooed for himself:
Her father has sold her for land and for pelf:
My steed, for whose equal the world they might search,
In mockery they borrow to bear her to church.

'Oh! there is one path through the forest so green,
Where thou and I only, my palfrey, have been:
We traversed it oft, when I rode to her bower
To tell my love tale through the rift of the tower.

'Thou know'st not my words, but thy instinct is good:
By the road to the church lies the path through the wood:
Thy instinct is good, and her love is as true:
Thou wilt see thy way homeward: dear palfrey, adieu.'

They feasted full late and full early they rose,
And church-ward they rode more than half in a doze:
The steed in an instant broke off from the throng,
And pierced the green path, which he bounded along.

In vain was pursuit, though some followed pell-mell:
Through bramble and thicket they floundered and fell.
On the backs of their coursers some dozed as before,
And missed not the bride till they reached the church-door.

* Braham said something like this to a Parliamentary Committee on Theatres, in 1832.

† Founded in *Le Vair Palefroi*: among the *Fabliaux* published by Barbazan.

The knight from his keep on the forest-bound gazed:
The drawbridge was down, the portcullis was raised:
And true to his hope came the palfrey amain,
With his only loved lady, who checked not the rein.

The drawbridge went up: the portcullis went down:
The chaplain was ready with bell, book, and gown:
The wreck of the bride-train arrived at the gate:
The bride showed the ring, and they muttered 'Too late!'

'Not too late for a feast, though too late for a fray:
What's done can't be undone: make peace while you may:'
So spake the young knight, and the old ones complied,
And quaffed a deep health to the bridegroom and bride.

Mr. Falconer had listened to the ballad with evident pleasure. He turned to resume his place on the sofa, but finding it pre-occupied by the Doctor, he put on a look of disappointment, which seemed to the Doctor exceedingly comic.

'Surely,' thought the Doctor, 'he is not in love with the old maid.'

Miss Gryll gave up her place to a young lady who in her turn sang a ballad of a different character.

LOVE AND AGE

I played with you 'mid cowslips blowing,
When I was six and you were four;
When garlands weaving, flower-balls throwing,
Were pleasures soon to please no more.
Through groves and meads, o'er grass and heather,
With little playmates, to and fro,
We wandered hand in hand together;
But that was sixty years ago.

You grew a lovely roseate maiden,
And still our early love was strong;
Still with no care our days were laden,
They glided joyously along;
And I did love you, very dearly,
How dearly words want power to show;
I thought your heart was touched as nearly;
But that was fifty years ago.

Then other lovers came around you,
Your beauty grew from year to year,
And many a splendid circle found you
The centre of its glittering sphere.
I saw you then, first vows forsaking,
On rank and wealth your hand bestow;
Oh, then I thought my heart was breaking,—
But that was forty years ago.

And I lived on, to wed another:
No cause she gave me to repine;
And when I heard you were a mother,
I did not wish the children mine.
My own young flock, in fair progression,
Made up a pleasant Christmas row:
My joy in them was past expression;—
But that was thirty years ago.

You grew a matron plump and comely,
You dwelt in fashion's brightest blaze;
My earthly lot was far more homely;
But I too had my festal days.
No merrier eyes have ever glistened
Around the hearth-stone's wintry glow,
Than when my youngest child was christened:—
But that was twenty years ago.

Time passed. My eldest girl was married,
And I am now a grandsire grey;
One pet of four years old I've carried
Among the wild-flowered meads to play.
In our old fields of childish pleasure,
Where now, as then, the cowslips blow,
She fills her basket's ample measure,—
And that is not ten years ago.

But though first love's impassioned blindness
Has passed away in colder light,
I still have thought of you with kindness,
And shall do, till our last good-night.
The ever-rolling silent hours
Will bring a time we shall not know,
When our young days of gathering flowers
Will be an hundred years ago.

MISS ILEX

That is a melancholy song. But of how many first loves is it the true tale? And how many are far less happy?

THE REVEREND DOCTOR OPIMIAN

It is simple and well sung, with a distinctness of articulation not often heard.

MISS ILEX

That young lady's voice is a perfect contralto. It is singularly beautiful, and I applaud her for keeping within her natural compass, and not destroying her voice by forcing it upwards, as too many do.

THE REVEREND DOCTOR OPIMIAN

Forcing, forcing, seems to be the rule of life. A young lady who forces her voice into *altissimo*, and a young gentleman who forces his mind into a receptable for a chaos of crudities, are pretty much on a par. Both do ill, were, if they were contented with attainments within the limits of natural taste and natural capacity, they might both do well. As to the poor young men, many of them become mere crammed fowls, with the same result as Hermogenes, who, after astonishing the world with his attainments at seventeen, came to a sudden end at the age of twenty-five, and spent the rest of a long life in hopeless imbecility.*

MISS ILEX

The poor young men can scarcely help themselves. They are not held qualified for a profession unless they have overloaded their understanding with things of no use in it—incongruous things too, which could never be combined into the pursuits of natural taste.

THE REVEREND DOCTOR OPIMIAN

Very true. Brindley would not have passed as a canal-maker, nor Edward Williams† as a bridge-builder. I saw the other day some examination papers which would have infallibly excluded Marlborough from the army and Nelson from the navy. I doubt if Haydn would have passed as a composer before a committee of lords like one of his pupils, who insisted on demonstrating to him that he was continually sinning against the rules of counterpoint; on which Haydn said to him, 'I thought I was to teach you, but it seems you are to teach me, and I do not want a preceptor,' and thereon he wished his lordship a good morning. Fancy Watt being

* Donaldson's *History of Greek Literature*, vol. iii. p. 156.

† The builder of Pont-y-Pryd.

asked, how much Joan of Naples got for Avignon when she sold it to Pope Clement the Sixth, and being held unfit for an engineer because he could not tell.

MISS ILEX

That is an odd question, Doctor. But how much did she get for it?

THE REVEREND DOCTOR OPIMIAN

Nothing. He promised ninety thousand golden florins, but he did not pay one of them: and that, I suppose, is the profound sense of the question. It is true he paid her after a fashion, in his own peculiar coin. He absolved her of the murder of her first husband, and perhaps he thought that was worth the money. But how many of our legislators could answer the question? Is it not strange that candidates for seats in parliament should not be subjected to competitive examination? Plato and Persius* would furnish good hints for it. I should like to see honourable gentlemen having to answer such questions as are deemed necessary tests for government clerks, before they would be held qualified candidates for seats in the legislature. That would be something like a reform in the parliament. Oh, that it were so, and I were the examiner! Ha, ha, ha, what a comedy!

The Doctor's hearty laugh was contagious, and Miss Ilex joined in it. Mr. MacBorrowdale came up.

MR. MACBORROWDALE

You are as merry as if you had discovered the object of Jack of Dover's quest.

THE REVEREND DOCTOR OPIMIAN

Something very like it. We have an honourale gentleman under competitive examination for a degree in legislative wisdom.

MR. MACBORROWDALE

Truly, that is fooling competition to the top of its bent.

THE REVEREND DOCTOR OPIMIAN

Competitive examination for clerks, and none for legislators—is not this an anomaly? Ask the honourable member for Muckborough on what acquisitions in history and mental and moral philosophy he founds his claim of competence to make laws for the

* PLATO: *Alcibiades* i; PERSIUS: *Sat.* iv.

nation? He can only tell you that he has been chosen as the most conspicuous Grub among the Money-grubs of his borough to be the representative of all that is sordid, selfish, hard-hearted, unintellectual, and antipatriotic, which are the distinguishing qualities of the majority among them. Ask a candidate for a clerkship what are his qualifications? He may answer, 'All that are requisite—reading, writing and arithmetic.' 'Nonsense,' says the questioner. 'Do you know the number of miles in direct distance from Timbuctoo to the top of Chimborazo?' 'I do not,' says the candidate. 'Then you will not do for a clerk,' says the competitive examiner. Does Moneygrub of Muckborough know? He does not; nor anything else. The clerk may be able to answer some of the questions put to him. Moneygrub could not answer one of them. But he is very fit for a legislator.

MR. MACBORROWDALE

Eh! but he is subjected to a pretty severe competitive examination of his own, by what they call a constituency, who just put him to the test in the art of conjuring, to see if he can shift money from his own pocket into theirs, without any inconvenient third party being aware of the transfer.

CHAPTER XVI

Amiam: che non ha tregua
Con gli anni umana vita, e si dilegua.
Amiam: che il sol si muore, e poi rinasce:
A noi sua breve luce
S'asconde, e il sonno eterna notte adduce.

TASSO: *Aminta.*

Love, while youth knows its prime,
For mortal life can make no truce with time.
Love: for the sun goes down to rise as bright:
To us his transient light
Is veiled, and sleep comes on with everlasting night.

LORD CURRYFIN was too much a man of the world to devote his attentions in society exclusively to one, and make them the subject of special remark. He left the inner drawing-room, and came up to the Doctor to ask him if he knew the young lady who had sung the last ballad. The Doctor knew her well. She was Miss Niphet, the only daughter of a gentleman of fortune, residing at a few miles distance.

LORD CURRYFIN

As I looked at her while she was singing, I thought of Southey's description of Laila's face in *Thalaba*:

A broad light floated o'er its marble paleness,
As the wind waved the fountain-fire.

Marble paleness suits her well. There is something statuesque in her whole appearance. I could not help thinking what an admirable Camilla she would make in Cimarosa's *Orazii*. Her features are singularly regular. They had not much play, but the expression of her voice was such as if she felt the full force of every sentiment she uttered.

THE REVEREND DOCTOR OPIMIAN

I consider her to be a person of very deep feeling, which she does not choose should appear on the surface. She is animated in conversation when she is led into it. Otherwise, she is silent and retiring, but obliging in the extreme; always ready to take part in anything that is going forward. She never needs, for example, being twice asked to sing. She is free from the vice which Horace ascribes to all singers, of not complying when asked, and never leaving off when they have once begun. If this be a general rule, she is an exception to it.

LORD CURRYFIN

I rather wonder she does not tinge her cheeks with a slight touch of artificial red, just as much as would give her a sort of blush-rose complexion.

MISS ILEX

You will not wonder when you know her better. The artificial, the false, in any degree, however little, is impossible to her. She does not show all she thinks and feels, but what she does show is truth itself.

LORD CURRYFIN

And what part is she to take in the Aristophanic comedy?

THE REVEREND DOCTOR OPIMIAN

She is to be the leader of the chorus.

LORD CURRYFIN

I have not seen her at the rehearsals.

THE REVEREND DOCTOR OPIMIAN

So far, her place has been supplied. You will see her at the next.

In the meantime Mr. Falconer had gone into the inner drawing-room, sat down by Miss Gryll, and entered into conversation with her. The Doctor observed them from a distance, but with all the opportunity he had had for observation, he was still undetermined in his opinion of the impression they might have made on each other.

'It is well,' he said to himself, 'that Miss Ilex is an old maid. If she were as young as Morgana, I think she would win our young friend's heart. Her mind is evidently much to his mind. But so would Morgana's be, if she could speak it as freely. She does not—why not? To him at any rate. She seems under no restraint with Lord Curryfin. A good omen, perhaps. I never saw a couple so formed for each other.

Heaven help me! I cannot help harping on that string. After all, the Vestals are the obstacle.'

Lord Curryfin, see Miss Niphet sitting alone at the side of the room, changed his place, sate down by her, and entered into conversation on the topics of the day, novels, operas, pictures, and various phenomena of London life. She kept up the ball with him very smartly. She was every winter—May a ıd June—in London, mixed much in society, and saw everytning that was to be seen. Lord Curryfin, with all his Protean accomplishments, could not start a subject on which she had not something to say. But she originated nothing. He spoke, and she answered. One thing he remarked as singular, that though she spoke with knowledge of many things, she did not speak as with taste or distaste of any. The world seemed to flow under her observation without even ruffling the surface of her interior thoughts. This perplexed his versatile lordship. He thought the young lady would be a subject worth studying: it was clear that she was a character. So far so well. He felt that he should not rest satisfied till he was able to define it.

The theatre made rapid progress. The walls were completed. The building was roofed in. The stage portion was so far finished as to allow Mr. Pallet to devote every morning to the scenery. The comedy was completed. The music was composed. The rehearsals went on with vigour, but for the present in the drawing-rooms.

Miss Niphet, returning one morning from a walk before breakfast, went into the theatre to see its progress, and found Lord Curryfin swinging over the stage on a seat suspended by long ropes from above the visible scene. He did not see her. He was looking upwards, not as one indulging in an idle pastime, but as one absorbed in serious meditation. All at once the seat was drawn up, and he disappeared in the blue canvas that represented the sky. She was not aware that gynmastics were to form any portion of the projected entertainment, and went away, associating the idea of his lordship, as many had done before, with something like a feeling of the ludicrous.

Miss Niphet was not much given to laughter, but whenever she looked at Lord Curryfin during breakfast she could not quite suppress a smile which hovered on her lips, and which was even the more forced on her by the contrast between his pantomimic disappearance and his quiet courtesy and remarkably good manners in company. The lines of Dryden—

> A man so various, that he seemed to be
> Not one, but all mankind's epitome,

—passed through her mind as she looked at him.

Lord Curryfin noticed the suppressed smile, but did not apprehend that it had any relation to himself. He thought some graceful facetiousness had presented itself to the mind of the young lady, and that she was amusing herself with her own fancy. It was, however, to him another touch of character, that lighted up her statuesque countenance with a new and peculiar beauty. By degrees her features resumed their accustomed undisturbed serenity. Lord Curryfin felt satisfied that in that aspect he had somewhere seen something like her, and after revolving a series of recollections, he remembered that it was a statue of Melpomene.

There was in the park a large lake, encircled with varieties of woodland, and by its side was a pavilion to which Miss Niphet often resorted to read in an afternoon. And at no great distance from it was the boat-house, to which Lord Curryfin often resorted for a boat, to row or sail on the water. Passing the pavilion in the afternoon, he saw the young lady, and entering into conversation, ascertained what had so amused her in the morning. He told her he had been trying—severally by himself, and collectively with the workmen—the strength of the suspending lines for the descent of the Chorus of Clouds in the Aristophanic comedy. She said she had been very ungrateful to laugh at the result of his solicitude for the safety of herself and her young friends. He said that in having moved her to smile, even at his expense, he considered himself amply repaid.

From this time they often met in the pavilion, that is to say, he often found her reading there on his way to a boat, and stopped awhile to converse with her. They had always plenty to say, and it resulted that he was always sorry to leave her, and she was always sorry to part with him. By degrees the feeling of the ludicrous ceased to be the predominant sentiment which she associated with him. *L'amour vient sans qu'on y pense.*

The days shortened, and all things were sufficiently advanced to admit of rehearsals in the theatre. The hours from twelve to two—from noon to luncheon—were devoted to this pleasant pastime. At luncheon there was much merriment over the recollections of the morning's work, and after luncheon there was walking in the park, rowing or sailing on the lake, riding or driving in the adjacent country, archery in a spacious field, and in bad weather billiards, reading in the library, music in the drawing-rooms, battledoor and shuttlecock in the hall; in short, all the methods of passing time agreeably which are available to good company, when there are ample means and space for their exercise; to say nothing of making love, which Lord Curryfin did with all delicacy and discretion—directly to Miss Gryll as he had begun, and indirectly to Miss Niphet, for whom he felt an involuntary and almost unconscious admiration. He had begun to apprehend that with the former he had a dangerous rival in the Hermit of the Folly, and he thought the latter had sufficient charms to console even Orlando for the loss of Angelica. In short, Miss Gryll had first made him think of marriage, and whenever he thought his hopes were dim in that quarter, he found an antidote to despair in the contemplation of the statue-like damsel.

Mr. Falconer took more and more pleasure in Miss Gryll's society, but he did not declare himself. He was more than once on the point of doing so, but the images of the Seven Sisters rose before him, and he suspended the intention. On these occasions he always went home for a day or two to fortify his resolution against his heart. Thus he passed his time between the Grange and the Tower, 'letting I dare not wait upon I would.'

Miss Gryll had listened to Lord Curryfin. She had neither encouraged nor discouraged him. She thought him the most amusing person she had ever known. She liked his temper, his acquirements, and his manners. She could not divest herself of that feeling of the ludicrous which everybody seemed to associate with him; but she thought the chances of life presented little hope of a happier marriage than a woman who would fall in with his tastes and pursuits—which, notwithstanding their tincture of absurdity, were entertaining and even amiable—might hope for with him. Therefore, she would not say No, though, when she thought of Mr. Falconer, she could not say Yes.

Lord Curryfin invented a new sail of infallible safety, which resulted, like most similar inventions, in capsizing the inventor on the first trial. Miss Niphet, going one afternoon, later than usual, to her accustomed pavilion, found his lordship scrambling up the bank, and his boat, keel upwards, at some little distance in the lake. For a moment her usual self-command forsook her. She held out both her hands to assist him up the bank, and as soon as he stood on dry land, dripping like a Triton in trousers, she exclaimed in such a tone as he had never before heard, 'Oh! my dear lord!' Then, as if conscious of her momentary aberration, she blushed with a deeper tinge than that of the artificial rose which he had once thought might improve her complexion. She attempted to withdraw her hands, but he squeezed them both ardently, and exclaimed in his turn, like a lover in a tragedy.

'Surely, till now I never looked on beauty.'

She was on the point of saying, 'Surely, before now you have looked on Miss Gryll,' but she checked herself. She was content to receive the speech as a sudden ebullition of gratitude for sympathy, and disengaging her hands, she insisted on his returning immediately to the house to change his 'dank and dripping weeds.'

As soon as he was out of sight she went to the boat-house, to summon the men who had charge of it to the scene of the accident. Putting off in another boat, they brought the capsized vessel to land, and hung up the sail to dry. She returned in the evening, and finding the sail dry, she set it on fire. Lord Curryfin, coming down to look after his tackle, found the young lady meditating over the tinder. She said to him,

'That sail will never put you under the water again.'

He was touched by this singular development of solicitude for his preservation, but could not help saying something in praise of his invention, giving a demonstration of the infallibility of the principle, with several scientific causes of error in working out the practice. He had no doubt it would be all right on another experiment. Seeing that her looks expressed unfeigned alarm at this announcement, he assured her that her kind interest in his safety was sufficient to prevent his trying his invention again. They walked back together to the house, and in the course of conversation she said to him,

'The last time I saw the words Infallible Safety, they were painted on the back of a stagecoach, which in one of our sumer tours we saw lying by the side of the road, with its top in a ditch and its wheels in the air.'

The young lady was still a mystery to Lord Curryfin.

'Sometimes,' he said to himself, 'I could almost fancy Melpomene in love with me. But I have seldom seen her laugh, and when she has done so now and then, it has usually been at me. That is not much like love. Her last remark was anything but a compliment to my inventive genius.'

CHAPTER XVII

> O gran contrasto in giovenil pensiero,
> Desir di laude, ed impeto d'amore!
>
> ARIOSTO: *c*.25.
>
> How great a strife in youthful minds can raise
> Impulse of love, and keen desire of praise.

LORD CURRYFIN, amongst his multifarious acquirements, had taken lessons from the great horse-tamer, and thought himself as well qualified as his master to subdue any animal of the species, however vicious. It was therefore with great pleasure he heard that there was a singularly refractory specimen in Mr. Gryll's stables. The next morning after hearing this, he rose early, and took his troublesome charge in hand. After some preliminary management he proceeded to gallop him round and round a large open space in the park, which was visible from the house. Miss Niphet, always an early riser, and having just prepared for a walk, saw him from her chamber window engaged in this perilous exercise, and though she knew nothing of the peculiar character of his recalcitrant disciple, she saw by its shakings, kickings, and plungings, that it was exerting all its energies to get rid of its rider. At last it made a sudden dash into the wood, and disappeared among the trees.

It was to the young lady a matter of implicit certainty that some disaster would ensue. She pictured to herself all the contingencies of accidents; being thrown to the ground and kicked by the horse's hoofs, being dashed against a tree, or suspended, like Absalom, by the hair. She hurried down and hastened towards the wood, from which, just as she reached it, the rider and horse emerged at full speed as before. But as soon as Lord Curryfin saw Miss Niphet, he took a graceful wheel round, and brought the horse to a stand by her side; for by this time he had mastered the animal, and brought it to the condition of Sir Walter's hunter in Wordsworth:—

> Weak as a lamb the hour that it is yeaned,
> And foaming like a mountain cataract.*

* Hartleap Well.

She did not attempt to dissemble that she had come to look for him, but said,

'I expected to find you killed.'

He said, 'You see all my experiments are not failures. I have been more fortunate with the horse than the sail.'

At this moment one of the keepers appeared at a little distance. Lord Curryfin beckoned to him, and asked him to take the horse to the stables. The keeper looked with some amazement, and exclaimed,

'Why, this is the horse that nobody could manage!'

'You will manage him easily enough now,' said Lord Curryfin.

So it appeared; and the keeper took charge of him, not altogether without misgiving.

Miss Niphet's feelings had been over-excited, the more so from the severity with which she was accustomed to repress them. The energy which had thus far upheld her, suddenly gave way. She sate down on a fallen tree, and burst into tears. Lord Curryfin sate down by her, and took her hand. She allowed him to retain it awhile; but all at once snatched it from him and sped towards the house over the grass, with the swiftness and lightness of Virgil's Camilla, leaving his lordship as much astonished at her movements as the Volscian crowd, *attonitis inhians animis*,* had been at those of her prototype. He could not help thinking, 'Few women run gracefully; but she runs like another Atalanta.'

When the party met at breakfast, Miss Niphet was in her place, looking more like a statue than ever, with, if possible, more of marble paleness. Lord Curryfin's morning exploit, of which the story had soon found its way from the stable to the hall, was the chief subject of conversation. He had received a large share of what he had always so much desired—applause and admiration; but now he thought he would willingly sacrifice all he had ever received in that line, to see even the shadow of a smile, or the expression of a sentiment of any kind, on the impassive face of Melpomene. She left the room when she rose from the breakfast-table, appeared at the rehearsal, and went through her part as

* Gaping with wondering minds.

usual; sate down at luncheon, and departed as soon as it was over. She answered, as she had always done, everything that was said to her, frankly, and to the purpose; and also, as usual, she originated nothing.

In the afternoon Lord Curryfin went down to the pavilion. She was not there. He wandered about the grounds in all directions, and returned several times to the pavilion, always in vain. At last he sate down in the pavilion, and fell into a meditation. He asked himself how it could be, that having begun by making love to Miss Gryll, having indeed gone too far to recede unless the young lady absolved him, he was now evidently in a transition state towards a more absorbing and violent passion, for a person who, with all her frankness, was incomprehensible, and whose snowy exterior seemed to cover a volcanic firc, which she struggled to repress, and was angry with herself when she did not thoroughly succeed in so doing. If he were quite free he would do his part towards the solution of the mystery, by making a direct and formal proposal to her. As a preliminary to this, he might press Miss Gryll for an answer. All he had yet obtained from her was, 'Wait till we are better acquainted.' He was in a dilemma between Morgana and Melpomene. It had not entered into his thoughts that Morgana was in love with him; but he thought it nevertheless very probable that she was in a fair way to become so, and that even as it was she liked him well enough to accept him. On the other hand, he could not divest himself of the idea that Melpomene was in love with him. It was true, all the sympathy she had yet shown might have arisen from the excitement of strong feelings, at the real or supposed peril of a person with whom she was in the habit of daily intercourse. It might be so. Still the sympathy was very impassioned; though, but for his rashness in self-exposure to danger, he might never have known it. A few days ago, he would not press Miss Gryll for an answer, because he feared it might be a negative. Now he would not, because he was at least not in haste for an affirmative. But supposing it were a negative, what certainty had he that a negative from Morgana would not be followed by a negative from Melpomene? Then his heart would be at sea without rudder or compass. We shall leave him awhile to the contemplation of his perplexities.

As his thoughts were divided, so were Morgana's. If Mr. Falconer should propose to her, she felt she could accept him without hesitation. She saw clearly the tendency of his feelings towards her. She saw, at the same time, that he strove to the utmost against them in behalf of his old associations, though, with all his endeavours, he could not suppress them in her presence. So there was the lover who did not propose, and who would have been preferred; and there was the lover who had proposed, and who, if it had been clear that the former chance was hopeless, would not have been lightly given up.

If her heart had been as much interested in Lord Curryfin as it was in Mr. Falconer, she would quickly have detected a diminution in the ardour of his pursuit; but so far as she might have noticed any difference in his conduct, she ascribed it only to deference to her recommendation to 'wait till they were better acquainted.' The longer and the more quietly he waited, the better it seemed to please her. It was not on him, but on Mr. Falconer, that the eyes of her observance were fixed. She would have given Lord Curryfin his liberty instantly if she had thought he wished it.

Mr. Falconer also had his own dilemma, between his new love and his old affections. Whenever the first seemed likely to gain the ascendancy, the latter rose in their turn, like Antaeus from earth, with renovated strength. And he kept up their force by always revisiting the Tower, when the contest seemed doubtful.

Thus, Lord Curryfin and Mr. Falconer were rivals, with a new phase of rivalry. In some of their variations of feeling, each wished the other success; the latter, because he struggled against a spell that grew more and more difficult to be resisted; the former, because he had been suddenly overpowered by the same kind of light that had shone from the statue of Pygmalion. Thus their rivalry, such as it was, was entirely without animosity, and in no way disturbed the harmony of the Aristophanic party.

The only person concerned in these complications whose thoughts and feelings were undivided, was Miss Niphet. She had begun by laughing at Lord Curryfin, and had ended by forming a decided partiality for him. She contended against the feeling; she was aware of his intentions towards Miss Gryll; and she would perhaps have achieved a conquest over herself, if her sympathies had not been kept in a continual fever by the rashness with which he exposed himself to accidents by flood and field. At the same time, as she was more interested in observing Morgana than Morgana was in observing her, she readily perceived the latter's predilection for Mr. Falconer, and the gradual folding around him of the enchanted net. These observations, and the manifest progressive concentration of Lord Curryfin's affections on herself, showed her that she was not in the way of inflicting any very severe wound on her young friend's feelings, or encouraging a tendency to absolute hopelessness in her own.

Lord Curryfin was pursuing his meditations in the pavilion, when the young lady, whom he had sought there in vain, presented herself before him in great agitation. He started up to meet her, and held out both his hands. She took them both, held them a moment, disengaged them, and sate down at a little distance, which he immediately reduced to nothing. He then expressed his disappointment at not having previously found her in the pavilion, and his delight at seeing her now. After a pause, she said: 'I felt so much disturbed in the morning, that I should have devoted the whole day to recovering calmness of thought, but for something I have just heard. My maid tells me that you are going to try that horrid horse in harness, and in a newly-invented high phaeton of your own, and that the grooms say they would not drive that horse in any carriage, nor any horse in that carriage, and that you have a double chance of breaking your neck. I have disregarded all other feelings to entreat you to give up your intention.'

Lord Curryfin assured her that he felt too confident in his power over horses, and in the safety of his new invention, to admit the possibility of danger: but that it was a very small sacrifice to her to restrict himself to tame horses and low carriages, or to abstinence from all horses and carriages, if she desired it.

'And from sailing-boats,' she added.

'And from sailing-boats,' he answered.

'And from balloons,' she said.

'And from balloons,' he answered. 'But what made you think of balloons?'

'Because,' she said, 'they are dangerous, and you are inquiring and adventurous.'

'To tell you the truth,' he said, 'I have been up in a balloon. I thought it the most charming excursion I ever made. I have thought of going up again. I have invented a valve——'

'Oh heavens!' she exclaimed. 'But I have your promise touching horses, and carriages, and sails, and balloons.'

'You have,' he said. 'It shall be strictly adhered to.'

She rose to return to the house. But this time he would not part with her, and they returned together.

Thus prohibited by an authority to which he yielded implicit obedience, from trying further experiments at the risk of his neck, he restricted his inventive faculty to safer channels, and determined that the structure he was superintending should reproduce, as far as possible, all the peculiarities of the Athenian Theatre. Amongst other things, he studied attentively the subject of the *êcheia*, or sonorous vases, which, in that vast theatre, propagated and clarified sound; and though in its smaller representative they were not needed, he thought it still possible that they might produce an agreeable effect. But with all the assistance of the Reverend Doctor Opimian, he found it difficult to arrive at a clear idea of their construction, or even of their principle; for the statement of Vitruvius, that they gave an accordant resonance in the fourth, the fifth, and the octave, seemed incompatible with the idea of changes of key, and not easily reconcileable with the doctrine of Harmonics. At last he made up his mind that they had no reference to key, but solely to pitch, modified by duly proportioned magnitude and distance; he therefore set to work assiduously, got a number of vases made, ascertained that they would give a resonance of some kind, and had them disposed at proper intervals, round the audience part of the building. This being done, the party assembled, some as audience, some as performers, to judge of the effect. The first burst of choral music produced a resonance, like the sound produced by sea-shells when placed against the ear, only many times multiplied, and growing like the sound of a gong: it was the exaggerated concentration of the symphony of a lime-grove full of cockchaffers,* on a fine evening in the early summer. The experiment was then tried with single voices: the hum was less in itself, but greater in proportion. It was then tried with speaking: the result was the same: a powerful and perpetual hum, not resonant peculiarly to the diatessaron, the diapente, or the diapason, but making a new variety of continuous fundamental bass.

'I am satisfied,' said Lord Curryfin, 'the art of making these vases is as hopelessly lost as that of making mummies.' Miss Niphet encouraged him to persevere. She said:

'You have produced a decided resonance: the only thing is to subdue it, which you may perhaps effect by diminishing the number and enlarging the intervals of the vases.'

He determined to act on the suggestion, and she felt that, for some little time at least, she had kept him out of mischief. But whenever anything was said or sung in the theatre, it was necessary, for the time, to remove the *êcheia*.

* The drone of the cockchaffer, as he wheels by you in drowsy hum, sounds his corno di bassetto on F below the line.—GARDINER's *Music of Nature*.

CHAPTER XVIII

> Si, Mimnermus uti censet, sine amore jocisque
> Nil est jucundum, vivas in amore jocisque.
>
> HOR. *Epist.* I. vi. 65, 66.
>
> If, as Mimnermus held, nought else can move
> Your soul to pleasure, live in sports and love.

THE theatre was completed, and was found to be, without the êcheia, a fine vehicle of sound. It was tried, not only in the morning rehearsals, but occasionally, and chiefly on afternoons of bad weather, by recitations, and even lectures; for though some of the party attached no value to that mode of dogmatic instruction, yet with the majority, and especially with the young ladies, it was decidedly in favour.

One rainy afternoon Lord Curryfin was entreated to deliver in the theatre his lecture on Fish. He readily complied, and succeeded in amusing his audience more, and instructing them as much, as any of his more pretentious brother lecturers could have done. We shall not report the lecture, but we refer those who may be curious on the subject to the next meeting of the Pantopragmatic Society, under the presidency of Lord Facing-both-ways, and the vice-presidency of Lord Michin Malicho.

At intervals in similar afternoons of bad weather some others of the party were requested to favour the company with lectures or recitations in the theatre. Mr. Minim delivered a lecture on music, Mr. Pallet on painting; Mr. Falconer, though not used to lecturing, got up one on domestic life in the Homeric age. Even Mr. Gryll took his turn, and expounded the Epicurean philosophy. Mr. MacBorrowdale, who had no objection to lectures before dinner, delivered one on all the affairs of the world—foreign and domestic, moral, political, and literary. In the course of it he touched on Reform. 'The stone which Lord Michin Malicho—who was the Gracchus of the last Reform, and is the Sisyphus of the present—has been so laboriously pushing up hill, is for the present deposited at the bottom in the Limbo of Vanity. If it should ever surmount the summit and run down on the other side, it will infallibly roll over and annihilate the franchise of the educated classes; for it would not be worth their while to cross the road to exercise it against the rabble preponderance which would then have been created. Thirty years ago, Lord Michin Malicho had several cogent arguments in favour of Reform. One was, that the people were roaring for it, and that therefore they must have it. He has now in its favour the no less cogent argument, that the people do not care about it, and that the less it is asked for the greater will be the grace of the boon. On the former occasion the out-of-door logic was irresistible. Burning houses, throwing dead cats and cabbage-stumps into carriages, and other varieties of the same system of didactics, demonstrated the fitness of those who practised them to have representatives in Parliament. So they got their representatives, and many think Parliament would have been better without them. My father was a stanch Reformer. In his neighbourhood in London was the place of assembly of a Knowledge-is-Power Club. The members, at the close of their meetings, collected mending-stones from the road, and broke the windows to the right and left of their line of march. They had a flag on which was inscribed THE POWER OF PUBLIC OPINION. Whenever the enlightened assembly met, my father closed his shutters, but, closing within, they did not protect the glass. One morning he picked up, from where it had fallen between the window and the shutter, a very large, and consequently very demonstrative, specimen of dialectical granite. He preserved it carefully, and mounted it on a handsome pedestal, inscribed with THE POWER OF PUBLIC OPINION. He placed it on the middle of his library mantelpiece, and the daily contemplation of it cured him of his passion for Reform. During the rest of his life he never talked, as he had used to do, of "the people:" he always said "the rabble," and delighted in quoting every passage of *Hudibras* in which the rabble-rout is treated as he had come to conclude it ought to be. He made this piece of granite the nucleus of many political disquisitions. It is still in my possession, and I look on it with veneration as my principal tutor, for it had certainly a large share in the elements of my education. If, which does not seem likely, another reform lunacy should arise in my time, I shall take care to close my shutters against THE POWER OF PUBLIC OPINION.'

The Reverend Doctor Opimian being called on to contribute his share to these diversions of rainy afternoons, said:

'The sort of prose lectures which I am accustomed to deliver would not be exactly appropriate to the present time and place. I will therefore recite to you some verses, which I made some time since, on what appeared to me a striking specimen of absurdity on the part of the advisers of royalty here—the bestowing the honours of knighthood, which is a purely Christian institution, on Jews and Paynim; very worthy persons in themselves, and entitled to any mark of respect befitting their class, but not to one strictly and exclusively Christian; money-dealers, too, of all callings the most antipathetic to that of a true knight. The contast impressed itself on me as I was reading a poem of the twelfth century, by Hues de Tabaret—*L'Ordène de Chevalerie*—and I endeavoured to express the contrast in the manner and form following:

'A NEW ORDER OF CHIVALRY

I

Sir Moses, Sir Aaron, Sir Jamramajee,
Two stock-jobbing Jews, and a shroffing Parsee,
Have girt on the armour of old Chivalrie,
And, instead of the Red Cross, have hoisted Balls Three.

Now fancy our Sovereign, so gracious and bland,
With the sword of Saint George in her royal right hand,
Instructing this trio of marvellous Knights
In the mystical meanings of Chivalry's rites.

"You have come from the bath, all in milk-white array,
To show you have washed worldly feelings away,
And, pure as your vestments from secular stain,
Renounce sordid passions and seekings for gain.

"This scarf of deep red o'er your vestments I throw,
In token, that down them your life-blood shall flow,
Ere Chivalry's honour, or Christendom's faith,
Shall meet, through your failure, or peril or scaith.

"These slippers of silk, of the colour of earth,
Are in sign of remembrance of whence you had birth;
That from earth you have sprung, and to earth you return,
But stand for the faith, life immortal to earn.

"This blow of the sword, on your shoulder-blades true,
Is the mandate of homage, where homage is due,
And the sign, that your swords from the scabbard shall fly,
When 'St. George and the Right' is the rallying cry.

"This belt of white silk, which no speck has defaced,
Is the sign of a bosom with purity graced,
And binds you to prove, whatsoever betides,
Of damsels distressed the friends, champions, and guides.

"These spurs of pure gold are the symbols which say,
As your steeds obey them, you the Church shall obey,
And speed at her bidding, through country and town,
To strike, with your faulchions, her enemies down."

II

Now fancy these Knights, when the speech they have heard,
As they stand, scarfed, shoed, shoulder-dubbed, belted and spurred,
With the cross-handled sword duly sheathed on the thigh,
Thus simply and candidly making reply:

"By your majesty's grace we have risen up Knights,
But we feel little relish for frays and for fights:
There are heroes enough, full of spirit and fire,
Always ready to shoot and be shot at for hire.

"True, with bulls and with bears we have battled our cause;
But the bulls have no horns, and the bears have no paws;
And the mightiest blow which we ever have struck,
Has achieved but the glory of laming a duck.*

* In Stock Exchange slang, Bulls are speculators for a rise, Bears for a fall. A lame duck is a man who cannot pay his differences, and is said to waddle off. The patriotism of the money-market is well touched by Ponsard, in his comedy *La Bourse:* Acte IV. Scène 3:—

ALFRED

Quand nous sommes vainqueurs, dire qu'on a baissé!
Si nous étions battus, on aurait donc haussé?

DELATOUR

On a craint qu'un succès, si brillant pour la France,
De la paix qu'on rêvait n'éloignât l'espérance.

ALFRED

Cette Bourse, morbleu! n'as donc rien dans le cœur!
Ventre affamé n'a point d'oreilles ... pour l'honneur!
Aussi je ne veux plus jouer—qu'après ma noce—
Et j'attends Waterloo pour me mettre à la hausse.

"With two nations in arms, friends impartial to both,
To raise each a loan we shall be nothing loth;
We will lend them the pay, to fit men for the fray;
But shall keep ourselves carefully out of the way.

"We have small taste for championing maids in distress:
For State we care little: for Church we care less:
To Premium and Bonus our homage we plight:
'Percentage!' we cry: and 'A fig for the right!'

"'Twixt Saint George and the Dragon, we settle it thus:
Which has scrip above par, is the Hero for us:
For a turn in the market, the Dragon's red gorge
Shall have our free welcome to swallow Saint George."

Now God save our Queen, and if aught should occur,
To peril the crown, or the safety of her,
God send that the leader, who faces the foe,
May have more of King Richard than Moses and Co.'

CHAPTER XIX

TRINCQ est ung mot panomphée, célébré et entendu de toutes nations, et nous signifie, BEUUEZ. Et ici maintenons que non rire, ains boyre est le propre de l'homme. Je ne dy boyre simplement et absolument, car aussy bien boyvent les bestes; je dy boyre vin bon et fraiz.—RABELAIS: l. v. c. 45.

SOME guests remained. Some departed and returned. Among these was Mr. MacBorrowdale. One day after dinner, on one of his reappearances, Lord Curryfin said to him:—

'Well, Mr. MacBorrowdale, in your recent observations, have you found anything likely to satisfy Jack of Dover, if he were prosecuting his inquiry among us?'

MR. MACBORROWDALE

Troth, no, my lord. I think, if he were among us, he would give up the search as hopeless. He found it so in his own day, and he would find it still more so now. Jack was both merry and wise. We have less mirth in practice; and we have more wisdom in pretension, which Jack would not have admitted.

THE REVEREND DOCTOR OPIMIAN

He would have found it like Juvenal's search for patriotic virtue, when Catiline was everywhere, and Brutus and Cato were nowhere.*

LORD CURRYFIN

Well, among us, if Jack did not find his superior, or even his equal, he would not have been at a loss for company to his mind. There is enough mirth for those who choose to enjoy it, and wisdom too, perhaps as much as he would have cared for. We ought to have more wisdom, as we have clearly more science.

THE REVEREND DOCTOR OPIMIAN

Science is one thing, and wisdom is another. Science is an edged tool, with which men play like children, and cut their own fingers.

* Et Catilinam
Quocumque in populo videas, quocumque sub axe:
Sed nec Brutus erit, Bruti nec avunculus usquam.
JUV. *Sat.* xiv. 41–43.

If you look at the results which science has brought in its train, you will find them to consist almost wholly in elements of mischief. See how much belongs to the word Explosion alone, of which the ancients knew nothing. Explosions of powder-mills and powder-magazines; of coal-gas in mines and in houses; of high-pressure engines in ships and boats and factories. See the complications and refinements of modes of destruction, in revolvers and rifles and shells and rockets and cannon. See collisions and wrecks and every mode of disaster by land and by sea, resulting chiefly from the insanity for speed, in those who for the most part have nothing to do at the end of the race, which they run as if they were so many Mercuries, speeding with messages from Jupiter. Look at our scientific drainage, which turns refuse into poison. Look at the subsoil of London, whenever it is turned up to the air, converted by gas leakage into one mass of pestilent blackness, in which no vegetation can flourish, and above which, with the rapid growth of the ever-growing nuisance, no living thing will breathe with impunity. Look at our scientific machinery, which has destroyed domestic manufacture, which has substituted rottenness for strength in the thing made, and physical degradation in crowded towns for healthy and comfortable country life in the makers. The day would fail, if I should attempt to enumerate the evils which science has inflicted on mankind. I almost think it is the ultimate destiny of science to exterminate the human race.

LORD CURRYFIN

You have gone over a wide field, which we might exhaust a good bin of claret in fully discussing. But surely the facility of motion over the face of the earth and sea is both pleasant and profitable. We may now see the world with little expenditure of labour or time.

THE REVEREND DOCTOR OPIMIAN

You may be whisked over it, but you do not see it. You go from one great town to another, where manners and customs are not even now essentially different, and with this facility of intercourse become progressively less and less so. The intermediate country—which you never see, unless there is a show mountain or waterfall or ruin, for which there is a station, and to which you go as you would to any other exhibition—the intermediate country contains all that is really worth seeing, to enable you to judge of the various characteristics of men and the diversified objects of nature.

LORD CURRYFIN

You can suspend your journey if you please, and see the intermediate country if you prefer it.

THE REVEREND DOCTOR OPIMIAN

But who does prefer it? You travel round the world by a handbook, as you do round an exhibition-room by a catalogue.

MR. MACBORROWDALE

Not to say, that in the intermediate country you are punished by bad inns and bad wine; of which I confess myself intolerant. I knew an unfortunate French tourist, who had made the round of Switzerland, and had but one expression for every stage of his journey: *Mauvaise auberge!*

LORD CURRYFIN

Well, then, what say you to the electric telegraph, by which you converse at the distance of thousands of miles? Even across the Atlantic, as no doubt we shall yet do.

MR. GRYLL

Some of us have already heard the Doctor's opinion on that subject.

THE REVEREND DOCTOR OPIMIAN

I have no wish to expedite communication with the Americans. If we could apply the power of electrical repulsion to preserve us from ever hearing anything more of them, I should think that we had for once derived a benefit from science.

MR. GRYLL

Your love for the Americans, Doctor, seems something like that of Cicero's friend Marius for the Greeks. He would not take the nearest road to his villa, because it was called the Greek-road.* Perhaps if your nearest way home were called the American-road, you would make a circuit to avoid it.

* Non enim te puto Græcos ludos desiderare: præsertim quum Græcos ita non ames, ut ne ad villam quidem tuam viâ Græcâ ire soleas.—CICERO: *Ep. ad Div.* vii. 1.

THE REVEREND DOCTOR OPIMIAN

I am happy to say I am not put to the test. Magnetism, galvanism, electricity, are 'one form of many names.'* Without magnetism we should never have discovered America; to which we are indebted for nothing but evil; diseases in the worst forms that can afflict humanity, and slavery in the worst form in which slavery can exist. The Old World had the sugar-cane and the cotton-plant, though it did not so misuse them. Then, what good have we got from America? What good of any kind, from the whole continent and its islands, from the Esquimaux to Patagonia?

MR. GRYLL

Newfoundland salt fish, Doctor.

THE REVEREND DOCTOR OPIMIAN

That is something, but it does not turn the scale.

MR. GRYLL

If they have given us no good, we have given them none.

THE REVEREND DOCTOR OPIMIAN

We have given them wine and classical literature; but I am afraid Bacchus and Minerva have equally

Scattered their bounty upon barren ground.

On the other hand, we have given the red men rum, which has been the chief instrument of their perdition. On the whole, our intercourse with America has been little else than an interchange of vices and diseases.

LORD CURRYFIN

Do you count it nothing to have substituted civilized for savage men?

THE REVEREND DOCTOR OPIMIAN

Civilized. The word requires definition. But looking into futurity, it seems to me that the ultimate tendency of the change is to substitute the worse for the better race; the Negro for the Red Indian. The Red Indian will not work for a master. No ill-usage will make him. Herein, he is the noblest specimen of humanity that

* *Πολλῶν ὀνομάτων μορφὴ μία.*—ÆSCHYLUS: *Prometheus.*

ever walked the earth. Therefore, the white man exterminates his race. But the time will come, when by mere force of numbers, the black race will predominate, and exterminate the white. And thus the worse race will be substituted for the better, even as it is in Saint Domingo, where the Negro has taken the place of the Caraib. The change is clearly for the worse.

LORD CURRYFIN

You imply, that in the meantime the white race is better than the red.

THE REVEREND DOCTOR OPIMIAN

I leave that as an open question. But I hold, as some have done before me, that the human mind degenerates in America, and that the superiority, such as it is, of the white race is only kept up by intercourse with Europe. Look at the atrocities in their ships. Look at their Congress and their Courts of Justice; debaters in the first; suitors, even advocates, sometimes judges, in the second, settling their arguments with pistol and dagger. Look at their extensions of slavery, and their revivals of the slave-trade, now covertly, soon to be openly. If it were possible that the two worlds could be absolutely dissevered for a century, I think a new Columbus would find nothing in America but savages.

LORD CURRYFIN

You look at America, Doctor, through your hatred of slavery. You must remember that we introduced it when they were our colonists. It is not so easily got rid of. Its abolition by France exterminated the white race in Saint Domingo, as the white race had exterminated the red. Its abolition by England ruined our West Indian colonies.

THE REVEREND DOCTOR OPIMIAN

Yes, in conjunction with the direct encouragement of foreign slave labour, given by our friends of liberty under the pretext of free trade. It is a mockery to keep up a squadron for suppressing the slave-trade on the one hand, while on the other hand we encourage it to an extent that counteracts in a tenfold degree the apparent power of suppression. It is a clear case of false pretension.

MR. GRYLL

You know, Doctor, the Old World had slavery throughout its entire extent; under the Patriarchs, the Greeks, the Romans;

everywhere, in short. Cicero thought our island not likely to produce anything worth having excepting slaves;* and of those none skilled, as some slaves were, in letters and music, but all utterly destitute of both. And in the Old World the slaves were of the same race with the masters. The Negroes are an inferior race, not fit, I am afraid, for anything else.

THE REVEREND DOCTOR OPIMIAN

Not fit, perhaps, for anything else belonging to what we call civilized life. Very fit to live on little, and wear nothing, in Africa; where it would have been a blessing to themselves and the rest of the world if they had been left unmolested; if they had had a Friar Bacon to surround their entire continent with a wall of brass.

MR. FALCONER

I am not sure, Doctor, that in many instances, even yet, the white slavery of our factories is not worse than the black slavery of America. We have done much to amend it, and shall do no more. Still much remains to be done.

THE REVEREND DOCTOR OPIMIAN

And will be done, I hope and believe. The Americans do nothing to amend their system. On the contrary, they do all they can to make bad worse. Whatever excuse there may be for maintaining slavery where it exists, there can be none for extending it into new territories; none for reviving the African slave-trade. These are the crying sins of America. Our white slavery, so far as it goes, is so far worse, that it is the degradation of a better race. But if it be not redressed, as I trust it will be, it will work out its own retribution. And so it is of all the oppressions that are done under the sun. Though all men but the red men will work for a master, they will not fight for an oppressor in the day of his need. Thus gigantic empires have crumbled into dust at the first touch of an invader's footstep. For petty, as for great oppressions, there is a day of

* Etiam illud jam cognitum est, neque argenti scripulum esse ullum in illâ insulâ, neque ullam spem prædæ, nisi ex mancipiis: ex quibus nullos puto te literis aut musicis eruditos expectare.—CICERO *ad Atticum:* iv. 16.

A hope is expressed by Pomponius Mela, l. iii. c. 6 (he wrote under Claudius), that, by the success of the Roman arms, the island and its savage inhabitants would soon be better known. It is amusing enough to peruse such passages in the midst of London.—GIBBON: c. i.

retribution growing out of themselves. It is often long in coming. *Ut sit magna, tamen certe lenta ira Deorum est.** But it comes.

> Raro antecedentem scelestum
> Deseruit pede Pœna claudo.†

LORD CURRYFIN

I will not say, Doctor, 'I've seen, and sure I ought to know.' But I have been in America, and I have found there, what many others will testify, a very numerous class of persons, who hold opinions very like your own: persons who altogether keep aloof from public life, because they consider it abandoned to the rabble; but who are as refined, as enlightened, as full of sympathy for all that tends to justice and liberty, as any whom you may most approve amongst ourselves.

THE REVEREND DOCTOR OPIMIAN

Of that I have no doubt. But I look to public acts and public men.

LORD CURRYFIN

I should much like to know what Mr. MacBorrowdale thinks of all this.

MR. MACBORROWDALE

Troth, my lord, I think we have strayed far away from the good company we began with. We have lost sight of Jack of Dover. But the discussion had one bright feature. It did not interfere with it, rather promoted, the circulation of the bottle: for every man who spoke pushed it on with as much energy as he spoke with, and those who were silent swallowed the wine and the opinion together, as if they relished them both.

THE REVEREND DOCTOR OPIMIAN

So far, discussion may find favour. In my own experience, I have found it very absorbent of claret. But I do not think it otherwise an incongruity after dinner, provided it be carried on, as our disquisitions have always been, with frankness and good humour. Consider how much instruction has been conveyed to us in the form of

* The anger of the Gods, though great, is slow.

† The foot of Punishment, though lame,
O'ertakes at last preceding Wrong.

conversations at banquets, by Plato and Xenophon and Plutarch. I read nothing with more pleasure than their *Symposia*: to say nothing of Athenæus, whose work is one long banquet.

MR. MACBORROWDALE

Nay, I do not object to conversation on any subject. I object to after-dinner lectures. I have had some unfortunate experiences. I have found what began in conversation end in a lecture. I have on different occasions met several men, who were in that respect all alike. Once started they never stopped. The rest of the good company, or rather the rest, which without them would have been good company, was no company. No one could get in a word. They went on with one unvarying stream of monotonous desolating sound. This makes me tremble when a discussion begins. I sit in fear of a lecture.

LORD CURRYFIN

Well, you and I have lectured, but never after dinner. We do it when we have promised it, and when those who are present expect it. After dinner, I agree with you, it is the most doleful blight that can fall on human enjoyment.

MR. MACBORROWDALE

I will give you one or two examples of these post-prandial inflictions. One was a great Indian reformer. He did not open his mouth till he had had about a bottle and a half of wine. Then he burst on us with a declamation, on all that was wrong in India, and its remedy. He began in the Punjaub, travelled to Calcutta, went southward, got into the Temple of Juggernaut, went southward again, and after holding forth more than an hour, paused for a moment. The man who sate next him attempted to speak: but the orater clapped him on the arm, and said: 'Excuse me: now I come to Madras.' On which his neighbour jumped up and vanished. Another went on in the same way about currency. His first hour's talking carried him just through the Restriction Act of ninety-seven. As we had then more than half a century before us, I took my departure. But these were two whom topography and chronology would have brought to a close. The bore of all bores was the third. His subject had no beginning, middle, nor end. It was education. Never was such a journey through the desert of mind: the Great Sahara of intellect. The very recollection makes me thirsty.

THE REVEREND DOCTOR OPIMIAN

If all the nonsense which, in the last quarter of a century, has been talked on all other subjects, were thrown into one scale, and all that has been talked on the subject of education alone were thrown into the other, I think the latter would preponderate.

LORD CURRYFIN

We have had through the whole period some fine specimens of nonsense on other subjects: for instance, with a single exception, Political Economy.

MR. MACBORROWDALE

I understand your lordship's politeness as excepting the present company. You need not except me. I am 'free to confess,' as they say 'in another place,' that I have talked a great deal of nonsense on that subject myself.

LORD CURRYFIN

Then, we have had latterly a mighty mass on the Purification of the Thames.

THE REVEREND DOCTOR OPIMIAN

Allowing full weight to the two last-named ingredients, they are not more than a counterpoise to Competitive Examination, which is also a recent exotic belonging to education.

LORD CURRYFIN

Patronage, it used to be alleged, considered only the fitness of the place for the man, not the fitness of the man for the place. It was desirable to reverse this.

THE REVEREND DOCTOR OPIMIAN

True: but

Dum vitant stulti vitium, in contraria currunt.*

Questions, which can only be answered by the parrotings of a memory, crammed to disease with all sorts of heterogeneous diet, can form no test of genius, taste, judgment, or natural capacity.

* When fools would from one vice take flight,
They rush into its opposite.
HOR. *Sat.* i, 2, 24.

Competitive Examination takes for its *norma*: 'It is better to learn many things ill than one thing well;' or rather: 'It is better to learn to gabble about everything than to understand anything.' This is not the way to discover the wood of which Mercuries are made. I have been told that this precious scheme has been borrowed from China: a pretty fountain-head for moral and political improvement: and if so, I may say, after Petronius: 'This windy and monstrous loquacity has lately found its way to us from Asia, and like a pestilential star has blighted the minds of youth, otherwise rising to greatness.'*

LORD CURRYFIN

There is something to be said on behalf of applying the same tests, addressing the same questions, to everybody.

THE REVEREND DOCTOR OPIMIAN

I shall be glad to hear what can be said on that behalf.

LORD CURRYFIN (*after a pause*)

'Mass,' as the second grave-digger says in *Hamlet*, 'I cannot tell.'

A chorus of laughter dissolved the sitting.

* Nuper ventosa isthæc et enormis loquacitas Athenas ex Asiâ commigravit, animosque juvenum, ad magna surgentes, veluti pestilenti quodam sidere afflavit.

CHAPTER XX

Les violences qu'on se fait pour s'empêcher d'aimer sont souvent plus cruelles que les rigueurs de ce qu'on aime.—LA ROCHEFOUCAULD.

THE winter set in early. December began with intense frost. Mr. Falconer, one afternoon, entering the inner drawing-room, found Miss Gryll alone. She was reading, and on the entrance of her visitor laid down her book. He hoped he had not interrupted her in an agreeable occupation. 'To observe romantic method,' we shall give what passed between them with the Christian names of the speakers.

MORGANA

I am only reading what I have often read before, *Orlando Innamorato*; and I was at the moment occupied with a passage about the enchantress from whom my name was borrowed. You are aware that enchantresses are in great favour here.

ALGERNON

Circe and Gryllus and your name sufficiently show that. And not your name only, but—— I should like to see the passage, and should be still better pleased if you would read it to me.

MORGANA

It is where Orlando, who had left Morgana sleeping by the fountain, returns to seek the enchanted key, by which alone he can liberate his friends.

> Il Conte, che d'intrare havea gran voglia,
> Subitamente al fonte ritornava:
> Quivi trovò Morgana, che con gioglia
> Danzava intorno, e danzando cantava.
> Nè più leggier si move al vento foglia
> Come ella sanza sosta si voltava,
> Mirando hora a la terra ed hora al sole;
> Ed al suo canto usava tal parole:
>
> 'Qualonque cerca al mondo haver thesoro,
> Over diletto, o segue onore e stato,
> Ponga la mano a questa chioma d'oro,
> Ch'io porto in fronte, e quel farò beato.

Ma quando ha il destro a far cotal lavoro,
Non prenda indugio, che'l tempo passato
Più non ritorna, e non si trova mai;
Ed io mi volto, e lui lascio con guai.'

Così cantava d'intorno girando
La bella Fata a quella fresca fonte:
Ma come gionto vide il Conte Orlando,
Subitamente rivoltò la fronte:
Il prato e la fontana abbandonando,
Prese il viaggio suo verso d'un monte,
Qual chiudea la valletta picciolina:
Quivi fuggendo Morgana cammina.*

* BOJARDO: l. ii. c. 8. *Ed. Vinegia:* 1544.

With earnest wish to pass the enchanted gate,
Orlando to the fount again advanced,
And found Morgana, all with joy elate,
Dancing around, and singing as she danced.
As lightly moved and twirled the lovely Fate
As to the breeze the lightest foliage glanced,
With looks alternate to the earth and sky,
She thus gave out her words of witchery:

Let him, who seeks unbounded wealth to hold,
Or joy, or honour, or terrestrial state,
Seize with his hand this lock of purest gold,
That crowns my brow, and blest shall be his fate.
But when time serves, behoves him to be bold,
Nor even a moment's pause interpolate:
The chance once lost he never finds again:
I turn, and leave him to lament in vain.'

Thus sang the lovely Fate in bowery shade,
Circling in joy around the crystal fount;
But when within the solitary glade
Glittered the armour of the approaching Count,
She sprang upon her feet, as one dismayed,
And took her way towards a lofty mount,
That rose the valley's narrow length to bound:
Thither Morgana sped along the ground.

I have translated *Fata,* Fate. It is usually translated Fairy. But the idea differs essentially from ours of a fairy. Amongst other things, there is no *Fato*, no Oberon to the Titania. It does not, indeed, correspond with our usual idea of Fate, but it is more easily distinguished as a class; for our old acquaintances the Fates are an inseparable three. The Italian *Fata* is independent of her sisters. They are enchantresses; but they differ from other enchantresses in being immortal. They are beautiful too, and their

ALGERNON

I remember the passage well. The beautiful *Fata,* dancing and singing by the fountain, presents a delightful picture.

MORGANA

Then, you know, Orlando, who had missed his opportunity of seizing the golden forelock while she was sleeping, pursues her a long while in vain through rocky deserts, *La Penitenza* following him with a scourge. The same idea was afterwards happily worked out by Machiavelli in his *Capitolo dell' Occasione.*

ALGERNON

You are fond of Italian literature? You read the language beautifully. I observe you have read from the original poem, and not from Berni's *rifacciamento.*

MORGANA

I prefer the original. It is more simple, and more in earnest. Berni's playfulness is very pleasant, and his exordiums are charming; and in many instances he has improved the poetry. Still, I think, he has less than the original of what are to me the great charms of poetry, truth and simplicity. Even the greater antiquity of style has its peculiar appropriateness to the subject. And Bojardo seems to have more faith in his narrative than Berni. I go on with him with ready credulity, where Berni's pleasantry interposes a doubt.

ALGERNON

You think that in narratives, however wild and romantic, the poet should write as if he fully believed in the truth of his own story.

MORGANA

I do; and I think so in reference to all narratives, not to poetry only. What a dry skeleton is the History of the early ages of Rome, told by one who believes nothing that the Romans believed. Religion pervades every step of the early Roman History; and in a great degree down at least to the Empire; but because their religion is not our religion, we pass over the supernatural part of the matter

beauty is immortal: always in Bojardo. He would not have turned Alcina into an old woman, as Ariosto did; which I must always consider a dreadful blemish on the many charms of the *Orlando Furioso.*

in silence, or advert to it in a spirit of contemptuous incredulity. We do not give it its proper place, nor present it in its proper colours, as a cause in the production of great effects. Therefore, I like to read Livy, and I do not like to read Niebuhr.

ALGERNON

May I ask if you read Latin?

MORGANA

I do; sufficiently to derive great pleasure from it. Perhaps, after this confession, you will not wonder that I am a spinster.

ALGERNON

So far, that I think it would tend to make you fastidious in your choice. Not that you would be less sought by any who would be worthy your attention. For I am told, you have had many suitors, and have rejected them all in succession. And have you not still many, and among them one very devoted lover, who would bring you title as well as fortune? A very amiable person, too, though not without a comic side to his character.

MORGANA

I do not well know. He so far differs from all my preceding suitors, that in every one of them I found the presence of some quality that displeased me, or the absence of some which would have pleased me: the want, in the one way or the other, of that entire congeniality in taste and feeling, which I think essential to happiness in marriage. He has so strong a desire of pleasing, and such power of acquisition and assimilation, that I think a woman truly attached to him might mould him to her mind. Still, I can scarcely tell why, he does not complete my idealities. They say, Love is his own avenger; and perhaps I shall be punished by finding my idealities realized in one who will not care for me.

ALGERNON

I take that to be impossible.

Morgana blushed, held down her head, and made no reply. Algernon looked at her with silent admiration. A new light seemed to break in on him. Though he had had so many opportunities of forming a judgment on the point, it seemed to strike him for the first time with irresistible conviction that he had never before heard such a sweet voice, nor seen such an expressive and intelligent countenance. And in this way they continued like two figures in a *tableau vivant*, till the entrance of other parties broke the spell which had thus fixed them in their positions.

A few minutes more, and their destinies might have been irrevocably fixed. But the interruption gave Mr. Falconer the opportunity of returning again to his Tower, to consider, in the presence of the seven sisters, whether he should not be in the position of a Roman, who was reduced to the dilemma of migrating without his household deities, or of suffering his local deities to migrate without him; and whether he could sit comfortably on either of the horns of this dilemma. He felt that he could not. On the other hand, could he bear to see the fascinating Morgana metamorphosed into Lady Curryfin? The time had been when he had half wished it, as the means of restoring him to liberty. He felt now, that when in her society he could not bear the idea; but he still thought, that in the midst of his domestic deities he might become reconciled to it.

He did not care for horses, nor keep any for his own use. But as time and weather were not always favourable to walking, he had provided for himself a comfortable travelling-chariot, without a box to intercept the view, in which, with post-horses after the fashion of the olden time, he performed occasional migrations. He found this vehicle of great use in moving to and fro between the Grange and the Tower; for then, with all his philosophy, Impatience was always his companion: Impatience on his way to the Grange, to pass into the full attraction of the powerful spell by which he was drawn like the fated ship to the magnetic rock in the *Arabian Nights*: Impatience on his way to the Tower, to find himself again in the 'Regions mild of pure and serene air,' in which the seven sisters seemed to dwell, like Milton's ethereal spirits 'Before the starry threshold of Jove's court.' Here was everything to soothe, nothing to irritate or disturb him: nothing on the spot: but it was with him, as it is with many, perhaps with all: the two great enemies of tranquillity, Hope and Remembrance, would still intrude: not like a bubble and a spectre, as in the beautiful lines of Coleridge:* for the remembrance of Morgana was not a spectre, and the hope of her love, which he cherished in spite of himself, was not a bubble: but their forces were not less disturbing, even in the presence of his earliest and most long and deeply cherished associations.

He did not allow his impatience to require that the horses should be put to extraordinary speed. He found something tranquillizing in the movement of a postilion in a smart jacket, vibrating on one horse upwards and downwards, with one invariable regulated motion like the crosshead of a side-lever steam-engine, and holding the whip quietly arched over the neck of the other. The mechanical monotony of the movement seemed less in contrast than in harmony with the profound stillness of the wintry forest: the leafless branches heavy with rime frost and glittering in the sun: the deep repose of nature, broken now and then by the traversing of deer, or the flight of wild birds: highest and loudest among them the long lines of rooks: but for the greater part of the way one long deep silence, undisturbed but by the rolling of the wheels and the iron tinkling of the hoofs on the frozen ground. By degrees he fell into a reverie, and meditated on his last dialogue with Morgana.

'It is a curious coincidence,' he thought, 'that she should have been dwelling on a passage, in which her namesake enchantress inflicted punishment on Orlando for having lost his opportunity. Did she associate Morgana with herself and Orlando with me? Did she intend a graceful hint to me not to lose *my* opportunity? I seemed in a fair way to seize the golden forelock, if we had not been interrupted. Do I regret

* Who late and lingering seeks thy shrine,
On him but seldom, Power divine,
Thy spirit rests. Satiety,
And sloth, poor counterfeits of thee,
Mock the tired worldling. Idle Hope,
And dire Remembrance, interlope,
And vex the feverish slumbers of the mind:
The bubble floats before: the spectre stalks behind.
COLERIDGE's *Ode to Tranquillity*.

that I did not? That is just what I cannot determine. Yet it would be more fitting, that whatever I may do should be done calmly, deliberately, philosophically, than suddenly, passionately, impulsively. One thing is clear to me. It is now or never: this or none. The world does not contain a second Morgana, at least not of mortal race. Well: the opportunity will return. So far, I am not in the predicament in which we left Orlando. I may yet ward off the scourge of *La Penitenza*.'

But his arrival at home, and the sight of the seven sisters, who had all come to the hall-door to greet him, turned his thoughts for awhile into another channel.

He dined at his usual hour, and his two Hebes alternately filled his glass with Madeira. After which the sisters played and sang to him in the drawing-room; and when he had retired to his chamber, had looked on the many portraitures of his Virgin Saint, and had thought by how many charms of life he was surrounded, he composed himself to rest with the reflection: 'I am here like Rasselas in the Happy Valley: and I can now fully appreciate the force of that beautiful chapter: *The wants of him who wants nothing*.'

CHAPTER XXI

Ubi lepos, joci, risus, ebrietas decent,
Gratiæ, decor, hilaritas, atque delectatio,
Qui quærit alia his, malum videtur quærere.
PLAUTUS: *In Pseudolo.*

Where sport, mirth, wine, joy, grace, conspire to please,
He seeks but ill who seeks aught else than these.

THE frost continued. The lake was covered over with solid ice. This became the chief scene of afternoon amusement, and Lord Curryfin carried off the honours of the skating. In the dead of the night, there came across his memory a ridiculous stave:

There's Mr. Tait, he cuts an eight,
He cannot cut a nine:

and he determined on trying if he could not outdo Mr. Tait. He thought it would be best to try his experiment without witnesses: and having more than an hour's daylight before breakfast, he devoted that portion of the morning to his purpose. But cutting a nine by itself baffled his skill, and treated him to two or three tumbles, which however did not abate his ardour. At length he bethought him of cutting a nine between two eights, and by shifting his feet rapidly at the points of difficulty, striking in and out of the nine to and from the eights on each side. In this he succeeded, and exhibiting his achievement in the afternoon, adorned the surface of the ice with successions of 898, till they amounted to as many sextillions, with their homogeneous sequences. He then enclosed the line with an oval, and returned to the bank through an admiring circle, who, if they had been as numerous as the spectators at the Olympic games, would have greeted him with as loud shouts of triumph as saluted Epharmostus of Opus.*

Among the spectators on the bank were Miss Niphet and

* Διήρχετο κύκλον ὅσσᾳ βοᾷ.—PIND. *Olymp.* ix.
With what a clamour he passed through the circle.

Mr. MacBorrowdale, standing side by side. While Lord Curryfin was cutting his sextillions, Mr. MacBorrowdale said: 'There is a young gentleman who is capable of anything, and who would shine in any pursuit, if he would keep to it. He shines as it is, in almost everything he takes in hand in private society: there is genius even in his failures, as in the case of the theatrical vases; but the world is a field of strong competition, and affords eminence to few in any sphere of exertion, and to those few rarely but in one.'

MISS NIPHET

Before I knew him, I never heard of him but as a lecturer on Fish: and to that he seems to limit his public ambition. In private life, his chief aim seems to be that of pleasing his company. Of course, you do not attach much value to his present pursuit. You see no utility in it.

MR. MACBORROWDALE

On the contrary, I see great utility in it. I am for a healthy mind in a healthy body: the first can scarcely be without the last, and the last can scarcely be without good exercise in pure air. In this way, there is nothing better than skating. I should be very glad to cut eights and nines with his lordship: but the only figure I should cut, would be that of as many feet as would measure my own length on the ice.

Lord Curryfin on his return to land, thought it his duty first to accost Miss Gryll, who was looking on by the side of Miss Ilex. He asked her if she ever skated? She answered in the negative. 'I have tried it,' she said, 'but unsuccessfully. I admire it extremely, and regret my inability to participate in it.' He then went up to Miss Niphet, and asked her the same question. She answered: 'I have skated often in our grounds at home.' 'Then why not now?' he asked. She answered: 'I have never done it before so many witnesses.' 'But what is the objection?' he asked. 'None that I know of,' she answered. 'Then,' he said, 'as I have done or left undone some things to please you, will you do this one thing to please me?' 'Certainly,' she replied: adding to herself: 'I will do anything in my power to please you.'

She equipped herself expeditiously, and started before he was well aware. She was half round the lake before he came up with her. She then took a second start, and completed the circle before he came up with her again. He saw that she was an Atalanta on ice as on turf. He placed himself by her side, slipped her arm through his, and they started together on a second round, which they completed arm-in-arm. By this time the blush-rose bloom which had so charmed him on a former occasion again mantled on her cheeks, though from a different cause, for it was now only the glow of healthful exercise; but he could not help exclaiming, 'I now see why and with what tints the Athenians coloured their statues.'

'Is it clear,' she asked, 'that they did so?'

'I have doubted it before,' he answered, 'but I am now certain that they did.'

In the meantime Miss Gryll, Miss Ilex, and the Reverend Doctor Opimian had been watching their movements from the bank.

MISS ILEX

I have seen much graceful motion in dancing, in private society and on the Italian stage; and some in skating before to-day; but anything so graceful as that double-gliding over the ice by those two remarkably handsome young persons, I certainly never saw before.

MISS GRYLL

Lord Curryfin is unquestionably handsome, and Miss Niphet, especially with that glow on her cheeks, is as beautiful a young woman as imagination can paint. They move as if impelled by a single will. It is impossible not to admire them both.

THE REVEREND DOCTOR OPIMIAN

They remind me of the mythological fiction, that Jupiter made men and women in pairs, like the Siamese twins; but in this way they grew so powerful and presumptuous, that he cut them in two; and now the main business of each half is to look for the other; which is very rarely found, and hence so few marriages are happy. Here the two true halves seem to have met.

The Doctor looked at Miss Gryll, to see what impression this remark might make on her. He concluded that, if she thought seriously of Lord Curryfin, she would show some symptom of jealousy of Miss Niphet; but she did not. She merely said,

'I quite agree with you, Doctor. There is evidently great congeniality between them, even in their respective touches of eccentricity.'

But the Doctor's remark had suggested to her what she herself had failed to observe; Lord Curryfin's subsidence from ardour into deference, in his pursuit of herself. She had been so undividedly 'the cynosure of neighbouring eyes,' that she could scarcely believe in the possibility of even temporary eclipse. Her first impulse was to resign him to her young friend. But then appearances might be deceitful. Her own indifference might have turned his attentions into another channel, without his heart being turned with them. She had seen nothing to show that Miss Niphet's feelings were deeply engaged in the question. She was not a coquet; but she would still feel it as a mortification that her hitherto unquestioned supremacy should be passing from her. She had felt all along, that there was one cause which would lead her to a decided rejection of Lord Curryfin. But her Orlando had not seized the golden forelock; perhaps he never would. After having seemed on the point of doing so, he had disappeared, and not returned. He was now again within the links of the sevenfold chain, which had bound him from his earliest days. She herself, too, had had, perhaps had still, the chance of the golden forelock in another quarter. Might she not subject her after-life to repentance, if her first hope should fail her, when the second had been irrevocably thrown away? The more she contemplated the sacrifice, the greater it appeared. Possibly doubt had given preponderance to her thoughts of Mr. Falconer; and certainty had caused them to repose in the case of Lord Curryfin; but when doubt was thrown into the latter scale also, the balance became more even. She would still give him his liberty, if she believed that he wished it; for then her pride would settle the question; but she must have more conclusive evidence on the point than the Reverend Doctor's metaphorical deduction from a mythological fiction.

In the evening, while the party in the drawing-room were

amusing themselves in various ways, Mr. MacBorrowdale laid a drawing on the table, and said, 'Doctor, what should you take that to represent?'

THE REVEREND DOCTOR OPIMIAN

An unformed lump of I know not what.

MR. MACBORROWDALE

Not unformed. It is a flint formation of a very peculiar kind.

THE REVEREND DOCTOR OPIMIAN

Very peculiar, certainly. Who on earth can have amused himself with drawing a misshapen flint? There must be some riddle in it; some ænigma, as insoluble to me as *Aelia Laelia Crispis.**

Lord Curryfin, and others of the party, were successively asked their opinions. One of the young ladies guessed it to be the petrifaction of an antediluvian muscle. Lord Curryfin said petrifications were often siliceous, but never pure silex; which this purported to be. It gave him the idea of an ass's head; which, however, could not by any process have been turned into flint.

Conjecture being exhausted, Mr. MacBorrowdale said, 'It is a thing they call a Celt. The ass's head is somewhat germane to the matter. The Artium Societatis Syndicus Et Socii have determined that it is a weapon of war, evidently of human manufacture. It has been found, with many others like it, among bones of mammoths and other extinct animals, and is therefore held to prove, that men and mammoths were contemporaries.'

THE REVEREND DOCTOR OPIMIAN

A weapon of war? Had it a handle? Is there a hole for a handle?

MR. MACBORROWDALE

That does not appear.

* This ænigma has been the subject of many learned disquisitions. The reader, who is unacquainted with it, may find it under the article 'Ænigma' in the *Encyclopædia Britannica*; and probably in every other encyclopædia.

THE REVEREND DOCTOR OPIMIAN

These flints, and no other traces of men, among the bones of mammoths?

MR. MACBORROWDALE

None whatever.

THE REVEREND DOCTOR OPIMIAN

What do the Artium Societatis Syndicus Et Socii suppose to have become of the men, who produced these demonstrations of high aboriginal art?

MR. MACBORROWDALE

They think these finished specimens of skill in the art of chipping prove that the human race is of greater antiquity than has been previously supposed; and the fact, that there is no other relic to prove the position, they consider of no moment whatever.

THE REVEREND DOCTOR OPIMIAN

Ha! ha! ha! this beats the Elephant in the Moon,* which turned out to be a mouse in a telescope. But I can help them to an explanation of what became of these primæval men-of-arms. They were an ethereal race, and evaporated.

* See Butler's poem, with that title, in his *Miscellaneous Works*.

CHAPTER XXII

> Over the mountains,
>   And over the waves;
> Under the fountains,
>   And under the graves;
> Under floods that are deepest,
>   Which Neptune obey;
> Over rocks that are steepest,
>   Love will find out the way.
>
> Old Song in PERCY's *Reliques*.

HARRY HEDGEROW had volunteered to be Mr. Falconer's Mercury during his absences from the Tower, and to convey to him letters and any communications which the sisters might have to make. Riding at a good trot, on a horse more distinguished for strength than grace, he found the shortest days long enough for the purpose of going and returning, with an ample interval for the refreshment of himself and his horse. While discussing beef and ale in the servants' hall, he heard a good deal of the family news, and many comments on the visitors. From these he collected, that there were several young gentlemen especially remarkable for their attention to the young lady of the mansion: that among them were two who were more in her good graces than the others: that one of these was the young gentleman who lived in the Duke's Folly, and who was evidently the favourite: and that the other was a young lord, who was the life and soul of the company, but who seemed to be very much taken with another young lady, who had, at the risk of her own life, jumped into the water and picked him out, when he was nearly being drowned. This story had lost nothing in travelling. Harry, deducing from all this the conclusion most favourable to his own wishes, determined to take some steps for the advancement of his own love-suit, especially as he had obtained some allies, who were willing to march with him to conquest, like the Seven against Thebes.

The Reverend Doctor Opimian had finished his breakfast, and had just sat down in his library, when he was informed that some young men wished to see him. The Doctor was always accessible, and the visitors were introduced. He recognised his friend Harry Hedgerow, who was accompanied by six others. After respectful salutations on their part, and benevolent acceptance on his, Harry, as the only one previously known to the Doctor, became spokesman for the deputation.

HARRY HEDGEROW

You see, Sir, you gave me some comfort when I was breaking my heart; and now we are told that the young gentleman at the Folly is going to be married.

THE REVEREND DOCTOR OPIMIAN

Indeed! you are better informed than I am.

HARRY HEDGEROW

Why, it's in everybody's mouth. He passes half his time at Squire Gryll's, and they say it's all for the sake of the young lady that's there: she that was some days at the Folly; that I carried in, when she was hurt in the great storm. I am sure I hope it be true. For you said, if he married, and suitable parties proposed for her sisters, Miss Dorothy might listen to me. I have lived in the hope of that ever since. And here are six suitable parties to propose for her six sisters. That is the long and the short of it.

THE REVEREND DOCTOR OPIMIAN

The short of it, at any rate. You speak like a Spartan. You come to the point at once. But why do you come to me? I have no control over the fair damsels.

HARRY HEDGEROW

Why, no, Sir, but you are the greatest friend of the young gentleman. And if you could just say a word for us to him, you see, Sir.

THE REVEREND DOCTOR OPIMIAN

I see seven notes in the key of A minor, proposing to sound in harmony with the seven notes of the octave above; but I really do not see what I can do in the matter.

HARRY HEDGEROW

Indeed, Sir, if you could only ask the young gentleman if he would object to our proposing to the young ladies.

THE REVEREND DOCTOR OPIMIAN

Why not propose to them yourselves? You seem to be all creditable young men.

HARRY HEDGEROW

I have proposed to Miss Dorothy, you know, and she would not have me; and the rest are afraid. We are all something to do with the land and the woods; farmers, and foresters, and nurserymen, and all that. And we have all opened our hearts to one another. They don't pretend to look above us; but it seems somehow as if they did, and couldn't help it. They are so like young ladies. They daze us, like. Why, if they'd have us, they'd be all in reach of one another. Fancy what a family party there'd be at Christmas. We just want a good friend to put a good foot foremost for us; and if the young gentleman does marry, perhaps they may better themselves by doing likewise.

THE REVEREND DOCTOR OPIMIAN

And so you seven young friends have each a different favourite among the seven sisters?

HARRY HEDGEROW

Why, that's the beauty of it.

THE REVEREND DOCTOR OPIMIAN

The beauty of it? Perhaps it is. I suppose there is an agistor* among you.

HARRY HEDGEROW (*after looking at his companions, who all shook their heads*)

I am afraid not. Ought there to be? We don't know what it means.

THE REVEREND DOCTOR OPIMIAN

I thought that among so many foresters there might be an agistor. But it is not indispensable. Well, if the young gentleman is

* An agistor was a forest officer, who superintended the taking in of strange cattle to board and lodge, and accounted for the profit to the sovereign. I have read the word, but never heard it. I am inclined to think, that in modern times the duty was carried on under another name, or merged in the duties of another office.

going to be married, he will tell me of it. And when he does tell me, I will tell him of you. Have patience. It may all come right.

HARRY HEDGEROW

Thank ye, Sir. Thank ye, Sir, kindly.

Which being echoed in chorus by the other six, they took their departure, much marvelling what the Reverend Doctor could mean by an agistor.

'Upon my word,' said the Doctor to himself, 'a very good-looking, respectable set of young men. I do not know what the others may have to say for themselves. They behaved like a Greek chorus. They left their share of the dialogue to the coryphæus. He acquitted himself well, more like a Spartan than an Athenian, but none the worse for that. Brevity, in this case, is better than rhetoric. I really like that youth. How his imagination dwells on the family party at Christmas. When I first saw him, he was fancying how the presence of Miss Dorothy would gladden his father's heart at that season. Now he enlarges the circle, but it is still the same predominant idea. He has lost his mother. She must have been a good woman, and his early home must have been a happy one. The Christmas hearth would not be so uppermost in his thoughts if it had been otherwise. This speaks well for him and his. I myself think much of Christmas and all its associations. I always dine at home on Christmas-day, and measure the steps of my children's heads on the wall, and see how much higher each of them as risen, since the same time last year, in the scale of physical life. There are many poetical charms in the heraldings of Christmas. The halcyon builds its nest on the tranquil sea. 'The bird of dawning singeth all night long.' I have never verified either of these poetical facts. I am willing to take them for granted. I like the idea of the Yule log, the enormous block of wood, carefully selected long before, and preserved where it would be thoroughly dry, which burned on the old-fashioned hearth. It would not suit the stoves of our modern saloons. We could not burn it in our kitchens, where a small fire in the midst of a mass of black iron, roasts, and bakes, and boils, and steams, and broils, and fries, by a complicated apparatus, which, whatever may be its other virtues, leaves no space for a Christmas fire. I like the festoons of holly on the walls and windows; the dance under the mistletoe; the gigantic sausage; the baron of beef; the vast globe of plum-pudding, the true image of the earth, flattened at the poles; the tapping of the old October; the inexhaustible bowl of punch; the life and joy of the old hall, when the squire and his household and his neighbourhood were as one. I like the idea of what has gone, and I can still enjoy the reality of what remains. I have no doubt Harry's father burns the Yule log, and taps the old October. Perhaps, instead of the beef, he produces a fat pig roasted whole, like Eumaeus, the divine swineherd in the *Odyssey*. How Harry will burn the Yule log if he can realize this day-dream of himself and his six friends with the seven sisters! I shall make myself acquainted with the position and characters of these young suitors. To be sure, it is not my business, and I ought to recollect the words of Cicero: 'Est enim difficilis cura rerum alienarum: quamquam Terentianus ille Chremes humani nihil a se alienum putat.'* I hold with Chremes too. I am not without hope, from some symptoms I have lately seen, that rumour in the present case is in a fair way of being right; and if, with the accordance of the young gentleman as key-note, these two heptachords should harmonize into a double octave, I do not see why I may not take my part as fundamental bass.'

* It is a hard matter to take active concern in the affairs of others; although the Chremes of Terence thinks nothing human alien to himself.—*De Officiis:* i. 9.

CHAPTER XXIII

Ἔγνωκα δ' οὖν . . . . .
Τοὺς ζῶντας ὥσπερ εἰς πανήγυρίν τινα
Ἀφειμένους ἐκ τοῦ θανάτου καὶ τοῦ σκότους
Εἰς τὴν διατριβὴν εἰς τὸ φῶς τε τοῦθ' ὃ δή
Ὁρῶμεν· ὃς δ' ἄν πλεῖστα γελάσῃ καὶ πίῃ,
Καὶ τῆς Ἀφροδίτης ἀντιλάβηται τὸν χρόνον
Τοῦτον ὃν ἀφεῖται, καὶ τύχῃ γ' ἐράνου τινός
Πανηγυρίσας, ἥδιστ' ἀπῆλθεν οἴκαδε.

ALEXIS: *Tarantini.*

As men who leave their homes for public games,
We leave our native element of darkness
For life's brief light. And who has most of mirth,
And wine, and love, may, like a satisifed guest,
Return contented to the night he sprang from.

IN the mean time Mr. Falconer, after staying somewhat longer than usual at home, had returned to the Grange. He found much the same party as he had left: but he observed, or imagined, that Lord Curryfin was much more than previously in favour with Miss Gryll; that she paid him more marked attention, and watched his conduct to Miss Niphet with something more than curiosity.

Amongst the winter evening's amusements were two forms of quadrille: the old-fashioned game of cards, and the more recently fashionable dance. On these occasions, it was of course a carpet-dance. Now dancing had never been in Mr. Falconer's line, and though modern dancing, especially in quadrilles, is little more than walking, still in that 'little more' there is ample room for grace and elegance of motion. Herein Lord Curryfin outshone all the other young men in the circle. He endeavoured to be as indiscriminating as possible in inviting partners: but it was plain to curious observation, especially if a spice of jealousy mingled with the curiosity, that his favourite partner was Miss Niphet. When they occasionally danced a polka, the Reverend Doctor's mythological theory came out in full force. It seemed as if Nature had preordained that they should be inseparable, and the interior conviction of both, that so it ought to be, gave them an accordance of movement that seemed to emanate from the innermost mind. Sometimes, too, they danced the *Minuet de la Cour*. Having once done it, they had been often unanimously requested to repeat it. In this they had no competitors. Miss Gryll confined herself to quadrilles, and Mr. Falconer did not even propose to walk through one with her. When dancing brought into Miss Niphet's cheeks the blush-rose bloom, which had more than once before so charmed Lord Curryfin, it required little penetration to see, through his external decorum, the passionate admiration with which he regarded her. Mr. Falconer remarked it, and looking round to Miss Gryll, thought he saw the trace of a tear in her eye. It was a questionable glistening: jealousy construed it into a tear. But why should it be there? Was her mind turning to Lord Curryfin? and the more readily because of a newly-perceived obstacle? Had mortified vanity any share in it? No: this was beneath Morgana. Then why was it there? Was it anything like regret that, in respect of the young lord, she too had lost her opportunity? Was he himself blameless in the matter? He had been on the point of declaration, and she had been apparently on the point of acceptance: and instead of following up his advantage, he had been absent longer than usual. This was ill; but in the midst of the contending forces which severally acted on him, how could he make it well? So he sate still, tormenting himself.

In the meantime, Mr. Gryll had got up at a card-table in the outer, which was the smaller, drawing-room, a quadrille party of his own, consisting of himself, Miss Ilex, the Reverend Doctor Opimian, and Mr. MacBorrowdale.

MR. GRYLL

This is the only game of cards that ever pleased me. Once it was the great evening charm of the whole nation. Now, when cards are played at all, it has given place to whist, which in my young days was considered a dry, solemn, studious game, played in moody silence, only interruped by an occasional outbreak of dogmatism and ill-humour. Quadrille is not so absorbing but that we may talk and laugh over it, and yet is quite as interesting as anything of the kind has need to be.

MISS ILEX

I delight in quadrille. I am old enough to remember when, in mixed society in the country, it was played every evening by some of the party. But *Chaque age a ses plaisirs, son esprit, et ses mœurs.** It is one of the evils of growing old, that we do not easily habituate ourselves to changes of custom. The old, who sit still while the young dance and sing, may be permitted to regret the once always accessible cards, which, in their own young days, delighted the old of that generation: and not the old only.

THE REVEREND DOCTOR OPIMIAN

There are many causes for the diminished attraction of cards in evening society. Late dinners leave little evening. The old time for cards was the interval between tea and supper. Now there is no such interval, except here and there in out-of-the-way places, where, perhaps, quadrille and supper may still flourish as in the days of Queen Anne. Nothing was more common in country towns and villages, half a century ago, than parties meeting in succession at each other's houses, for tea, supper, and quadrille. How popular this game had been you may judge from Gay's ballad, which represents all classes as absorbed in quadrille.† Then the facility of locomotion dissipates, annihilates neighbourhood. People are not now the fixtures they used to be in their respective localities, finding their amusements within their own limited circle. Half the inhabitants of a country place are here to-day and gone to-morrow. Even of those, who are more what they call settled, the greater portion is less probably at home than whisking about the world. Then, again, where cards are played at all, whist is more consentaneous to

* Boileau.

† For example:

> When patients lie in piteous case,
>   In comes the apothecary,
> And to the Doctor cries, 'Alas!
>   *Non debes quadrillare.*'
> The patient dies without a pill;
> For why? The Doctor's at quadrille.
>
> Should France and Spain again grow loud,
>   The Muscovite grow louder,
> Britain, to curb her neighbours proud,
>   Would want both ball and powder;
> Must want both sword and gun to kill;
> For why? The General's at quadrille.

modern solemnity: there is more wiseacre-ism about it: in the same manner that this other sort of quadrille, in which people walk to and from one another with faces of exemplary gravity, has taken the place of the old-fashioned country dance. 'The merry dance I dearly love' would never suggest the idea of a quadrille, any more than 'merry England' would call up any image not drawn from ancient ballads and the old English drama.

MR. GRYLL

Well, Doctor, I intend to have a ball at Christmas, in which all modes of dancing shall have fair play, but country dances shall have their full share.

THE REVEREND DOCTOR OPIMIAN

I rejoice in the prospect. I shall be glad to see the young dancing as if they were young.

MISS ILEX

The variety of the game called tredrille—the Ombre of Pope's *Rape of the Lock*—is a pleasant game for three. Pope had many opportunities of seeing it played, yet he has not described it correctly: and I do not know that this has been observed.

THE REVEREND DOCTOR OPIMIAN

Indeed, I never observed it. I shall be glad to know how it is so.

MISS ILEX

Quadrille is played with forty cards: tredrille usually with thirty: sometimes, as in Pope's Ombre, with twenty-seven. In forty cards, the number of trumps is eleven in the black suits, twelve in the red:* in thirty, nine in all suits alike.† In twenty-seven, they cannot be more than nine in one suit, and eight in the other three. In Pope's Ombre spades are trumps, and the number is eleven: the number which they would be if the cards were forty. If you follow his description carefully, you will find it to be so.

MR. MACBORROWDALE

Why then, we can only say, as a great philosopher said on

* Nine cards in the black and ten in the red suits, in addition to the aces of spades and clubs, Spadille and Basto, which are trumps in all suits.

† Seven cards in each of the four suits in addition to Spadille and Basto.

another occasion: The description is sufficient 'to impose on the degree of attention with which poetry is read.'*

MISS ILEX

It is a pity it should be so. Truth to nature is essential to poetry. Few may perceive an inaccuracy: but to those who do, it causes a great diminution, if not a total destruction, of pleasure in perusal. Shakspeare never makes a flower blossom out of season. Wordsworth, Coleridge, and Southey are true to nature, in this and in all other respects: even in their wildest imaginings.

THE REVEREND DOCTOR OPIMIAN

Yet here is a combination, by one of our greatest poets, of flowers that never blossom in the same season:—

> Bring the rathe primrose, that forsaken dies,
> The tufted crow-toe, and pale jessamine,
> The white pink, and the pansie freakt with jet,
> The glowing violet,
> The musk rose, and the well-attired woodbine,
> With cowslips wan, that hang the pensive head,
> And every flower that sad embroidery wears:
> Bid amaranthus all his beauty shed,
> And daffadillies fill their cups with tears,
> To deck the laureat hearse where Lycid lies.

And at the same time he plucks the berries of the myrtle and the ivy.

MISS ILEX

Very beautiful, if not true to English seasons: but Milton might have thought himself justified in making this combination in Arcadia. Generally he is strictly accurate, to a degree that is in itself a beauty. For instance, in his address to the nightingale:

> Thee, chauntress, oft the woods among,
> I woo to hear thy even-song,
> And missing thee, I walk unseen,
> On the dry smooth-shaven green.

The song of the nightingale ceases about the time that the grass is mown.

* DUGALD STEWART, in the *Philosophy of the Human Mind*, I think; but I quote from memory.

THE REVEREND DOCTOR OPIMIAN

The old Greek poetry is always true to nature, and will bear any degree of critical analysis. I must say, I take no pleasure in poetry that will not.

MR. MACBORROWDALE

No poet is truer to nature than Burns, and no one less so than Moore. His imagery is almost always false. Here is a highly-applauded stanza, and very taking at first sight:

> The night-dew of heaven, though in silence it weeps,
> Shall brighten with verdure the sod where he sleeps;
> And the tear that we shed, though in secret it rolls,
> Shall long keep his memory green in our souls.

But it will not bear analysis. The dew is the cause of the verdure: but the tear is not the cause of the memory: the memory is the cause of the tear.

THE REVEREND DOCTOR OPIMIAN

There are inaccuracies more offensive to me than even false imagery. Here is one, in a song which I have often heard with displeasure. A young man goes up a mountain, and as he goes higher and higher, he repeats *Excelsior*: but *excelsior* is only taller in the comparison of things on a common basis, not higher as a detached object in the air. Jack's bean-stalk was *excelsior* the higher it grew: but Jack himself was no more *celsus* at the top than he had been at the bottom.

MR. MACBORROWDALE

I am afraid, Doctor, if you look for profound knowledge in popular poetry, you will often be disappointed.

THE REVEREND DOCTOR OPIMIAN

I do not look for profound knowledge. But I do expect that poets should understand what they talk of. Burns was not a scholar, but he was always master of his subject. All the scholarship of the world would not have produced *Tam O'Shanter*: but in the whole of that poem, there is not a false image nor a misused word. What do you suppose these lines represent?

> I turning saw, throned on a flowery rise,
>   One sitting on a crimson scarf unrolled:
> A queen, with swarthy cheeks and bold black eyes,
>   Brow-bound with burning gold.

MR. MACBORROWDALE

I should take it to be a description of the Queen of Bambo.

THE REVEREND DOCTOR OPIMIAN

Yet thus one of our most popular poets describes Cleopatra: and one of our most popular artists has illustrated the description by a portrait of a hideous grinning Æthiop. Moore led the way to this perversion by demonstrating, that the Ægyptian women must have been beautiful, because they were 'the countrywomen of Cleopatra.'* Here we have a sort of counter-demonstration, that Cleopatra must have been a fright, because she was the countrywoman of the Ægyptians. But Cleopatra was a Greek, the daughter of Ptolemy Auletes and a lady of Pontus. The Ptolemies were Greeks, and whoever will look at their genealogy, their coins, and their medals, will see how carefully they kept their pure Greek blood uncontaminated by African intermixture. Think of this description and this picture, applied to one who, Dio says—and all antiquity confirms him—was 'the most superlatively beautiful of women, splendid to see, and delightful to hear.'† For she was eminently accomplished: she spoke many languages with grace and facility. Her mind was as wonderful as her personal beauty. There is not a shadow of intellectual expression in that horrible portrait.

The conversation at the quadrille table was carried on with occasional pauses, and intermingled with the technicalities of the game.

Miss Gryll continued to alternate between joining in the quadrille dances and resuming her seat by the side of the room, where she was the object of great attention from some young gentlemen, who were glad to find her unattended by either Lord Curryfin or Mr. Falconer. Mr. Falconer continued to sit, as if he had been fixed to his seat, like Theseus. The more he reflected on his conduct, in disappearing at that critical point of time and staying away so long, the more he

* De Pauw, the great depreciator of everything Ægyptian, has, on the authority of a passage in Aelian, presumed to affix to the countrywomen of Cleopatra the stigma of complete and unredeemed ugliness.—MOORE's *Epicurean*, fifth note.

† *Περικαλλεστάτη γυναικῶν . . . . λαμπρά τε ἰδεῖν καὶ ἀκουσθῆναι οὖσα.*—DIO, xlii. 34.

felt that he had been guilty of an unjustifiable, and perhaps unpardonable, offence. He noticed with extreme discomposure the swarm of moths, as he called them to himself, who were fluttering in the light of her beauty: he would gladly have put them to flight; and this being out of the question, he would have been contented to take his place among them; but he dared not try the experiment.

Nevertheless, he would have been graciously received. The young lady was not cherishing any feeling of resentment against him. She understood, and made generous allowance for, his divided feelings. But his irresolution, if he were left to himself, was likely to be of long duration : and she meditated within herself the means of forcing him to a conclusion one way or the other.

CHAPTER XXIV

Δέρκεο τὴν νεᾶνιν, δέρκεο, κοῦρε·
Ἔγρεο, μὴ σε φύγῃ πέρδικος ἄγρα.
Ῥόδον ἀνθέων ἀνάσσει·
Ῥόδον ἐν κόραις Μυρίλλα.
ANACREON.

See, youth, the nymph who charms your eyes;
Watch, lest you lose the willing prize.
As queen of flowers the rose you own,
And her of maids the rose alone.

WHILE light, fire, mirth, and music were enlivening the party within the close-drawn curtains, without were moonless night and thickly-falling snow; and the morning opened on one vast expanse of white, mantling alike the lawns and the trees, and weighing down the wide-spreading branches. Lord Curryfin, determined not to be baulked of his skating, sallied forth immediately after breakfast, collected a body of labourers, and swept clear an ample surface of ice, a path to it from the house, and a promenade on the bank. Here he and Miss Niphet amused themselves in the afternoon, in company with a small number of the party, and in the presence of about the usual number of spectators. Mr. Falconer was there, and contented himself with looking on.

Lord Curryfin proposed a reel, Miss Niphet acquiesced, but it was long before they found a third. At length one young gentleman, of the plump and rotund order, volunteered to supply the deficiency, and was soon deposited on the ice, where his partners in the ice-dance would have tumbled over him if they had not anticipated the result and given him a wide berth. One or two others followed, exhibiting several varieties in the art of falling ungracefully. At last the lord and the lady skated away on as large a circuit as the cleared ice permitted, and as they went he said to her,

'If you were the prize of skating, as Atalanta was of running, I should have good hope to carry you off against all competitors but yourself.'

She answered, 'Do not disturb my thoughts, or I shall slip.'

He said no more, but the words left their impression. They gave him as much encouragement as, under their peculiar circumstances, he could dare to wish for, or she could venture to intimate.

Mr. Falconer admired their 'poetry of motion' as much as all others had done. It suggested a remark which he would have liked to address to Miss Gryll, but he looked round for her in vain. He returned to the house in the hope that he might find her alone, and take the opportunity of making his peace.

He found her alone, but it seemed that he had no peace to make. She received him with a smile, and held out her hand to him, which he grasped fervently. He fancied that it trembled, but her features were composed. He then sate down at the table, on which the old edition of Bojardo was lying open as before. He said, 'You have not been down to the lake to see that wonderful skating.' She answered, 'I have seen it every day but this. The snow deters me today. But it is wonderful. Grace and skill can scarcely go beyond it.'

He wanted to apologize for the mode and duration of his departure and absence, but did not know how to begin. She gave him the occasion. She said, 'You have been longer absent than usual—from our rehearsals. But we are all tolerably perfect in our parts. But your absence was remarked—by some of the party. You seemed to be especially missed by Lord Curryfin. He asked the Reverend Doctor every morning if he thought you would return that day.'

ALGERNON

And what said the Doctor?

MORGANA

He usually said, 'I hope so.' But one morning he said something more specific.

ALGERNON

What was it?

MORGANA

I do not know that I ought to tell you.

ALGERNON

Oh, pray do.

MORGANA

He said, 'The chances are against it.' 'What are the odds?' said Lord Curryfin. 'Seven to one,' said the Doctor. 'It ought not to be so,' said Lord Curryfin, 'for here is a whole Greek chorus against seven vestals.' The Doctor said, 'I do not estimate the chances by the mere balance of numbers.'

ALGERNON

He might have said more as to the balance of numbers.

MORGANA

He might have said more, that the seven outweighed the one.

ALGERNON

He could not have said that.

MORGANA

It would be much for the one to say that the balance was even.

ALGERNON

But how if the absentee himself had been weighed against another in that one's own balance?

MORGANA

One to one promises at least more even weight.

ALGERNON

I would not have it so. Pray, forgive me.

MORGANA

Forgive you? For what?

ALGERNON

I wish to say, and I do not well know how, without seeming to assume what I have no right to assume, and then I must have double cause to ask your forgiveness.

MORGANA

Shall I imagine what you wish to say, and say it for you?

ALGERNON

You would relieve me infinitely, if you imagine justly.

MORGANA

You may begin by saying with Achilles,

> My mind is troubled, like a fountain stirred;
> And I myself see not the bottom of it.*

ALGERNON

I think I do see it more clearly.

MORGANA

You may next say, I live an enchanted life. I have been in danger of breaking the spell; it has once more bound me with sevenfold force; I was in danger of yielding to another attraction; I went a step too far in all but declaring it; I do not know how to make a decent retreat.

ALGERNON

Oh! no, no; nothing like that.

MORGANA

Then there is a third thing you may say; but before I say that for you, you must promise to make no reply, not even a monosyllable; and not to revert to the subject for four times seven days. You hesitate.

ALGERNON

It seems as if my fate were trembling in the balance.

MORGANA

You must give me the promise I have asked for.

ALGERNON

I do give it.

MORGANA

Repeat it then, word for word.

* *Troilus and Cressida*, act iii. scene 3.

ALGERNON

To listen to you in silence; not to say a syllable in reply; not to return to the subject for four times seven days.

MORGANA

Then you may say, I have fallen in love; very irrationally—(*he was about to exclaim, but she placed her finger on her lips*)—very irrationally; but I cannot help it. I fear I must yield to my destiny. I will try to free myself from all obstacles; I will, if I can, offer my hand where I have given my heart. And this I will do, if I ever do, at the end of four times seven days: if not then, never.

She placed her finger on her lips again, and immediately left the room; having first pointed to a passage in the open pages of *Orlando Innamorato*. She was gone before he was aware that she was going; but he turned to the book, and read the indicated passage. It was a part of the continuation of Orlando's adventure in the enchanted garden, when himself pursued and scourged by *La Penitenza*, he was pursuing the Fata Morgana over rugged rocks and through briary thickets.

Cosi diceva. Con molta roina
Sempre seguia Morgana il cavalliero:
Fiacca ogni bronco ed ogni mala spina,
Lasciando dietro a se largo il sentiero:
Ed a la Fata molto s' avicina
E già d'averla presa è il suo pensiero:
Ma quel pensiero è ben fallace e vano,
Però che presa anchor scappa di mano.

O quante volte gli dette di piglio,
Hora ne' panni ed hor nella persona:
Ma il vestimento, ch' è bianco e vermiglio,
Ne la speranza presto l'abbandona:
Pur una fiata rivoltando il ciglio,
Come Dio volse e la ventura bona,
Volgendo il viso quella Fata al Conte
Ei ben la prese al zuffo ne la fronte.

Allor cangiosse il tempo, e l'aria scura
Divenne chiara, e il ciel tutto sereno,
E l'aspro monte si fece pianura;
E dove prima fu di spine pieno,

Se coperse de fiori e de verdura:
E'l flagellar dell' altra venne meno:
La qual, con miglior viso che non suole,
Verso del Conte usava tal parole.

Attenti, cavalliero, a quella chioma. . . .*

'She must have anticipated my coming,' said the young gentleman to himself. 'She had opened the book at this

* BOJARDO, *Orlando Innamorato,* l. ii. c. 9. *Ed. di Vinegia.* 1544.

So spake Repentance. With the speed of fire
Orlando followed where the enchantress fled,
Rending and scattering tree and bush and briar,
And leaving wide the vestige of his tread.
Nearer he drew, with feet that could not tire,
And strong in hope to seize her as she sped.
How vain the hope! Her form he seemed to clasp,
But soon as seized, she vanished from his grasp.

How many times he laid his eager hand
On her bright form, or on her vesture fair;
But her white robes, and their vermilion band,
Deceived his touch, and passed away like air.
But once, as with a half-turned glance she scanned
Her foe—Heaven's will and happy chance were there—
No breath for pausing might the time allow—
He seized the golden forelock of her brow.

Then passed the gloom and tempest from the sky;
The air at once grew calm and all serene;
And where rude thorns had clothed the mountain high,
Was spread a plain, all flowers and vernal green.
Repentence ceased her scourge. Still standing nigh,
With placid looks, in her but rarely seen,
She said: 'Beware how yet the prize you lose;
The key of fortune few can wisely use.'

In the last stanza of the preceding translation, the seventh line is the essence of the stanza immediately following; the eighth is from a passage several stanzas forward, after Orlando has obtained the key, which was the object of his search:

Che mal se trova alcun sotto la Luna,
Ch'adopri ben la chiave de Fortuna.

The first two books of Bojardo's poem were published in 1486. The first complete edition was published in 1495.

The Venetian edition of 1544, from which I have cited this passage, and the preceding one in chapter XX., is the fifteenth and last complete Italian passage, and has left it to say to me for her—choose between love and repentance. Four times seven days! This is to ensure calm for the Christmas holidays. The term will pass over twelfth night. The lovers of old romance were subjected to a probation of seven years:—

> Seven long years I served thee, fair one,
> Seven long years my fee was scorn.

But here, perhaps, the case is reversed. She may have feared a probation of seven years for herself; and not

edition. The original work was superseded by the *Rifacciamenti* of Berni and Domenichi. Mr. Panizzi has rendered a great service to literature in reprinting the original. He collated all accessible editions. *Verum opere in longo fas est obrepere somnum.* He took for his standard, as I think, unfortunately, the Milanese edition of 1539. With all the care he bestowed on his task, he overlooked one fearful perversion in the concluding stanza, which in all editions but the Milanese reads thus:

> Mentre ch'io canto, ahimè Dio redentore,
> Veggio l'Italia tutta a fiamma e a foco,
> Per questi Galli, che con gran furore
> Vengon per disertar non so che loco.
> Però vi lascio in questo vano amore
> Di Fiordespina ardente a poco a poco:
> Un' altra fiata, se mi fia concesso,
> Racconterovi il tutto per espresso.

> Even while I sing, ah me, redeeming Heaven!
> I see all Italy in fire and flame,
> Raised by these Gauls, who, by great fury driven,
> Come with destruction for their end and aim.
> The maiden's heart, by vainest passion riven,
> Not now the rudely-broken song may claim;
> Some future day, if Fate auspicious prove,
> Shall end the tale of Fiordespina's love.

The Milanese edition of 1539 was a reprint of that of 1513, in which year the French, under Louis XII, had reconquered Milan. The Milanese editions read *valore* for *furore*.

It was no doubt in deference to the conquerors that the printer of 1513 made this substitution; but it utterly perverts the whole force of the passage. The French, under Charles VIII, invaded Italy in September, 1494, and the horror with which their devastations inspired Bojardo not only stopped the progress of his poem, but brought his life prematurely to a close. He died in December, 1494. The alteration of this single word changes almost into a compliment an expression of cordial detestation.

without reason. And what have I to expect if I let the four times seven days pass by? Why, then, I can read in her looks—and they are interpreted in the verses before me—I am assigned to repentance, without the hope of a third opportunity. She is not without a leaning towards Lord Curryfin. She thinks he is passing from her, and on the twenty-ninth day, or perhaps in the meantime, she will try to regain him. Of course she will succeed. What rivalry could stand against her? If her power over him is lessened, it is that she has not chosen to exert it. She has but to will it, and he is again her slave. Twenty-eight days! twenty-eight days of doubt and distraction.' And starting up, he walked out into the park, not choosing the swept path, but wading knee-deep in snow where it lay thickest in the glades. He was recalled to himself by sinking up to his shoulders in a hollow. He emerged with some difficulty, and retraced his steps to the house, thinking that, even in the midst of love's most dire perplexities, dry clothes and a good fire are better than a hole in the snow.

CHAPTER XXV

Μνηστῆρες δ' ὁμαδήσαν ἀνὰ μέγαρα σκιόεντα.
HOMERUS *in Odysseâ.*

The youthful suitors, playing each his part,
Stirred pleasing tumult in each fair one's heart.
Adapted—not translated.

HARRY HEDGEROW had found means on several occasions of delivering farm and forest produce at the Tower, to introduce his six friends to the sisters, giving all the young men in turn to understand that they must not think of Miss Dorothy; an injunction which, in the ordinary perverse course of events, might have led them all to think of no one else, and produced a complication very disagreeable for their introducer. It was not so, however. 'The beauty of it,' as Harry said to the Reverend Doctor, was that each had found a distinct favourite among the seven vestals. They had not, however, gone beyond giving pretty intelligible hints. They had not decidedly ventured to declare or propose. They left it to Harry to prosecute his suit to Miss Dorothy, purposing to step in on the rear of his success. They had severally the satisfaction of being assured by various handsome young gipsies, whose hands they had crossed with lucky shillings, that each of them was in love with a fair young woman, who was quite as much in love with him, and whom he would certainly marry before twelve months were over. And they went on their way rejoicing.

Now Harry was indefatigable in his suit, which he had unbounded liberty to plead; for Dorothy always listened to him complacently, though without departing from the answer she had originally given, that she and her sisters would not part with each other and their young master.

The sisters had not attached much importance to Mr. Falconer's absences; for on every occasion of his return, the predominant feeling he had seemed to express was that of extreme delight at being once more at home.

One day, while Mr. Falconer was at the Grange, receiving

admonition from *Orlando Innamorato*, Harry, having the pleasure to find Dorothy alone, pressed his suit as usual, was listened to as usual, and seemed likely to terminate without being more advanced than usual, excepting in so far as they both found a progressive pleasure, she in listening, and he in being listened to. There was to both a growing charm in thus 'dallying with the innocence of love,' and though she always said No with her lips, he began to read Yes in her eyes.

HARRY

Well, but Miss Dorothy, though you and your sisters will not leave your young master, suppose somebody should take him away from you, what would you say then?

DOROTHY

What do you mean, Master Harry?

HARRY

Why, suppose he should get married, Miss Dorothy?

DOROTHY

Married!

HARRY

How should you like to see a fine lady in the Tower, looking at you as much as to say, This is mine?

DOROTHY

I will tell you very candidly, I should not like it at all. But what makes you think of such a thing?

HARRY

You know where he is now?

DOROTHY

At Squire Gryll's, rehearsing a play for Christmas.

HARRY

And Squire Gryll's niece is a great beauty, and a great fortune.

DOROTHY

Squire Gryll's niece was here, and my sisters and myself saw a great deal of her. She is a very nice young lady; but he has seen great beauties and great fortunes before; he has always been indifferent to the beauties, and he does not care about fortune. I am sure he would not like to change his mode of life.

HARRY

Ah, Miss Dorothy! you don't know what it is to fall in love. It tears a man up by the roots, like a gale of wind.

DOROTHY

Is that your case, Master Harry?

HARRY

Indeed it is, Miss Dorothy. If you didn't speak kindly to me, I do not know what would become of me. But you always speak kindly to me, though you wont have me.

DOROTHY

I never said wont, Master Harry.

HARRY

No; but you always say can't, and that's the same as wont, so long as you don't.

DOROTHY

You are a very good young man, Master Harry. Everybody speaks well of you. And I am really pleased to think you are so partial to me. And if my young master and my sisters were married, and I were disposed to follow their example, I will tell you very truly, you are the only person I should think of, Master Harry.

Master Harry attempted to speak, but he felt choked in the attempt at utterance; and in default of words, he threw himself on his knees before his beloved, and clasped his hands together with a look of passionate imploring, which was rewarded by a benevolent smile. And they did not change their attitude till the entrance of one of the sisters startled them from their sympathetic reverie.

Harry having thus made a successful impression on one of the Theban gates, encouraged his six allies to carry on the siege of the others; for which they had ample opportunity, as the absences of the young gentleman became longer, and the rumours of an attachment between him and Miss Gryll obtained more ready belief.

CHAPTER XXVI

Οὐ χρὴ κακοῖσι θυμὸν ἐπιτρέπειν·
Προκόψομεν γὰρ οὐδὲν, ἀσάμενοι,
Ὦ Βακχί· φάρμακον δ' ἄριστον
Οἶνον ἐνεικαμένοις μεθύσθαι.
ALCAEUS.

Bacchis! 'tis vain to brood on care,
Since grief no remedy supplies;
Be ours the sparkling bowl to share,
And drown our sorrows as they rise.

MR. FALCONER saw no more of Miss Gryll till the party assembled in the drawing-rooms. She necessarily took the arm of Lord Curryfin for dinner, and it fell to the lot of Mr. Falconer to offer his to Miss Niphet, so that they sate at remote ends of the table, each wishing himself in the other's place; but Lord Curryfin paid all possible attention to his fair neighbour. Mr. Falconer could see that Miss Gryll's conversation with Lord Curryfin was very animated and joyous: too merry, perhaps, for love: but cordial to a degree that alarmed him. It was, however, clear by the general mirth at the head of the table, that nothing very confidential or sentimental was passing. Still, a young lady who had placed the destiny of her life on a point of brief suspense ought not to be so merry as Miss Gryll evidently was. He said little to Miss Niphet; and she, with her habit of originating nothing, sate in her normal state of statue-like placidity, listening to the conversation near her. She was on the left hand of Mr. Gryll. Miss Ilex was on his right, and on her right was the Reverend Doctor Opimian. These three kept up an animated dialogue. Mr. MacBorrowdale was in the middle of the table, and amused his two immediate fair neighbours with remarks appertaining to the matter immediately before them, the preparation and arrangement of a good dinner: remarks that would have done honour to Francatelli.

After a while, Mr. Falconer bethought him that he would

try to draw out Miss Niphet's opinion on the subject nearest his heart. He said to her: 'They are very merry at the head of the table.'

MISS NIPHET

I suppose Lord Curryfin is in the vein for amusing his company, and he generally succeeds in his social purposes.

MR. FALCONER

You lay stress on social, as if you thought him not successful in all his purposes.

MISS NIPHET

Not in all his inventions, for example. But in the promotion of social enjoyment he has few equals. Of course, it must be in congenial society. There is a power of being pleased, as well as the power of pleasing. With Miss Gryll and Lord Curryfin, both meet in both. No wonder that they amuse those around them.

MR. FALCONER

In whom there must also be a power of being pleased.

MISS NIPHET

Most of the guests here have it. If they had not they would scarcely be here. I have seen some dismal persons, any one of whom would be a kill-joy to a whole company. There are none such in this party. I have also seen a whole company all willing to be pleased, but all mute from not knowing what to say to each other: not knowing how to begin. Lord Curryfin would be a blessing to such a party. He would be the steel to their flint.

MR. FALCONER

Have you known him long?

MISS NIPHET

Only since I met him here.

MR. FALCONER

Have you heard that he is a suitor to Miss Gryll?

MISS NIPHET

I have heard so.

MR. FALCONER

Should you include the probability of his being accepted in your estimate of his social successes?

MISS NIPHET

Love affairs are under influences too capricious for the calculation of probabilities.

MR. FALCONER

Yet I should be very glad to hear your opinion. You know them both so well.

MISS NIPHET

I am disposed to indulge you, because I think it is not mere curiosity that makes you ask the question. Otherwise I should not be inclined to answer it. I do not think he will ever be the affianced lover of Morgana. Perhaps he might have been if he had persevered as he began. But he has been used to smiling audiences. He did not find the exact reciprocity he looked for. He fancied that it was, or would be, for another. I believe he was right.

MR. FALCONER

Yet you think he might have succeeded if he had persevered.

MISS NIPHET

I can scarcely think otherwise, seeing how much he has to recommend him.

MR. FALCONER

But he has not withdrawn.

MISS NIPHET

No, and will not. But she is too high-minded to hold him to a proposal not followed up as it commenced; even if she had not turned her thoughts elsewhere.

MR. FALCONER

Do you not think she could recal him to his first ardour if she exerted all her fascinations for the purpose?

MISS NIPHET

It may be so. I do not think she will try. (*She added, to herself*:) I do not think she would succeed.

Mr. Falconer did not feel sure she would not try: he thought he saw symptoms of her already doing so. In his opinion Morgana was, and must be, irresistible. But as he had thought his fair neighbour somewhat interested in the subject, he wondered at the apparent impassiveness with which she replied to his quesions.

In the meantime he found, as he had often done before, that the more his mind was troubled, the more Madeira he could drink without disordering his head.

CHAPTER XXVII

Il faut avoir aimé une fois en sa vie, non pour le moment où l'on aime, car on n'éprouve alors que des tourmens, des regrets, de la jalousie: mais peu à peu ce tourmens-là deviennent des souvenirs, qui charment notre arrière saison: . . . et quand vous verrez la vieillesse douce, facile et tolérante, vous pourrez dire comme Fontenelle: L'amour a passé par-là.

SCRIBE: *La Vieille.*

MISS GRYLL carefully avoided being alone with Mr. Falconer, in order not to give him an opportunity of speaking on the forbidden subject. She was confident that she had taken the only course which promised to relieve her from a life of intolerable suspense; but she wished to subject her conduct to dispassionate opinion, and she thought she could not submit it to a more calmly-judging person than her old spinster friend, Miss Ilex, who had, moreover, the great advantage of being a woman of the world. She therefore took an early opportunity of telling her what had passed between herself and Mr. Falconer, and asking her judgment on the point.

MISS ILEX

Why, my dear, if I thought there had been the slightest chance of his ever knowing his own mind sufficiently to come to the desired conclusion himself, I should have advised your giving him a little longer time; but as it is clear to me that he never would have done so, and as you are decidedly partial to him, I think you have taken the best course which was open to you. He had all but declared to you more than once before; but this 'all but' would have continued, and you would have sacrificed your life to him for nothing.

MISS GRYLL

But do you think you would in my case have done as I did?

MISS ILEX

No, my dear, I certainly should not; for, in a case very similar, I did not. It does not follow that I was right. On the contrary, I think you are right and I was wrong. You have shown true moral courage where it was most needed.

MISS GRYLL

I hope I have not revived any displeasing recollections.

MISS ILEX

No, my dear, no; the recollections are not displeasing. The day-dreams of youth, however fallacious, are a composite of pain and pleasure: for the sake of the latter the former is endured, nay, even cherished in memory.

MISS GRYLL

Hearing what I hear you were, seeing what I see you are, observing your invariable cheerfulness, I should not have thought it possible that you could have been crossed in love, as your words seem to imply.

MISS ILEX

I was, my dear, and have been foolish enough to be constant all my life to a single idea; and yet I would not part with this shadow for any attainable reality.

MISS GRYLL

If it were not opening the fountain of an ancient sorrow, I could wish to know the story, not from idle curiosity, but from my interest in you.

MISS ILEX

Indeed, my dear Morgana, it is very little of a story: but such as it is, I am willing to tell it you. I had the credit of being handsome and accomplished. I had several lovers; but my inner thoughts distinguished only one; and he, I think, had a decided preference for me, but it was a preference of present impression. If some Genius had commanded him to choose a wife from any company of which I was one, he would, I feel sure, have chosen me; but he was very much of an universal lover, and was always overcome by the smiles of present beauty. He was of a romantic turn of mind: he disliked and avoided the ordinary pursuits of young men: he delighted in the society of accomplished young women, and in that alone. It was the single link between him and the world. He would disappear for weeks at a time, wandering in forests, climbing mountains, and descending into the dingles of mountain-streams, with no other companion than a Newfoundland dog; a large black dog, with a white breast, four white paws, and a white tip to his tail: a beautiful affectionate dog: I often patted him on the head, and fed him with my hand. He knew me as well as Bajardo* knew Angelica.

Tears started into her eyes at the recollection of the dog. She paused for a moment.

MISS GRYLL

I see the remembrance is painful. Do not proceed.

MISS ILEX

No, my dear. I would not, if I could, forget that dog. Well, my young gentleman, as I have said, was a sort of universal lover, and made a sort of half-declaration to half the young women he knew: sincerely for the moment to all: but with more permanent earnestness, more constant return, to me than to any other. If I had met him with equal earnestness, if I could have said or implied to him in any way, 'Take me while you may, or think of me no more,' I am persuaded I should not now write myself spinster. But I wrapped myself up in reserve. I thought it fitting that all advances should come from him: that I should at most show nothing more than willingness to hear, not even the semblance of anxiety to receive them. So nothing came of our love but remembrance and regret. Another girl, whom I am sure he loved less, but who understood him better, acted towards him as I ought to have done, and became his wife. Therefore, my dear, I applaud your moral courage, and regret that I had it not when the occasion required it.

MISS GRYLL

My lover, if I may so call him, differs from yours in this: that he is not wandering in his habits, nor versatile in his affections.

* Rinaldo's horse: He had escaped from his master, and had repelled Sacripante with his heels:—

Indi va mansueto alla donzella,
Con umile sembiante e gesto umano:
Come intorno al padrone il can saltella,
Che sia due giorni o tre stato lontano.
Bajardo ancora avea memoria d'ella,
Che in Albracca il servia già di sua mano.
Orlando Furioso, c. i. s. 75.

MISS ILEX

The peculiar system of domestic affection, in which he was brought up and which his maturer years have confirmed, presents a greater obstacle to you than any which my lover's versatility presented to me, if I had known how to deal with it.

MISS GRYLL

But how was it, that, having so many admirers as you must have had, you still remained single?

MISS ILEX

Because I had fixed my heart on one who was not like any one else. If he had been one of a class, such as most persons in this world are, I might have replaced the first idea by another; but 'his soul was like a star, and dwelt apart.'

MISS GRYLL

A very erratic star, apparently. A comet, rather.

MISS ILEX

No. For the qualities which he loved and admired in the object of his temporary affection, existed more in his imagination than in her. She was only the frame-work of the picture of his fancy. He was true to his idea, though not to the exterior semblance on which he appended it, and to or from which he so readily transferred it. Unhappily for myself, he was more of a reality to me than I was to him.

MISS GRYLL

His marriage could scarcely have been a happy one. Did you ever meet him again?

MISS ILEX

Not of late years, but for a time occasionally in general society, which he very sparingly entered. Our intercourse was friendly; but he never knew, never imagined, how well I loved him, nor even, perhaps, that I had loved him at all. I had kept my secret only too well. He retained his wandering habits, disappearing from time to time, but always returning home. I believe he had no cause to complain of his wife. Yet I cannot help thinking that I could have fixed him and kept him at home. Your case is in many respects similar to mine; but the rivalry to me was in a wandering fancy: to you it is in fixed domestic affections. Still, you were in as much danger as I was of being the victim of an idea and a punctilio: and you have taken the only course to save you from it. I regret that I gave in to the punctilio: but I would not part with the idea. I find a charm in the recollection far preferable to

> The waveless calm, the slumber of the dead,

which weighs on the minds of those who have never loved, or never earnestly.

CHAPTER XXVIII

Non duco contentionis funem, dum constet inter nos, quod fere totus mundus exerceat histrioniam.—PETRONIUS ARBITER.

I do not draw the rope of contention,* while it is agreed amongst us, that almost the whole world practises acting.

All the world's a stage.—SHAKSPEARE.

En el teatro del mundo
Todos son representantes.—CALDERON.

Tous les comédiens ne sont pas au théâtre.—*French Proverb.*

RAIN came, and thaw, followed by drying wind. The roads were in good order for the visitors to the Aristophanic Comedy. The fifth day of Christmas was fixed for the performance. The theatre was brilliantly lighted, with spermaceti candles in glass chandeliers for the audience, and argand lamps for the stage. In addition to Mr. Gryll's own houseful of company, the beauty and fashion of the surrounding country, which comprised an extensive circle, adorned the semicircular seats; which, however, were not mere stone benches, but were backed, armed, and padded, into comfortable stalls. Lord Curryfin was in his glory, in the capacity of stage-manager.

The curtain rising, as there was no necessity for its being made to fall,† discovered the scene, which was on the London bank of the Thames, on the terrace of a mansion occupied by the Spirit-rapping Society, with an archway in the centre of the building, showing a street in the back-

* A metaphor apparently taken from persons pulling in opposite directions at each end of a rope. I cannot see, as some have done, that it has anything in common with Horace's *Tortum digna sequi potius quam ducere funem:* 'More worthy to follow than to lead the tightened cord:' which is a metaphor taken from a towing line, or any line acting in a similar manner, where one draws and another is drawn. Horace applies it to money, which, he says, should be the slave, and not the master of its possessor.

† The Athenian theatre was open to the sky, and if the curtain had been made to fall it would have been folded up in mid air, destroying the effect of the scene. Being raised from below, it was invisible when not in use.

ground. Gryllus was lying asleep. Circe, standing over him, began the dialogue.

CIRCE

Wake, Gryllus, and arise in human form.

GRYLLUS

I have slept soundly, and had pleasant dreams.

CIRCE

I, too, have soundly slept. Divine how long.

GRYLLUS

Why, judging by the sun, some fourteen hours.

CIRCE

Three thousand years.

GRYLLUS

That is a nap, indeed.
But this is not your garden, nor your palace.
Where are we now?

CIRCE

Three thousand years ago,
This land was forest, and a bright pure river
Ran through it to and from the Ocean stream.
Now, through a wilderness of human forms,
And human dwellings, a polluted flood
Rolls up and down, charged with all earthly poisons,
Poisoning the air in turn.

GRYLLUS

I see vast masses
Of strange unnatural things.

CIRCE

Houses, and ships,
And boats, and chimneys vomiting black smoke,
Horses, and carriages of every form,
And restless bipeds, rushing here and there
For profit or for pleasure, as they phrase it.

GRYLLUS

Oh, Jupiter and Bacchus! what a crowd,
Flitting, like shadows without mind or purpose,
Such as Ulysses saw in Erebus.
But wherefore are we here?

CIRCE

There have arisen
Some mighty masters of the invisible world,
And these have summoned us.

GRYLLUS

With what design?

CIRCE

That they themselves must tell. Behold they come,
Carrying a mystic table, around which
They work their magic spells. Stand by, and mark.

Three spirit-rappers appeared, carrying a table, which they placed on one side of the stage:

1. Carefully the table place,
Let our gifted brother trace
A ring around the enchanted space.
2. Let him tow'rd the table point,
With his first fore-finger joint,
And, with memerized beginning,
Set the sentient oak-slab spinning.
3. Now it spins around, around,
Sending forth a murmuring sound,
By the initiate understood
As of spirits in the wood.
ALL. Once more Circe we invoke.

CIRCE

Here: not bound in ribs of oak,
Nor, from wooden disk revolving,
In strange sounds strange riddles solving,
But in native form appearing,
Plain to sight, as clear to hearing.

THE THREE

Thee with wonder we behold.
By thy hair of burning gold,
By thy face with radiance bright,
By thine eyes of beaming light,
We confess thee, mighty one,
For the daughter of the Sun.
On thy form we gaze appalled.

CIRCE

Gryllus, too, your summons called.

THE THREE

Him of yore thy powerful spell
Doomed in swinish shape to dwell:
Yet such life he reckoned then
Happier than the life of men.
Now, when carefully he ponders
All our scientific wonders,
Steam-driven myriads, all in motion,
On the land and on the ocean,
Going, for the sake of going,
Wheresoever waves are flowing,
Wheresoever winds are blowing;
Converse through the sea transmitted,
Swift as ever thought has flitted;
All the glories of our time,
Past the praise of loftiest rhyme;
Will he, seeing these, indeed,
Still retain his ancient creed,
Ranking, in his mental plan,
Life of beast o'er life of man?

CIRCE

Speak, Gryllus.

GRYLLUS

It is early yet to judge:
But all the novelties I yet have seen
Seem changes for the worse.

THE THREE

If we could show him
Our triumphs in succession, one by one,
'Twould surely change his judgment: and herein
How might'st thou aid us, Circe!

CIRCE

I will do so:
And calling down, like Socrates of yore,
The Clouds to aid us, they shall shadow forth,
In bright succession, all that they behold,
From air, on earth and sea. I wave my wand:
And lo! they come, even as they came in Athens,
Shining like virgins of ethereal life.

The Chorus of Clouds descended, and a dazzling array of female beauty was revealed by degrees through folds of misty gauze. They sang their first choral song:

CHORUS OF CLOUDS*

I

Clouds ever-flowing, conspicuously soaring,
From loud-rolling Ocean, whose stream† gave us birth,
To heights, whence we look over torrents down-pouring
To the deep quiet vales of the fruit-giving earth,—
As the broad eye of Æther, unwearied in brightness,
Dissolves our mist-veil in its glittering rays,
Our forms we reveal from its vapoury lightness,
In semblance immortal, with far-seeing gaze.

II

Shower-bearing Virgins, we seek not the regions
Whence Pallas, the Muses, and Bacchus have fled,
But the city, where Commerce embodies her legions,
And Mammon exalts his omnipotent head.
All joys of thought, feeling, and taste are before us,
Wherever the beams of his favour are warm:
Though transient full oft as the veil of our chorus,
Now golden with glory, now passing in storm.

Reformers, scientific, moral, educational, political, passed in succession, each answering a question of Gryllus. Gryllus observed, that so far from everything being better than it had been, it seemed that everything was wrong and wanted mending. The chorus sang its second song.

Seven competitive examiners entered with another table, and sate down on the opposite side of the stage to the spirit-rappers. They brought forward Hermogenes‡ as a crammed fowl to argue with Gryllus. Gryllus had the best of the argument; but the examiners adjudged the victory to Hermogenes. The chorus sangs its third song.

Circe, at the request of the spirit-rappers, whose power was limited to the production of sound, called up several visible spirits, all illustrious in their day, but all appearing as in the days of their early youth, 'before their renown was

* The first stanza is pretty closely adapted from the strophe of Aristophanes: Ἀέναοι Νεφέλαι. The second is only a distant imitation of the antistrophe: Παρθένοι ὀμβροφόροι.

† In Homer, and all the older poets, the ocean is a river surrounding the earth, and the seas are inlets from it.

‡ See Chapter XV. page 185.

around them.' They were all subjected to competitive examination, and were severally pronounced disqualified for the pursuit in which they had shone. At last came one whom Circe recommended to the examiners as a particularly promising youth. He was a candidate for military life. Every question relative to his profession he answered to the purpose. To every question not so relevant he replied, that he did not know and did not care. This drew on him a reprimand. He was pronounced disqualified, and ordered to join the rejected, who were ranged in a line along the back of the scene. A touch of Circe's wand changed them into their semblance of maturer years. Among them were Hannibal and Oliver Cromwell; and in the foreground was the last candidate, Richard Cœur-de-Lion. Richard flourished his battle-axe over the heads of the examiners who jumped up in great trepidation, overturned their table, tumbled over one another, and escaped as best they might in haste and terror. The heroes vanished. The chorus sang its fourth song.

CHORUS

I

As before the pike will fly
Dace and roach and such small fry;
As the leaf before the gale,
As the chaff beneath the flail;
As before the wolf the flocks,
As before the hounds the fox;
As before the cat the mouse,
As the rat from falling house;
As the fiend before the spell
Of holy water, book, and bell;
As the ghost from dawning day,—
So has fled, in gaunt dismay,
This septemvirate of quacks,
From the shadowy attacks
Of Cœur-de-Lion's battle-axe.

II

Could he in corporeal might,
Plain to feeling as to sight,
Rise again to solar light,
How his arm would put to flight
All the forms of Stygian night,

That round us rise in grim array,
Darkening the meridian day:
Bigotry, whose chief employ
Is embittering earthly joy;
Chaos, throned in pedant state,
Teaching echo how to prate;
And 'Ignorance, with looks profound,'
Not 'with eye that loves the ground,'
But stalking wide, with lofty crest,
In science's pretentious vest.

III

And now, great masters of the realms of shade,
To end the task which called us down from air
We shall present, in pictured show arrayed,
Of this your modern world the triumphs rare,
That Gryllus's benighted spirit
May wake to your transcendant merit,
And, with profoundest admiration thrilled,
He may with willing mind assume his place
In your steam-nursed, steam-borne, steam-killed,
And gas-enlightened race.

CIRCE

Speak, Gryllus, what you see.

GRYLLUS

I see the ocean,
And o'er its face ships passing wide and far;
Some with expanded sails before the breeze,
And some with neither sails nor oars, impelled
By some invisible power against the wind,
Scattering the spray before them. But of many
One is on fire, and one has struck on rocks
And melted in the waves like fallen snow.
Two crash together in the middle sea,
And go to pieces on the instant, leaving
No soul to tell the tale; and one is hurled
In fragments to the sky, strewing the deep
With death and wreck. I had rather live with Circe
Even as I was, than flit about the world
In those enchanted ships, which some Alastor
Must have devised as traps for mortal ruin.

CIRCE

Look yet again.

GRYLLUS

Now the whole scene is changed.
I see long trains of strange machines on wheels,
With one in front of each, puffing white smoke
From a black hollow column. Fast and far
They speed, like yellow leaves before the gale,
When autumn winds are strongest. Through their windows
I judge them thronged with people; but distinctly
Their speed forbids my seeing.

SPIRIT-RAPPER

This is one
Of the great glories of our modern time.
'Men are become as birds,' and skim like swallows
The surface of the world.

GRYLLUS

For what good end?

SPIRIT-RAPPER

The end is in itself—the end of skimming
The surface of the world.

GRYLLUS

If that be all,
I had rather sit in peace in my old home:
But while I look, two of them meet and clash,
And pile their way with ruin. One is rolled
Down a steep bank; one through a broken bridge
Is dashed into a flood. Dead, dying, wounded,
Are there as in a battle-field. Are these
Your modern triumphs? Jove preserve me from them.

SPIRIT-RAPPER

These ills are rare. Millions are borne in safety
Where one incurs mischance. Look yet again.

GRYLLUS

I see a mass of light brighter than that
Which burned in Circe's palace, and beneath it
A motley crew dancing to joyous music.
But from that light explosion comes, and flame;
And forth the dancers rush in haste and fear
From their wide-blazing hall.

SPIRIT-RAPPER

Oh, Circe! Circe!
Thou show'st him all the evil of our arts

In more than just proportion to the good.
Good without evil is not given to man.
Jove, from his urns dispensing good and ill,
Gives ill unmixed to some, and good and ill
Mingled to many—good unmixed to none.*
Our arts are good. The inevitable ill
That mixes with them, as with all things human,
Is as a drop of water in a goblet
Full of old wine.

GRYLLUS

More than one drop, I fear,
And those of bitter water.

CIRCE

There is yet
An ample field of scientific triumph:
What shall we show him next?

SPIRIT-RAPPER

Pause we awhile.
He is not in the mood to feel conviction
Of our superior greatness. He is all
For rural comfort and domestic ease,
But our impulsive days are all for moving:

* This is the true sense of the Homeric passage:—

Δοιοὶ γάρ τε πίθοι κατακείαται ἐν Διὸς οὔδει
Δώρων, οἷα δίδωσι, κακῶν, ἕτερος δὲ ἑάων·
Ὧι μὲν καμμίξας δῴη Ζεὺς τερπικέραυνος,
Ἄλλοτε μέν τε κακῷ ὅγε κύρεται, ἄλλοτε δ' ἐσθλῷ.
Ὧι δέ κε τῶν λυγρῶν δῳη, λωβητὸν ἔθηκε,
Καὶ ἑ κακὴ βούβρωστις ἐπὶ χθόνα δῖαν ἐλαυνει·
Φοιτᾷ δ' οὔτε θεοῖσι τετιμένος, οὔτε βροτοῖσιν.

HOMER: *Il.* xxiv.

There are only two distributions: good and ill mixed, and unmixed ill. None, as Heyne has observed, receive unmixed good. *Ex dolio bonorum nemo meracius accipit: hoc memorare omisit.* This sense is implied, not expressed. Pope missed it in his otherwise beautiful translation.

Two urns by Jove's high throne have ever stood,
The source of evil one, and one of good:
From thence the cup of mortal man he fills,
Blessings to these, to those distributes ills,
To most he mingles both: the wretch decreed
To taste the bad, unmixed, is curst indeed:
Pursued by wrongs, by meagre famine driven,
He wanders, outcast both of earth and heaven.

POPE.

Sometimes with some ulterior end, but still
For moving, moving, always. There is nothing
Common between us in our points of judgment.
He takes his stand upon tranquillity,
We ours upon excitement. There we place
The being, end, and aim of mortal life.
The many are with us: some few, perhaps,
With him. We put the question to the vote
By universal suffrage. Aid us, Circe!
On talismanic wings your spells can waft
The question and reply. Are we not wiser,
Happier, and better, than the men of old,
Of Homer's days, of Athens, and of Rome?

VOICES WITHOUT

Aye. No. Aye, aye. No. Aye, aye, aye, aye, aye.
We are the wisest race the earth has known,
The most advanced in all the arts of life,
In science, and in morals.

SPIRIT-RAPPER

The Ayes have it.
What is that wondrous sound, that seems like thunder,
Mixed with gigantic laughter?

CIRCE

It is Jupiter,
Who laughs at your presumption; half in anger,
And half in mockery. Now, my worthy masters,
You must in turn experience in yourselves
The mighty magic thus far tried on others.

The table turned slowly, and by degrees went on spinning with accelerated speed. The legs assumed motion, and it danced off the stage. The arms of the chairs put forth hands, and pinched the spirit-rappers, who sprang up and ran off, pursued by their chairs. This piece of mechanical pantomime was a triumph of Lord Curryfin's art, and afforded him ample satisfaction for the failure of his resonant vases.

CIRCE

Now, Gryllus, we may seek our ancient home
In my enchanted isle.

GRYLLUS

Not yet, not yet.
Good signs are toward of a joyous supper.
Therein the modern world may have its glory,
And I, like an impartial judge, am ready
To do it ample justice. But, perhaps,
As all we hitherto have seen are shadows,
So too may be the supper.

CIRCE

Fear not, Gryllus.
That you will find a sound reality,
To which the land and air, seas, lakes, and rivers,
Have sent their several tributes. Now, kind friends,
Who with your smiles have graciously rewarded
Our humble but most earnest aims to please,
And with your presence at our festal board
Will charm the winter midnight, Music gives
The signal: Welcome and old wine await you.

THE CHORUS

Shadows to-night have offered portraits true
Of many follies which the world enthral.
'Shadows we are, and shadows we pursue:'
But in the banquet's well-illumined hall,
Realities, delectable to all,
Invite you now our festal joy to share.
Could we our Attic prototype recal,
One compound word should give our bill of fare:*
But where our language fails, our hearts true welcome bear.

Miss Gryll was resplendent as Circe; and Miss Niphet, as leader of the Chorus, looked like Melpomene herself, slightly unbending her tragic severity into that solemn smile which characterized the chorus of the old comedy. The charm of the first acted irresistibly on Mr. Falconer. The second would have completed, if anything had been wanted to complete it, the conquest of Lord Curryfin.

The supper passed off joyously, and it was a late hour of the morning before the company dispersed.

* As at the end of the *Ecclesiazusæ*.

CHAPTER XXIX

Within the temple of my purer mind
One imaged form shall ever live enshrined,
And hear the vows, to first affection due,
Still breathed: for love, that ceases, ne'er was true.

LEYDEN's *Scenes of Infancy*.

AN interval of a week was interposed between the comedy and the intended ball. Mr. Falconer having no fancy for balls, and disturbed beyond endurance by the interdict which Miss Gryll had laid on him against speaking for four times seven days on the subject nearest his heart, having discharged with becoming self-command his share in the Aristophanic comedy, determined to pass his remaining days of probation in the Tower, where he found in the attentions of the seven sisters, not a perfect Nepenthe, but the only possible antidote to intense vexation of spirit. It is true, his two Hebes, pouring out his Madeira, approximated as nearly as anything could do to Helen's administration of the true Nepenthe. He might have sung of Madeira, as Redi's Bacchus sang of one of his favourite wines:—

Egli è il vero oro potabile,
Che mandar suole in esilio
Ogni male inrimediabile:
Egli è d'Elena il Nepente,
Che fa stare il mondo allegro,
Dai pensieri
Foschi e neri
Sempre sciolto, e sempre esente.*

Matters went on quietly at the Grange. One evening Mr. Gryll said to the Reverend Doctor Opimian—

I have heard you, Doctor, more than once, very eulogistic of hair as indispensable to beauty. What say you to the Bald Venus of the Romans—*Venus Calva*?

* REDI: *Bacco in Toscana*.

THE REVEREND DOCTOR OPIMIAN

Why, sir, if it were a question whether the Romans had any such deity, I would unhesitatingly maintain the *negatur*. Where do you find her?

MR GRYLL

In the first place, I find her in several dictionaries.

THE REVEREND DOCTOR OPIMIAN

A dictionary is nothing without an authority. You have no authority but that of one or two very late writers, and two or three old grammarians, who had found the word and guessed at its meaning. You do not find her in any genuine classic. A bald Venus! It is as manifest a contradiction in terms as hot ice or black snow.

LORD CURRYFIN

Yet I have certainly read, though I cannot at this moment say where, that there was in Rome a temple to *Venus Calva*, and that it was so dedicated in consequence of one of two circumstances: the first being, that through some divine anger the hair of the Roman women fell off, and that Ancus Martius set up a bald statue of his wife, which served as an expiation, for all the women recovered their hair, and the worship of the Bald Venus was instituted; the other being, that when Rome was taken by the Gauls, and when they had occupied the city and were besieging the Capitol, the besieged having no materials to make bowstrings, the women cut off their hair for the purpose, and after the war a statue of the Bald Venus was raised in honour of the women.

THE REVEREND DOCTOR OPIMIAN

I have seen the last story transferred to the time of the younger Maximin.* But when two or three explanations of which only one can possibly be true, are given of any real or supposed fact, we may safely conclude that all are false. These are ridiculous myths, founded on the misunderstanding of an obsolete word. Some hold that *Calva*, as applied to Venus, signifies Pure; but I hold with others that it signifies alluring, with a sense of deceit. You will find

* Julius Capitolinus: *Max. Jun.* c. 7.

the cognate verbs, *calvo* and *calvor*, active,* passive,† and deponent,‡ in Servius, Plautus, and Sallust. Nobody pretends that the Greeks had a bald Venus. The *Venus Calva* of the Romans was the *Aphroditê Doliê* of the Greeks.§ Beauty cannot co-exist with baldness; but it may and does co-exist with deceit. Homer makes deceitful allurement an essential element in the girdle of Venus.|| Sappho addresses her as craft-weaving Venus.** Why should I multiply examples when poetry so abounds with complaints of deceitful love, that I will be bound every one of this company could, without a moment's hesitation, find a quotation in point?—Miss Gryll to begin with.

MISS GRYLL

Oh, Doctor, with every one who has a memory for poetry, it must be *l'embarras de richesses*. We could occupy the time till midnight in going round and round on the subject. We should soon come to an end with instances of truth and constancy.

THE REVEREND DOCTOR OPIMIAN

Not so soon, perhaps. If we were to go on accumulating examples, I think I could find you a Penelope for a Helen, a Fiordiligi for an Angelica, an Imogene for a Calista, a Sacripant for a Rinaldo, a Romeo for an Angelo, to nearly the end of the chapter.

* Est et Venus Calva, ob hanc causam, quod cum Galli Capitolium obsiderent, et deessent funes Romanis ad tormenta facienda, Prima Domitia crinem suum, post cæteræ matronæ, imitatæ eam, exsecuerunt, unde facta tormenta; et post bellum statua Veneri hoc nomine collocata est: licet alii Calvam Venerem quasi puram tradant: *alii Calvam, quod corda calviat, id est, fallat atque eludat.* Quidam dicunt, porrigine olim capillos cecidisse fœminis, et Ancum regem suæ uxori statuam Calvam posuisse, quod constitit piaculo; nam mox omnibus fœminis caplli renati sunt: unde institutum ut Calva Venus coleretur.—SERVIUS ad *Aen.* i. 720.

The substance of this passage is given in the text.

† Contra ille *calvi* ratus.—SALLUST: *Hist.* iii.
Thinking himself to be deceitfully allured.

‡ Nam ubi domi sola sum, sopor manus *calvitur.* – PLAUTUS *in Casinâ.*
For when I am at home alone, sleep alluringly deceives my hands.

§ *'Αφροδίτη Δολίη.*

|| *Πάρφασις, ἥτ' ἔκλεψε νόον πύκα περ φρονεόντων.*—*Il.* xiv. 217.

** *Παῖ Διὸς δολοπλόκε.*

I will not say quite, for I am afraid at the end of the catalogue the numbers of the unfaithful would predominate.

MISS ILEX

Do you think, Doctor, you would find many examples of love that is one, and once for all; love never transferred from its first object to a second?

THE REVEREND DOCTOR OPIMIAN

Plato holds that such is the essence of love, and poetry and romance present it in many instances.

MISS ILEX

And the contrary in many more.

THE REVEREND DOCTOR OPIMIAN

If we look indeed, into the realities of life, as they offer themselves to us in our own experience, in history, in biography, we shall find few instances of constancy to first love; but it would be possible to compile a volume of illustrious examples of love which, though it may have previously ranged, is at last fixed in single, unchanging constancy. Even Iñez de Castro was only the second love of Don Pedro of Portugal; yet what an instance is there of love enduring in the innermost heart as if it had been engraved on marble.

MISS GRYLL

What is that story, Doctor? I know it but imperfectly.

THE REVEREND DOCTOR OPIMIAN

Iñez de Castro was the daughter, singularly beautiful and accomplished, of a Castilian nobleman, attached to the Court of Alphonso the Fourth of Portugal. When very young, she became the favourite and devoted friend of Constance, the wife of the young Prince Don Pedro. The Princess died early, and the grief of Iñez touched the heart of Pedro, who found no consolation but in her society. Thence grew love, which resulted in secret marriage. Pedro and Iñez lived in seclusion at Coimbra, perfectly happy in each other, and in two children who were born to them, till three of Alphonso's courtiers, moved by I know not what demon of mischief—for I never could discover an adequate motive—induced the king to attempt the dissolution of the marriage, and failing in this, to authorize them to murder Iñez during a brief absence of her husband. Pedro raised a rebellion, and desolated the estates of the assassins, who escaped, one into France, and two into Castille. Pedro laid down his arms on the entreaty of his mother, but would never again see his father, and lived with his two children in the strictest retirement in the scene of his ruined happiness. When Alphonso died, Pedro determined not to assume the crown till he had punished the assassins of his wife. The one who had taken refuge in France was dead; the others were given up by the King of Castille. They were put to death, their bodies were burned, and their ashes were scattered to the winds. He then proceeded to the ceremony of his coronation. The mortal form of Iñez, veiled and in royal robes, was enthroned by his side: he placed the queenly crown on her head, and commanded all present to do her homage. He raised in a monastery side by side, two tombs of white marble, one for her, one for himself. He visited the spot daily, and remained inconsolable till he rejoined her in death. This is the true history, which has been sadly perverted by fiction.

MISS ILEX

There is, indeed, something grand in that long-enduring constancy: something terribly impressive in that veiled spectral image of robed and crowned majesty. You have given this, Doctor, as an instance that the first love is not necessarily the strongest, and this, no doubt, is frequently true. Even Romeo had loved Rosalind before he saw Juliet. But love which can be so superseded, is scarcely love. It is acquiescence in a semblance: acquiescence, which may pass for love through the entire space of life, if the latent sympathy should never meet its perfect counterpart.

THE REVEREND DOCTOR OPIMIAN

Which it very seldom does; but acquiescence in the semblance is rarely enduring, and hence there are few examples of life-long constancy. But I hold with Plato that true love is single, indivisible, unalterable.

MISS ILEX

In this sense, then, true love is first love: for the love which endures to the end of life, though it may be the second in semblance, is the first in reality.

The next morning Lord Curryfin said to Miss Niphet: 'You took no part in the conversation of last evening. You gave no opinion on the singleness and permanence of love.'

MISS NIPHET

I mistrust the experience of others and I have none of my own.

LORD CURRYFIN

Your experience, when it comes, cannot but confirm the theory. The love which once dwells on you can never turn to another.

MISS NIPHET

I do not know that I ought to wish to inspire such an attachment.

LORD CURRYFIN

Because you could not respond to it?

MISS NIPHET

On the contrary; because I think it possible I might respond to it too well.

She paused a moment, and then, afraid of trusting herself to carry on the dialogue, she said: 'Come into the hall and play at battledore and shuttlecock.'

He obeyed the order: but in the exercise her every movement developed some new grace, that maintained at its highest degree the intensity of his passionate admiration.

CHAPTER XXX

——dum fata sinunt, jungamus amores:
Mox veniet tenebris Mors adoperta caput.
Jam subrepet iners ætas, nec amare licebit,
Dicere nec cano blanditias capite.

TIBULLUS.

Let us, while Fate allows, in love combine,
Ere our last night its shade around us throw,
Or Age, slow-creeping, quench the fire divine,
And tender words befit not locks of snow.

THE shuttlecock had been some time on the wing, struck to and fro with unerring aim, and to all appearances would never have touched the ground, if Lord Curryfin had not seen, or fancied he saw, symptoms of fatigue on the part of his fair antagonist. He therefore, instead of returning the shuttlecock, struck it upward, caught it in his hand, and presented it to her, saying, 'I give in. The victory is yours.' She answered, 'The victory is yours, as it always is, in courtesy.'

She said this with a melancholy smile, more fascinating to him than the most radiant expression from another. She withdrew to the drawing-room, motioning to him not to follow.

In the drawing-room, she found Miss Gryll, who appeared to be reading; at any rate, a book was open before her.

MISS GRYLL

You did not see me just now, as I passed through the hall. You saw only two things: the shuttlecock, and your partner in the game.

MISS NIPHET

It is not possible to play, and see anything but the shuttlecock.

MISS GRYLL

And the hand that strikes it.

MISS NIPHET

That comes unavoidably into sight.

MISS GRYLL

My dear Alice, you are in love, and do not choose to confess it.

MISS NIPHET

I have no right to be in love with your suitor.

MISS GRYLL

He was my suitor, and has not renounced his pursuit: but he is your lover. I ought to have seen long ago, that from the moment his eyes rested on you, all else was nothing to him. With all that habit of the world, which enables men to conceal their feelings in society, with all his exertion to diffuse his attentions as much as possible among all the young ladies in his company, it must have been manifest to a careful observer, that when it came, as it seemed in ordinary course, to be your turn to be attended to, the expression of his features was changed from complacency and courtesy to delight and admiration. I could not have failed to see it, if I had not been occupied with other thoughts. Tell me candidly, do you not think it is so?

MISS NIPHET

Indeed, my dear Morgana, I did not designedly enter into rivalry with you; but I do think you conjecture rightly.

MISS GRYLL

And if he were free to offer himself to you, and if he did so offer himself, you would accept him?

MISS NIPHET

Assuredly I would.

MISS GRYLL

Then, when you next see him, he shall be free. I have set my happiness on another cast, and I will stand the hazard of the die.

MISS NIPHET

You are very generous, Morgana: for I do not think you give up what you do not value.

MISS GRYLL

No, indeed. I value him highly. So much so, that I have hesitated, and might have finally inclined to him, if I had not perceived his invincible preference of you. I am sorry, for your sake, and his, that I did not clearly perceive it sooner; but you see what it is to be spoiled by admirers. I did not think it possible that any one could be preferred to me. I ought to have thought it possible, but I had no experience in that direction. So now you see a striking specimen of mortified vanity.

MISS NIPHET

You have admirers in abundance, Morgana: more than have often fallen to the lot of the most attractive young women. And love is such a capricious thing, that to be the subject of it is no proof of superior merit. There are inexplicable affinities of sympathy, that make up an irresistible attraction, heaven knows how.

MISS GRYLL

And these inexplicable affinities Lord Curryfin has found in you, and you in him.

MISS NIPHET

He has never told me so.

MISS GRYLL

Not in words: but looks and actions have spoken for him. You have both struggled to conceal your feelings from others, perhaps even from yourselves. But you are both too ingenuous to dissemble successfully. You suit each other thoroughly: and I have no doubt you will find in each other the happiness I most cordially wish you.

Miss Gryll soon found an opportunity of conversing with Lord Curryfin, and began with him somewhat sportively: 'I have been thinking,' she said, 'of an old song which contains a morsel of good advice—

Be sure to be off with the old love,
Before you are on with the new.

You begin by making passionate love to me, and all at once you turn round to one of my young friends, and say, "Zephyrs whisper how I love you."'

LORD CURRYFIN

Oh, no! no, indeed. I have not said that, nor anything to the same effect.

MISS GRYLL

Well, if you have not exactly said it, you have implied it. You have looked it. You have felt it. You cannot conceal it. You cannot deny it. I give you notice, that, if I die for love of you, I shall haunt you.

LORD CURRYFIN

Ah! Miss Gryll, if you do not die till you die for love of me, you will be as immortal as Circe, whom you so divinely represented.

MISS GRYLL

You offered yourself to me, to have and to hold, for ever and aye. Suppose I claim you. Do not look so frightened. You deserve some punishment, but that would be too severe. But, to a certain extent, you belong to me, and I claim the right to transfer you. I shall make a present of you to Miss Niphet. So, according to the old rules of chivalry, I order you, as my captive by right, to present yourself before her, and tell her that you have come to receive her commands, and obey them to the letter. I expect she will keep you in chains for life. You do not look much alarmed at the prospect. Yet you must be aware, that you are a great criminal; and you have not a word to say in your own justification.

LORD CURRYFIN

Who could be insensible to charms like yours, if hope could have mingled with the contemplation? But there were several causes by which hope seemed forbidden, and therefore——

MISS GRYLL

And therefore when beauty, and hope, and sympathy shone under a more propitious star, you followed its guidance. You could not help yourself:

> What heart were his that could resist
> That melancholy smile?

I shall flatter myself that I might have kept you, if I had tried hard for it at first; but

> Il pentirsi da sesto nulla giova.

No doubt you might have said with the old song,

> I ne'er could any lustre see
> In eyes that would not look on me.

But you scarcely gave me time to look on you before you were gone. You see, however, like our own Mirror of Knighthood, I make the best of my evil fate, and

Cheer myself up with ends of verse,
And sayings of philosophers.

LORD CURRYFIN

I am glad to see you so merry; for even if your heart were more deeply touched by another than it ever could have been by me, I think I may say of you, in your own manner,

So light a heel
Will never wear the everlasting flint.

I hope and I believe you will always trip joyously over the surface of the world. You are the personification of L'Allegra.

MISS GRYLL

I do not know how that may be. But go now to the personification of La Pensierosa. If you do not turn her into a brighter Allegra than I am, you may say I have no knowledge of woman's heart.

It was not long after this dialogue that Lord Curryfin found an opportunity of speaking to Miss Niphet alone. He said, 'I am charged with a duty, such as was sometimes imposed on knights in the old days of chivalry. A lady, who claims me as her captive by right, has ordered me to kneel at your feet, to obey your commands, and to wear your chains, if you please to impose them.'

MISS NIPHET

To your kneeling, I say, Rise; for your obedience, I have no commands; for chains, I have none to impose.

LORD CURRYFIN

You have imposed them. I wear them already, inextricably, indissolubly.

MISS NIPHET

If I may say, with the witch in *Thalaba*,

Only she,
Who knit his bonds, can set him free.

I am prepared to unbind the bonds. Rise, my lord, rise.

LORD CURRYFIN

I will rise, if you give me your hand to lift me up.

MISS NIPHET

There it is. Now that it has helped you up, let it go.

LORD CURRYFIN

And do not call me, my lord.

MISS NIPHET

What shall I call you?

LORD CURRYFIN

Call me Richard, and let me call you Alice.

MISS NIPHET

That is a familiarity only sanctioned by longer intimacy than ours has been.

LORD CURRYFIN

Or closer?

MISS NIPHET

We have been very familiar friends during the brief term of our acquaintance. But let go my hand.

LORD CURRYFIN

I have set my heart on being allowed to call you Alice, and on your calling me Richard.

MISS NIPHET

It must not be so—at least, not yet.

LORD CURRYFIN

There is nothing I would not do to acquire the right.

MISS NIPHET

Nothing?

LORD CURRYFIN

Nothing.

MISS NIPHET

How thrives your suit with Miss Gryll?

LORD CURRYFIN

That is at an end. I have her permission—her command she calls it—to throw myself at your feet, and on your mercy.

MISS NIPHET

How did she take leave of you, crying or laughing?

LORD CURRYFIN

Why, if anything, laughing.

MISS NIPHET

Do you not feel mortified?

LORD CURRYFIN

I have another and deeper feeling, which predominates over any possible mortification.

MISS NIPHET

And that is—

LORD CURRYFIN

Can you doubt what it is!

MISS NIPHET

I will not pretend to doubt. I have for some time been well aware of your partiality for me.

LORD CURRYFIN

Partiality! Say love, adoration, absorpton of all feelings into one.

MISS NIPHET

Then you may call me Alice. But once more, let go my hand.

LORD CURRYFIN

My hand, is it not?

MISS NIPHET

Yours, when you claim it.

LORD CURRYFIN

Then thus I seal my claim.

He kissed her hand as respectfully as was consistent with

'masterless passion;' and she said to him, 'I will not dissemble. If I have had one wish stronger than another—strong enough to exclude all others—it has been for the day when you might be free to say to me what you have now said. Am I too frank with you?'

LORD CURRYFIN

Oh, heaven, no! I drink in your words as a stream from paradise.

He sealed his claim again, but this time it was on her lips. The rose again mantled on her cheeks, but the blush was heightened to damask. She withdrew herself from his arms, saying 'Once for all, till you have an indisputable right.'

CHAPTER XXXI

Sic erimus cuncti, postquam nos auferet Orcus:
Ergo vivamus, dum licet esse bene.

So must we be, when ends our mortal day:
Then let us live, while yet live well we may.
TRIMALCHIO, *with the silver skeleton: in* PETRONIUS: c. 34.

TWELFTH Night was the night of the ball. The folding-doors of the drawing-rooms, which occupied their entire breadth, were thrown wide open. The larger room was appropriated to grown dancers; the smaller to children, who came in some force, and were placed within the magnetic attraction of an enormous twelfthcake, which stood in a decorated recess. The carpets had been taken up, and the floors were painted with forms in chalk* by skilful artists, under the superintendence of Mr. Pallet. The library, separated from all the apartments by ante-chambers with double doors, was assigned, with an arrangement of whist-tables, to such of the elder portion of the party as might prefer that mode of amusement to being mere spectators of the dancing. Mr. Gryll, with Miss Ilex, Mr. MacBorrowdale, and the Reverend Doctor Opimian, established his own quadrille party in a corner of the smaller drawing-room, where they could at once play and talk, and enjoy the enjoyment of the young. Lord Curryfin was Master of the Ceremonies.

After two or three preliminary dances, to give time for the arrival of the whole of the company, the twelfthcake was divided. The characters were drawn exclusively among the children, and the little king and queen were duly crowned, placed on a theatrical throne, and paraded in state round both drawing-rooms, to their own great delight and that of their little associates. Then the ball was supposed to com-

* These all wear out of me, like forms, with chalk
Painted on rich men's floors, for one feast-night:

says WORDSWORTH, of 'chance acquaintance' in his neighbourhood.—*Miscellaneous Sonnets*, No. 39.

mence, and was by general desire opened with a minuet by Miss Niphet and Lord Curryfin. Then came the alternations of quadrilles and country dances, interspersed with occasional waltzes and polkas. So the ball went merrily, with, as usual, abundant love-making in mute signs and in *sotto voce* parlance.

Lord Curryfin, having brought his own love-making to a satisfactory close, was in exuberant spirits, sometimes joining in the dance, sometimes—in his official capacity—taking the round of the rooms to see that everything was going on to everybody's satisfaction. He could not fail to observe that his proffered partnership in the dance, though always graciously, was not so ambitiously accepted as before he had disposed of himself for life. A day had sufficed to ask and obtain the consent of Miss Niphet's father, who now sate on the side the larger drawing-room, looking with pride and delight on his daughter, and with cordial gratification on her choice; and when it was once, as it was at once, known that Miss Niphet was to be Lady Curryfin, his lordship passed into the class of married men, and was no longer the object of that solicitous attention which he had received as an undrawn prize in the lottery of marriage, while it was probable that somebody would have him, and nobody knew who.

The absence of Mr. Falconer was remarked by several young ladies, to whom it appeared that Miss Gryll had lost her two most favoured lovers at once. However, as she had still many others, it was not yet a decided case for sympathy. Of course she had no lack of partners, and whatever might have been her internal anxiety, she was not the least gay among the joyous assembly.

Lord Curryfin, in his circuit of the apartments, paused at the quadrille-table, and said, 'You have been absent two or three days, Mr. MacBorrowdale—what news have you brought from London?

MR. MACBORROWDALE

Not much, my lord. Tables turn as usual, and the ghost-trade appears to be thriving: for instead of being merely audible, the ghosts are becoming tangible, and shake hands under the tables with living wiseacres, who solemnly attest the fact. Civilized men ill-use their wives; the wives revenge themselves in their own way, and the Divorce Court has business enough on its hands to employ it twenty years at its present rate of progression. Commercial bubbles burst, and high-pressure boilers blow up, and mountebanks of all descriptions flourish on public credulity. Everywhere there are wars and rumours of wars. The Peace Society has wound up its affairs in the Insolvent Court of Prophecy. A great tribulation is coming on the earth, and Apollyon in person is to be perpetual dictator of all the nations. There is, to be sure, one piece of news in your line, but it will be no news to you. There is a meeting of the Pantopragmatic Society, under the presidency of Lord Facing-both-ways, who has opened it with a long speech, philanthropically designed as an elaborate exercise in fallacies, for the benefit of young rhetoricians. The society has divided its work into departments, which are to meddle with everything, from the highest to the lowest—from a voice in legislation to a finger in Jack Horner's pie. I looked for a department of Fish, with your lordship's name at the head of it; but I did not find it. It would be a fine department. It would divide itself naturally into three classes—living fish, fossil fish, and fish in the frying-pan.

LORD CURRYFIN

I assure you, Mr. MacBorrowdale, all this seems as ridiculous now to me as it does to you. The third class of fish is all that I shall trouble myself with in future, and that only at the tables of myself and my friends.

MR. GRYLL

I wonder the Pantopragmatics have not a department of cookery; a female department, to teach young wives how to keep their husbands at home, by giving them as good dinners as they can get abroad, especially at clubs. Those antidomestic institutions receive their chief encouragement from the total ignorance of cookery on the part of young wives: for in this, as in all other arts of life, it is not sufficient to order what shall be done: it is necessary to know how it ought to be done. This is a matter of more importance to social well-being, than nine-tenths of the subjects the Pantopragmatics meddle with.

THE REVEREND DOCTOR OPIMIAN

And therefore I rejoice that they do not meddle with it. A dinner, prepared from a New Art of Cookery, concocted under their auspices, would be more comical and more uneatable than the Roman dinner in Peregrine Pickle. Let young ladies learn cookery by all means: but let them learn under any other tuition than that of the Pantopragmatic Society.

MR. GRYLL

As for the tribulation coming on the earth, I am afraid there is some ground to expect it, without looking for its foreshadowing exclusively to the Apocalypse. Niebuhr, who did not draw his opinions from prophecy, rejoiced that his career was coming to a close, for he thought we were on the eve of a darker middle age.

THE REVEREND DOCTOR OPIMIAN

He had not before his eyes the astounding march of intellect, drumming and trumpeting science from city to city. But I am afraid that sort of obstreperous science only gives people the novel 'use of their eyes to see the way of blindness.'*

> Truths which, from action's paths retired,
> My silent search in vain required,†

I am not likely to find in the successive gabblings of a dozen lecturers of Babel.

MR. GRYLL

If you could so find them, they would be of little avail against the new irruption of Goths and Vandals, which must have been in the apprehension of Niebuhr. There are Vandals on northern thrones, anxious for nothing so much as to extinguish truth and liberty wherever they show themselves—Vandals in the bosom of society everywhere, even amongst ourselves, in multitudes, with precisely

* *Gaoler.*—For look you, sir: you know not which way you shall go.
Posthumus.—Yes, indeed, do I, fellow.
Gaoler.—Your death has eyes in's head, then: I have not seen him so pictured. ...
Posthumus.—I tell thee, fellow, there are none want eyes to direct them the way I am going, but such as wink, and will not use them.
Gaoler.—What an infinite mock is this, that a man should have the best use of eyes to see the way of blindness!

Cymbeline: Act v, Scene iv.

† COLLINS: *Ode on the Manners.*

the same aim, only more disguised by knaves, and less understood by dupes.

THE REVEREND DOCTOR OPIMIAN

And, you may add, Vandals dominating over society throughout half America, who deal with free speech and even the suspicion of free thought, just as the Inquisition dealt with them, only substituting Lynch law and the gallows for a different mockery of justice, ending in fire and faggot.

MR. GRYLL

I confine my view to Europe. I dread northern monarchy, and southern anarchy; and rabble brutality amongst ourselves, smothered and repressed for the present, but always ready to break out into inextinguishable flame, like hidden fire under treacherous ashes.*

MR. MACBORROWDALE

In the meantime, we are all pretty comfortable: and sufficient for the day is the evil thereof; which in our case, so far as I can see, happens to be precisely none.

MISS ILEX

Lord Curryfin seems to be of that opinion, for he has flitted away from the discussion, and is going down a country-dance with Miss Niphet.

THE REVEREND DOCTOR OPIMIAN

He has chosen his time well. He takes care to be her last partner before supper, that he may hand her to the table. But do you observe, how her tragic severity has passed away? She was always pleasant to look on, but it was often like contemplating ideal beauty in an animated statue. Now she is the image of perfect happiness, and irradiates all around her.

MISS ILEX

How can it be otherwise? The present and the future are all brightness to her. She cannot but reflect their radiance.

* ——incedis per ignes
Suppositos cineri doloso.
HOR.: *Carm.* l. ii. 1.

Now came the supper, which, as all present had dined early, was unaffectedly welcomed and enjoyed. Lord Curryfin looked carefully to the comfort of his idol, but was unremitting in his attentions to her fair neighbours. After supper, dancing was resumed, with an apparent resolution in the greater portion of the company not to go home till morning. Mr. Gryll, Mr. MacBorrowdale, the Reverend Doctor Opimian, and two or three elders of the party, not having had their usual allowance of wine after their early dinner, remained at the supper table over a bowl of punch, which had been provided in ample quantity, and, in the intervals of dancing, circulated, amongst other refreshments, round the sides of the ball-room, where it was gratefully accepted by the gentlemen, and not absolutely disregarded even by the young ladies. This may be conceded on occasion, without admitting Goldoni's facetious position, that a woman, masked and silent, may be known to be English by her acceptance of punch.*

* Lord Runebif, in Venice, meets Rosaura, who is masked, before a *bottega di caffè*. She makes him a curtsey in the English fashion.

Milord.—Madama, molto compita, volete caffè?
Rosaura.—(*Fa cenno di no.*)
Milord.—Cioccolata?
Rosaura.—(*Fa cenno di no.*)
Milord.—Volete ponce?
Rosaura.—(*Fa cenno di sì.*)
Milord.—Oh è Inglese.

La Vedova Scaltra, A. 3, S. 10.

He does not offer her tea, which, as a more English drink than either coffee or chocolate, might have entered into rivalry with punch: especially if, as Goldoni represented in another comedy, the English were in the habit of drinking it, not with milk, but with arrack. Lord Arthur calls on his friend Lord Bonfil in the middle of the day, and Lord Bonfil offers him tea, which is placed on the table with sugar and arrack. While they are drinking it, Lord Coubrech enters.

Bonfil.—Favorite, bevete con noi.
Coubrech.—Il tè non si rifiuta.
Artur.—È bevanda salutifera.
Bonfil.—Volete rak?
Coubrech.—Sì, rak.
Bonfil.—Ecco, vi servo.

Pamela Fancuilla, A. I. S. 15.

CHAPTER XXXII

Ὑμεῖς δε, πρέσβεις, χαίρετ', ἐν κακοῖς ὅμως
Ψυχῇ διδόντες ἡδονὴν καθ' ἡμέραν,
Ὡς τοῖς θανοῦσι πλοῦτος οὐδὲν ὠφελεῖ.

The Ghost of Darius to the Chorus, in the Persæ of ÆSCHYLUS.

Farewell, old friends: and even if ills surround you,
Seize every joy the passing day can bring,
For wealth affords no pleasure to the dead.

DOROTHY had begun to hope that Harry's news might be true, but even Harry's sanguineness began to give way: the pertinacity with which the young master remained at home, threw a damp on their expectations. But having once fairly started, in the way of making love on the one side and responding to it on the other, they could not but continue as they had begun, and she permitted him to go on building castles in the air, in which the Christmas of the ensuing year was arrayed in the brightest apparel of fire and festival.

Harry, walking home one afternoon, met the Reverend Doctor Opimian, who was on his way to the Tower, where he purposed to dine and pass the night. Mr. Falconer's absence from the ball had surprised him, especially as Lord Curryfin's rivalry had ceased, and he could imagine no good cause for his not returning to the Grange. The Doctor held out his hand to Harry, who returned the grasp most cordially. The Doctor asked him, 'How he and his six young friends were prospering in their siege of the hearts of the seven sisters.'

HARRY HEDGEROW

Why, sir, so far as the young ladies are concerned, we have no cause to complain. But we can't make out the young gentleman. He used to sit and read all the morning, at the top of the Tower. Now he goes up the stairs, and after a little while he comes down again, and walks into the forest. Then he goes up stairs again, and down again, and out again. Something must be come to him, and the only thing we can think of is, that he is crossed in love. And he never gives me a letter or a message to the Grange. So putting all that together, we haven't a merry Christmas, you see, sir.

THE REVEREND DOCTOR OPIMIAN

I see, still harping on a merry Christmas. Let us hope that the next may make amends.

HARRY HEDGEROW

Have they a merry Christmas at the Grange, sir?

THE REVEREND DOCTOR OPIMIAN

Very merry.

HARRY HEDGEROW

Then there's nobody crossed in love there, sir.

THE REVEREND DOCTOR OPIMIAN

That is more than I can say. I cannot answer for others. I am not, and never was, if that is any comfort to you.

HARRY HEDGEROW

It is a comfort to me to see you, and hear the sound of your voice, sir. It always does me good.

THE REVEREND DOCTOR OPIMIAN

Why then, my young friend, you are most heartily welcome to see and hear me whenever you please, if you will come over to the Vicarage. And you will always find a piece of cold roast beef and a tankard of good ale; and just now a shield of brawn. There is some comfort in them.

HARRY HEDGEROW

Ah! thank ye, sir. They are comfortable things in their way. But it isn't for them I should come.

THE REVEREND DOCTOR OPIMIAN

I believe you, my young friend. But a man fights best when he has a good basis of old English fare to stand on, against all opposing forces, whether of body or mind. Come and see me. And whatever happens in this world, never let it spoil your dinner.

HARRY HEDGEROW

That's father's advice, sir. But it wont always do. When he lost mother, that spoiled his dinner for many a day. He has never been the same man since, though he bears up as well as he can. But if I could take Miss Dorothy home to him, I'm sure that would all but make him young again. And if he had a little Harry to dandle next Christmas, wouldn't he give him the first spoonful out of the marrow-bone!

THE REVEREND DOCTOR OPIMIAN

I doubt if that would be good food for little Harry, notwithstanding it was Hector's way of feeding Astyanax.* But we may postpone the discussion of his diet till he makes his appearance. In the meantime, live in hope; but live on beef and ale.

The Doctor again shook him heartily by the hand, and Harry took his leave.

The Doctor walked on, soliloquising as usual. 'This young man's father has lost a good wife, and has never been the same man since. If he had had a bad wife, he would have felt it as a happy release. This life has strange compensations. It helps to show the truth of Juvenal's remark, that the gods alone know what is good for us.† Now, here again is my friend at the Tower. If he had not, as I am sure he has, the love of Morgana, he would console himself with his Vestals. If he had not their sisterly affection, he would rejoice in the love of Morgana, but having both the love and the affection, he is between two counter-attractions, either of which would make him happy, and both together make him miserable. Who can say which is best for him? or for them? or for Morgana herself? I almost wish the light of her favour had shone on Lord Curryfin. That chance has passed from her; and she will not easily find such another. Perhaps she might have held him in her bonds, if she had been so disposed. But Miss Niphet is a glorious girl, and there is a great charm in such perfect reciprocity. Jupiter himself, as I have before had occasion to remark, must have pre-arranged their consentaneity. The young lord went on some time, adhering, as he supposed, to his first pursuit, and falling unconsciously and inextricably into the second: and the young lady went on, devoting her whole heart and soul to him, not clearly perhaps knowing it herself, but certainly not suspecting that

* *Il.* xxii. vv. 500, 501.

† Juvenal: *Sat.* x. v. 346, sqq.

any one else could dive into the heart of her mystery. And now they both seem surprised that nobody seems surprised at their sudden appearance in the character of affianced lovers. His is another example of strange compensation; for if Morgana had accepted him on his first offer, Miss Niphet would not have thought of him; but she found him a waif and stray, a flotsome on the waters of love, and landed him at her feet without art or stratagem. Artlessness and simplicity triumphed, where the deepest design would have failed. I do not know if she had any compensation to look for; but if she had, she has found it; for never was a man with more qualities for domestic happiness, and not Pedro of Portugal himself was more overwhelmingly in love. When I first knew him, I saw only the comic side of his character: he has a serious one too, and not the least agreeable part of it: but the comic still shows itself. I cannot well define whether his exuberant good-humour is contagious, and makes me laugh by anticipation as soon as I fall into his company, or whether it is impossible to think of him, gravely lecturing on Fish, as a member of the Pantopragmatic Society, without perceiving a ludicrous contrast between his pleasant social face and the unpleasant social impertinence of those would-be meddlers with everything. It is true, he has renounced that folly; but it is not so easy to dissociate him from the recollection. No matter: if I laugh, he laughs with me: if he laughs, I laugh with him. "Laugh when you can" is a good maxim: between well-disposed sympathies a very little cause strikes out the fire of merriment—

> As long liveth the merry man, they say,
> As doth the sorry man, and longer by a day.

And a day so acquired is a day worth having. But then—

> Another sayd sawe doth men advise,
> That they be together both merry and wise.*

Very good doctrine, and fit to be kept in mind: but there is

* These two quotations are from the oldest comedy in the English language: *Ralph Roister Doister*, 1566. Republished by the Shakspeare Society, 1847.

much good laughter without much wisdom, and yet with no harm in it.'

The Doctor was approaching the Tower when he met Mr. Falconer, who had made one of his feverish exits from it, and was walking at double his usual speed. He turned back with the Doctor, who having declined taking anything before dinner but a glass of wine and a biscuit, they went up together to the library.

They conversed only on literary subjects. The Doctor, though Miss Gryll was uppermost in his mind, determined not to originate a word respecting her, and Mr. Falconer, though she was also his predominant idea, felt that it was only over a bottle of Madeira he could unbosom himself freely to the Doctor.

The Doctor asked, 'What he had been reading of late?' He said, 'I have tried many things, but I have always returned to *Orlando Innamorato*. There it is on the table, an old edition of the original poem.' The Doctor said, 'I have seen an old edition, something like this, on the drawing-room table at the Grange.' He was about to say something touching sympathy in taste, but he checked himself in time. The two younger sisters brought in lights. 'I observe,' said the Doctor, 'that your handmaids always move in pairs. My hot water for dressing is always brought by two inseparables, whom it seems profanation to call housemaids.'

MR. FALCONER

It is always so on my side of the house, that not a breath of scandal may touch their reputation. If you were to live here from January to December, with a houseful of company, neither you, nor I, nor any of my friends, would see one of them alone for a single minute.

THE REVEREND DOCTOR OPIMIAN

I approve the rule. I would stake my life on the conviction that these sisters are

> Pure as the new-fall'n snow,
> When never yet the sullying sun
> Has seen its purity,
> Nor the warm zephyr touched and tainted it.*

* SOUTHEY: *Thalaba.*

But as the world is constituted, the most perfect virtue needs to be guarded from suspicion. I cannot, however, associate your habits with a houseful of company.

MR. FALCONER

There must be sympathies enough in the world to make up society for all tastes: more difficult to find in some cases than in others; but still always within the possibility of being found. I contemplated, when I arranged this house, the frequent presence of a select party. The Aristophanic comedy and its adjuncts brought me into pleasant company elsewhere. I have postponed the purpose, not abandoned it.

Several thoughts passed through the Doctor's mind. He was almost tempted to speak them. 'How beautiful was Miss Gryll in Circe; how charmingly she acted. What was a select party without women? And how could a bachelor invite them?' But this would be touching a string which he had determined not to be the first to strike. So, *apropos* of the Aristophanic comedy, he took down Aristophanes, and said, 'What a high idea of Athenian comedy is given by this single line, in which the poet opines "The bringing out of comedy to be the most difficult of all arts."* It would not seem to be a difficult art now-a-days, seeing how much new comedy is nightly produced in London, and still more in Paris, which, whatever may be its literary value, amuses its audiences as much as Aristophanes amused the Athenians.'

MR. FALCONER

There is this difference, that though both audiences may be equally amused, the Athenians felt they had something to be proud of in the poet, which our audiences can scarcely feel, as far as novelties are concerned. And as to the atrocious outrages on taste and feeling perpetrated under the name of burlesques, I should be astonished if even those who laugh at them could look back on their amusement with any other feeling than that of being most heartily ashamed of the author, the theatre, and themselves.

When the dinner was over, and a bottle of claret had been placed by the side of the Doctor, and a bottle of Madeira by

* *Κωμῳδοδιασκαλίαν εἶναι χαλεπώτατον ἔργον ἁπάντων.*

Equites.

the side of his host, who had not been sparing during dinner of his favourite beverage, which had been to him for some days, like ale to the Captain and his friends in Beaumont and Fletcher,* almost 'his eating and his drinking solely,' the Doctor said, 'I am glad to perceive that you keep up your practice of having a good dinner; though I am at the same time sorry to see that you have not done your old justice to it.'

MR. FALCONER

A great philosopher had seven friends, one of whom dined with him in succession on each day of the week. He directed, amongst his last dispositions, that during six months after his death the establishment of his house should be kept on the same footing, and that a dinner should be daily provided for himself and his single guest of the day, who was to be entreated to dine there in memory of him, with one of his executors (both philosophers) to represent him in doing the honours of the table alternately.

THE REVEREND DOCTOR OPIMIAN

I am happy to see that the honours of your table are done by yourself, and not by an executor, administrator, or assign. The honours are done admirably, but the old justice on your side is wanting. I do not, however, clearly see what the *feralis cœna* of guest and executor has to do with the dinner of two living men.

MR. FALCONER

Ah, Doctor, you should say one living man and a ghost. I am only the ghost of myself. I do the honours of my departed conviviality.

THE REVEREND DOCTOR OPIMIAN

I thought something was wrong; but whatever it may be, take Horace's advice—'Alleviate every ill with wine and song, the sweet consolations of deforming anxiety.'†

MR. FALCONER

I do, Doctor. Madeira, and the music of the Seven Sisters, are my

* Ale is their eating and their drinking solely.
Scornful Lady, Act iv. Scene ii.

† Illic omne malum vino cantuque levato,
Deformis ægrimoniæ dulcibus alloquiis.
Epod. 13.

consolations, and great ones; but they do not go down to the hidden care that gnaws at the deepest fibres of the heart, like Ratatosk at the roots of the Ash of Ygdrasil.

THE REVEREND DOCTOR OPIMIAN

In the Scandinavian mythology: one of the most poetical of all mythologies. I have a great respect for Odin and Thor. Their adventures have always delighted me; and the system was admirably adapted to foster the high spirit of a military people. Lucan has a fine passage on the subject.*

The Doctor repeated the passage of Lucan with great emphasis. This was not what Mr. Falconer wanted. He had wished that the Doctor should inquire into the cause of his trouble; but independently of the Doctor's determination to ask no questions, and to let his young friend originate his own disclosures, the unlucky metaphor had carried the Doctor into one of his old fields, and if it had not been that he awaited the confidence, which he felt sure his host would spontaneously repose in him, the Scandinavian mythology would have formed his subject for the evening. He paused, therefore, and went on quietly sipping his claret.

Mr. Falconer could restrain himself no longer, and without preface or note of preparation, he communicated to the Doctor all that had passed between Miss Gryll and himself, not omitting a single word of the passages of Bojardo, which were indelibly impressed on his memory.

THE REVEREND DOCTOR OPIMIAN

I cannot see what there is to afflict you in all this. You are in love with Miss Gryll. She is disposed to receive you favourably. What more would you wish in that quarter?

MR. FALCONER

No more in that quarter, but the seven sisters are as sisters to me. If I had seven real sisters, the relationship would subsist, and marriage would not interfere with it; but, be a woman as amiable, as liberal, as indulgent, as confiding as she may, she could not treat the unreal, as she would the real tie.

* *Pharsalia,* l. i. vv. 458–462.

THE REVEREND DOCTOR OPIMIAN

I admit, it is not to be expected. Still there is one way out of the difficulty. And that is by seeing all the seven happily married.

MR. FALCONER

All the seven married! Surely that is impossible.

THE REVEREND DOCTOR OPIMIAN

Not so impossible as you apprehend.

The Doctor thought it a favourable opportunity to tell the story of the seven suitors, and was especially panegyrical on Harry Hedgerow, observing, that if the maxim *Noscitur à sociis* might be reversed, and a man's companions judged by himself, it would be a sufficient recommendation of the other six; whom, moreover, the result of his inquiries had given him ample reason to think well of. Mr. Falconer received with pleasure at Christmas, a communication which at the Midsummer preceding, would have given him infinite pain. It struck him all at once, that, as he had dined so ill, he would have some partridges for supper, his larder being always well stocked with game. They were presented accordingly, after the usual music in the drawing-room, and the Doctor, though he had dined well, considered himself bound in courtesy to assist in their disposal; when recollecting how he had wound up the night of the ball, he volunteered to brew a bowl of punch, over which they sate till a late hour, discoursing of many things, but chiefly of Morgana.

CHAPTER XXXIII

Ἦ σοφὸς, ἦ σοφὸς ἦν,
Ὃς πρῶτος ἐν γνώμᾳ τόδ' ἐβάστασε,
Καὶ γλώσσᾳ διεμυθολόγησεν,
Ὡς τὸ κηδεῦσαι καθ' ἑαυτὸν ἀριστεύει μακρῷ·
Καὶ μήτε τῶν πλούτῳ διαθρυπτομένων,
Μήτε τῶν γέννᾳ μεγαλυνομένων,
Ὄντα χερνήταν ἐραστεῦσαι γάμων.
ÆSCHYLUS: *Prometheus.*

Oh! wise was he, the first who taught
This lesson of observant thought,
That equal fates alone may dress
The bowers of nuptial happiness;
That never, where ancestral pride
Inflames, or affluence rolls its tide,
Should love's ill-omened bonds entwine
The offspring of an humbler line.

MR. FALCONER, the next morning, after the Doctor had set out on his return walk, departed from his usual practice of not seeing one of the sisters alone, and requested that Dorothy would come to him in the drawing-room. She appeared before him, blushing and trembling.

'Sit down,' he said, 'dear Dorothy; I have something to say to you and your sisters; but I have reasons for saying it first to you. It is probable, at any rate possible, that I shall very soon marry, and perhaps, in that case, you may be disposed to do the same. And I am told, that one of the best young men I have ever known is dying for love of you.'

'He is a good young man, that is certain,' said Dorothy; then becoming suddenly conscious of how much she had undesignedly admitted, she blushed deeper than before. And by way of mending the matter, she said, 'but I am not dying for love of him.'

'I dare say you are not,' said Mr. Falconer; 'you have no cause to be so, as you are sure of him, and only your consent is wanting.'

'And yours,' said Dorothy, 'and that of my sisters;

especially my elder sisters; indeed, they ought to set the example.'

'I am not sure of that,' said Mr. Falconer. 'So far, if I understand rightly, they have followed yours. It was your lover's indefatigable devotion that brought together suitors to them all. As to my consent, that you shall certainly have. So the next time you see Master Harry, send him to me.'

'He is here now,' said Dorothy.

'Then ask him to come in,' said Mr. Falconer.

And Dorothy retired in some confusion. But her lips could not contradict her heart. Harry appeared.

MR. FALCONER

So, Harry, you have been making love in my house, without asking my leave.

HARRY HEDGEROW

I couldn't help making love, sir; and I didn't ask your leave, because I thought I shouldn't get it.

MR. FALCONER

Candid, as usual, Harry. But do you think Dorothy would make a good farmer's wife?

HARRY HEDGEROW

I think, sir, she is so good, and so clever, and so ready and willing to turn her hand to anything, that she would be a fit wife for anybody, from a lord downwards. But it may be most for her own happiness to keep in the class in which she was born.

MR. FALCONER

She is not very pretty, you know.

HARRY HEDGEROW

Not pretty, sir! If she isn't a beauty, I don't know who is.

MR. FALCONER

Well, no doubt she is a handsome girl.

HARRY HEDGEROW

Handsome is not the thing, sir. She's beautiful.

MR. FALCONER

Well, Harry, she is beautiful, if that will please you.

HARRY HEDGEROW

It does please me, sir. I ought to have known you were joking when you said she was not pretty.

MR. FALCONER

But, you know, she has no fortune.

HARRY HEDGEROW

I don't want fortune. I want her, and nothing else, and nobody else.

MR. FALCONER

But I cannot consent to her marrying without a fortune of her own.

HARRY HEDGEROW

Why, then, I'll give her one beforehand. Father has saved some money, and she shall have that. We'll settle it on her, as the lawyers say.

MR. FALCONER

You are a thoroughly good fellow, Harry, and I really wish Dorothy joy of her choice; but that is not what I meant. She must bring you a fortune, not take one from you; and you must not refuse it.

Harry repeated that he did not want fortune; and Mr. Falconer repeated that, so far as depended on him, he should not have Dorothy without one. It was not an arduous matter to bring to an amicable settlement.

The affair of Harry and Dorothy being thus satisfactorily arranged, the other six were adjusted with little difficulty; and Mr. Falconer returned with a light heart to the Grange, where he presented himself at dinner on the twenty-seventh day of his probation.

He found much the same party as before; for though some of them absented themselves for a while, they could not resist Mr. Gryll's earnest entreaties to return. He was cordially welcomed by all, and with a gracious smile from Morgana.

CHAPTER XXXIV

Jane. . . . . We'll draw round
The fire, and grandmamma perhaps will tell us
One of her stories.
Harry. Aye, dear grandmamma!
A pretty story! something dismal now!
A bloody murder.
Jane. Or about a ghost.
SOUTHEY: *The Grandmother's Tale.*

IN the evening Miss Gryll said to the Doctor,

'We have passed Christmas without a ghost story. This is not as it should be. One evening at least of Christmas ought to be devoted to *merveilleuses histoires racontées autour du foyer*; which Chateaubriand enumerates among the peculiar enjoyments of those *qui n'ont pas quitté leur pays natal*. You must have plenty of ghosts in Greek and Latin, Doctor.'

THE REVEREND DOCTOR OPIMIAN

No doubt. All literature abounds with ghosts. But there are not many classical ghosts that would make a Christmas tale, according to the received notion of a ghost story. The ghost of Patroclus in Homer, of Darius in Æschylus, of Polydorus in Euripides, are fine poetical ghosts: but none of them would make a ghost story. I can only call to mind one such story in Greek: but even that, as it has been turned into ballads by Goethe in the *Bride of Corinth*, and by Lewis in the *Gay Gold Ring*,* would not be new to anyone here.

* Lewis says, in a note on the *Gay Gold Ring*:—'I once read in some Grecian author, whose name I have forgotten, the story which suggested to me the outline of the foregoing ballad. It was as follows: a young man arriving at the house of a friend, to whose daughter he was betrothed, was informed that some weeks had passed since death had deprived him of his intended bride. Never having seen her, he soon reconciled himself to her loss, especially as, during his stay at his friend's house, a young lady was kind enough to visit him every night in his chamber, whence she retired at daybreak, always carrying with her some valuable present from her lover. This intercourse continued till accident showed the young man the picture of his deceased bride, and he recognised, with horror, the features of his nocturnal visitor. The young lady's tomb being opened, he found in it the

There are some classical tales of wonder, not ghost stories, but suitable Christmas tales. There are two in Petronius, which I once amused myself by translating as closely as possible to the originals, and, if you please, I will relate them as I remember them. For I hold with Chaucer:

> Whoso shall telle a tale after a man,
> He moste reherse, as nigh as ever he can,
> Everich word, if it be in his charge,
> All speke he never so rudely and so large:
> Or elles he moste tellen his tale untrewe,
> Or feinen things, or finden wordes newe.*

This proposal being received with an unanimous 'By all means, Doctor,' the Doctor went on:

These stories are told at the feast of Trimalchio: the first by Niceros, a freedman, one of the guests:

'While I was yet serving, we lived in a narrow street where now is the house of Gavilla. There, as it pleased the gods, I fell in love with the wife of Terentius, the tavern-keeper—Melissa Tarentiana—many of you knew her, a most beautiful kiss-thrower.'

MISS GRYLL

That is an odd term, Doctor.

various presents which his liberality had bestowed on his unknown *innamorata*.'—M. G. LEWIS: *Tales of Wonder*, v. i. p. 99.

The Greek author here alluded to was Phlegon, whom some assign to the age of Augustus, and others, more correctly, to that of Hadrian. He wrote a treatise, *Περὶ θαυμασίων*: *On Wonderful Things*. The first, in what remains of the treatise, is the story in question, and the beginning of the story is lost. There is no picture in the case. The lover and his nocturnal visitor had interchanged presents, and the parents recognised those which had belonged to their daughter: a gold ring, and a neckerchief. They surprised their daughter on her third nightly visit, and she said to them:—'Oh, mother and father! how unjustly have you envied me the passing three days with your guest under my paternal roof. Now deeply will you lament your curiosity. I return to my destined place: for not without divine will came I hither.' Having spoken thus, she fell immediately dead. The tomb was opened, and they found an iron ring and a gilt cup, which she had received from her lover: who, in grief and horror, put an end to his life. It appears to be implied, that, if the third night had passed like the two preceding, she would have regained her life, and been restored to her parents and bridegroom.

* *Canterbury Tales*, vv. 733–738.

THE REVEREND DOCTOR OPIMIAN

It relates, I imagine, to some graceful gesture of pantomimic dancing: for beautiful hostesses were often accomplished dancers. Virgil's Copa, which, by the way, is only half panegyrical, gives us, nevertheless, a pleasant picture in this kind. It seems to have been one of the great attractions of a Roman tavern: and the host, in looking out for a wife, was probably much influenced by her possession of this accomplishment. The dancing, probably, was of that kind which the moderns call *demi-caractère,* and was performed in picturesque costume. . . .

The Doctor would have gone off in a dissertation on dancing hostesses; but Miss Gryll recalled him to the story, which he continued, in the words of Niceros:

'But, by Hercules, mine was pure love; her manners charmed me, and her friendliness. If I wanted money, if she had earned an *as*, she gave me a *semis*. If I had money, I gave it into her keeping. Never was woman more trustworthy. Her husband died at a farm, which they possessed in the country. I left no means untried to visit her in her distress; for friends are shown in adversity. It so happened, that my master had gone to Capua, to dispose of some cast-off finery. Seizing the opportunity, I persuaded a guest of ours to accompany me to the fifth milestone. He was a soldier, strong as Pluto. We set off before cock-crow; the moon shone like day; we passed through a line of tombs. My man began some ceremonies before the pillars. I sate down, singing, and counting the stars. Then, as I looked round to my comrade, he stripped himself, and laid his clothes by the wayside. My heart was in my nose: I could no more move than a dead man. But he walked three times round his clothes, and was suddenly changed into a wolf. Do not think I am jesting. No man's patrimony would tempt me to lie. But, as I had begun to say, as soon as he was changed into a wolf, he set up a long howl, and fled into the woods. I remained a while bewildered; then I approached to take up his clothes, but they were turned into stone. Who was dying of fear but I? But I drew my sword, and went on cutting shadows till I arrived at the farm. I entered the narrow way. The life was half boiled out of me; perspiration ran down me like a torrent: my eyes were dead. I could scarcely come to myself. My Melissa began to wonder why I walked so late; "and if you had come sooner," she said, "you might at least have helped us; for a wolf entered the farm and fell on the sheep, tearing them, and leaving them all bleeding. He escaped; but with cause to remember us; for our man drove a spear through his neck." When I heard these things, I could not think of sleep; but hurried homeward with the dawn; and when I came to the place where the clothes had been turned into stone, I found nothing but blood. When I reached home, my soldier was in bed, lying like an ox, and a surgeon was dressing his neck. I felt that he was a turnskin, and I could never after taste bread with him, not if you would have killed me. Let those who doubt of such things look into them. If I lie, may the wrath of all your Genii fall on me.'

This story being told, Trimalchio, the lord of the feast, after giving his implicit adhesion to it, and affirming the indisputable veracity of Niceros, relates another, as a fact of his own experience.

'While yet I wore long hair, for from a boy I led a Chian life,* our little Iphis, the delight of the family, died; by Hercules, a pearl; quick, beautiful, one of ten thousand. While, therefore, his unhappy mother was weeping for him, and we all were plunged in sorrow, suddenly witches came in pursuit of him, as dogs, you may suppose, of a hare. We had then in the house a Cappadocian, tall, brave to audacity, capable of lifting up an angry bull. He boldly, with a drawn sword, rushed out through the gate, having his left hand carefully wrapped up, and drove his sword through a woman's bosom; here as it were; safe be what I touch! We heard a groan; but, assuredly I will not lie, we did not see the women. But our stout fellow returning, threw himself into bed, and all his body was livid, as if he had been beaten with whips; for the evil hand had touched him. We closed the gate, and resumed our watch over the dead; but when the mother went to embrace the body of her son, she touched it, and found it was only a figure, of which all the interior was straw, no heart, nothing. The witches had stolen away the boy, and left in his place a straw-stuffed image. I ask you—it is impossible not—to believe, that there are women with more than mortal knowledge, nocturnal women, who can make that which is uppermost downmost. But our tall hero after this was never again of his own colour; indeed, after a few days, he died raving.'

'We wondered and believed,' says a guest who heard the story, 'and kissing the table, we implored the nocturnals to keep themselves to themselves, while we were returning from supper.'

* Free boys wore long hair. A Chian life is a delicate and luxurious life. Trimalchio implies that, though he began life as a slave, he was a pet in the household, and was treated as if he had been free.

MISS GRYLL

Those are pleasant stories, Doctor; and the peculiar style of the narrators testifies to their faith in their own marvels. Still, as you say, they are not ghost stories.

LORD CURRYFIN

Shakspeare's are glorious ghosts, and would make good stories, if they were not so familiarly known. There is a ghost much to my mind in Beaumont and Fletcher's *Lover's Progress*. Cleander has a beautiful wife, Calista, and a friend, Lisander. Calista and Lisander love each other, *en tout bien, tout honneur*. Lisander, in self-defence and in fair fight, kills a court favourite, and is obliged to conceal himself in the country. Cleander and Dorilaus, Calista's father, travel in search of him. They pass the night at a country inn. The jovial host had been long known to Cleander, who had extolled him to Dorilaus; but on inquiring for him they find he has been dead three weeks. They call for more wine, dismiss their attendants, and sit up alone, chatting of various things, and, among others, of mine host, whose skill on the lute and in singing is remembered and commended by Cleander. While they are talking, a lute is struck within; followed by a song, beginning

'Tis late and cold, stir up the fire,—
Sit close, and draw the table nigher:
Be merry, and drink wine that's old.

And ending:—

Welcome, welcome, shall go round,
And I shall smile, though underground.

And when the song ceases, the host's ghost enters. They ask him why he appears? He answers to wait once more on Cleander, and to entreat a courtesy:—

——to see my body buried
In holy ground: for now I lie unhallowed,
By the clerk's fault: let my new grave be made
Amongst good fellows, that have died before me,
And merry hosts of my kind.

Cleander promises that it shall be done; and Dorilaus, who is a merry old gentleman throughout the play, adds:—

And forty stoops of wine drank at thy funeral.

Cleander asks him:—

Is't in your power, some hours before my death,
To give me warning?

The host replies:—

I cannot tell you truly:
But if I can, so much on earth I loved you,
I will appear again.

In a subsequent scene, the ghost forewarns him, and he is soon after assassinated: not premeditatedly, but as an accident in the working out, by subordinate characters, of a plot to bring into question the purity of Calista's love for Lisander.

MISS ILEX

In my young days ghosts were so popular, that the first question asked about any new play was, Is there a ghost in it? The *Castle Spectre* had set this fashion. It was one of the first plays I saw, when I was a very little girl. The opening of the folding-doors disclosing the illuminated oratory; the extreme beauty of the actress who personated the ghost; the solemn music to which she moved slowly forward to give a silent blessing to her kneeling daughter; and the chorus of female voices chanting *Jubilate*; made an impression on me which no other scene of the kind has ever made. That is my ghost, but I have no ghost-story worth telling.

MR. FALCONER

There are many stories in which the supernatural is only apparent, and is finally explained. But some of these, especially the novels of Brockden Brown, carry the principle of terror to its utmost limits. What can be more appalling than his *Wieland*? It is one of the few tales in which the final explanation of the apparently supernatural does not destroy or diminish the original effect.

MISS GRYLL

Generally, I do not like that explaining away. I can accord a ready faith to the supernatural in all its forms, as I do to the adventures of Ulysses and Orlando. I should be sorry to see the enchantments of Circe expounded into sleights of hand.

THE REVEREND DOCTOR OPIMIAN

I agree with you, Miss Gryll. I do not like to find a ghost, which has frightened me through two volumes, turned into a Cock-lane ghost in the third.

MISS GRYLL

We are talking about ghosts, but we have not a ghost story. I want a ghost story.

MISS NIPHET

I will try to tell you one, which I remember imperfectly. It relates, as many such stories do, to a buried treasure. An old miser had an only daughter; he denied himself everything, but he educated her well, and treated her becomingly. He had accumulated a treasure, which he designed for her, but could not bear the thought of parting with it, and died without disclosing the place of its concealment. The daughter had a lover, not absolutely poor, nor much removed from it. He farmed a little land of his own. When her father died, and she was left destitute and friendless, he married her, and they endeavoured by economy and industry to make up for the deficiencies of fortune. The young husband had an aunt, with whom they sometimes passed a day of festival, and Christmas Day especially. They were returning home late at night on one of these occasions; snow was on the ground; the moon was in the first quarter, and nearly setting. Crossing a field, they paused a moment to look on the beauty of the starry sky; and when they again turned their eyes to the ground, they saw a shadow on the snow; it was too long to have any distinct outline; but no substantial form was there to throw it. The young wife clung trembling to the arm of her husband. The moon set, and the shadow disappeared. New Year's Day came, and they passed it at the aunt's. On their return the moon was full, and high in heaven. They crossed the same field, not without hesitation and fear. In the same spot as before, they again saw the shadow; it was that of a man in a large loose wrapper, and a high-peaked hat. They recognised the outline of the old miser. The husband sustained his nearly fainting wife; as their eyes were irresistibly fixed on it, it began to move, but a cloud came over the moon, and they lost sight of it. The next night was bright, and the wife had summoned all her courage to follow out the mystery; they returned to the spot at the same hour; the shadow again fell on the snow, and again it began to move, and glided away slowly over the surface of the snow. They followed it fearfully. At length it stopped on a small mound in another field of their own farm. They walked round and round it, but it moved no more. The husband entreated his wife to remain, while he sought a stick to mark the place. When she was alone, the shadow spread out its arms as in the act of benediction, and vanished. The husband found her extended on the snow; he raised her in his arms; she recovered, and they walked home. He returned in the morning with pick-axe and spade, cleared away the snow, broke into the ground, and found a pot of gold, which was unquestionably their own. And then, with the usual end of a nurse's tale, 'they lived happily all the rest of their lives'.

MISS ILEX

Your story, though differing in all other respects, reminds me of a ballad in which there is a shadow on the snow,

> Around it, and round, he had ventured to go,
> But no form that had life threw that stamp on the snow.*

MR. GRYLL

In these instances, the shadow has an outline, without a visible form to throw it. I remember a striking instance of shadows without distinguishable forms. A young chevalier was riding through a forest of pines, in which he had before met with fearful adventures, when a strange voice called on him to stop. He did not stop, and the stranger jumped up behind him. He tried to look back, but could not turn his head. They emerged into a glade, where he hoped to see in the moonlight the outline of the unwelcome form. But 'unaccountable shadows fell around, unstamped with delineations of themselves'.†

MISS GRYLL

Well, Mr. MacBorrowdale, have you no ghost story for us?

MR. MACBORROWDALE

In faith, Miss Gryll, ghosts are not much in my line: the main business of my life has been among the driest matters of fact: but I will tell you a tale of a bogle, which I remember from my boyish days.

There was a party of witches and warlocks assembled in the refectory of a ruined abbey, intending to have a merry supper, if they could get the materials. They had no money, and they had for servant a poor bogle, who had been lent to them by his Satanic majesty, on condition that he should provide their supper if he could; but without buying or stealing. They had a roaring fire, with nothing to roast, and a large stone table, with nothing on it but broken dishes and empty mugs. So the fire-light shone on an

* MISS BANNERMAN's *Tales of Superstition and Chivalry.*

† *The Three Brothers*, vol. iv. p. 193.

uncouth set of long hungry faces. Whether there was among them 'ae winsome wench and wawlie'* is more than I can say; but most probably there was, or the bogle would scarcely have been so zealous in the cause. Still he was late on his quest. The friars of a still flourishing abbey were making preparations for a festal day, and had despatched a man with a cart to the nearest town, to bring them a supply of good things. He was driving back his cart well loaded with beef, and poultry, and ham; and a supply of choice rolls for which a goodwife in the town was famous; and a new arrival of rare old wine, a special present to the Abbot from some great lord. The bogle having smelt out the prize, presented himself before the carter in the form of a sailor with a wooden leg, imploring charity. The carter said he had nothing for him, and the sailor seemed to go on his way. He re-appeared in various forms, always soliciting charity, more and more importunately every time, and always receiving the same denial. At last he appeared as an old woman, leaning on a stick, who was more pertinacious in her entreaties than the preceding semblances; and the carter, after asseverating with an oath, that a whole ship-load of beggars must have been wrecked that night on the coast, reiterated that he had nothing for her. 'Only the smallest coin, master,' said the old woman. 'I have no coin,' said the carter. 'Just a wee bite and sup of something,' said the old woman; 'you are scarcely going about without something to eat and drink; something comfortable for yourself. Just look in the cart: I am sure you will find something good.' 'Something, something, something,' said the carter; 'if there is anything fit to eat or drink in the cart, I wish a bogle may fly away with it.' 'Thank you,' said the bogle, and changed himself into a shape which laid the carter on his back, with his heels in the air. The bogle made lawful prize of the contents of the cart. The refectory was soon fragrant with the odour of roast, and the old wine flowed briskly, to the great joy of the assembly, who passed the night in feasting, singing, and dancing, and toasting Old Nick.

MISS GRYLL

And now, Mr, Falconer, you who live in an old tower, among old

* But Tam kend what was what fu' brawlie:
There was ae winsome wench and wawlie,
That night enlisted in the core,
Lang after kend on Carrick shore.
Tam O'Shanter.

books, and are deep in the legends of saints, surely you must have a ghost-story to tell us.

MR. FALCONER

Not exactly a ghost-story, Miss Gryll, but there is a legend which took my fancy, and which I turned into a ballad. If you permit me, I will repeat it.

The permission being willingly granted, Mr. Falconer closed the series of fireside marvels by reciting

THE LEGEND OF SAINT LAURA

Saint Laura, in her sleep of death,
Preserves beneath the tomb
—'Tis willed where what is will must be—*
In incorruptibility
Her beauty and her bloom.

So pure her maiden life had been,
So free from earthly stain,
'Twas fixed in fate by Heaven's own Queen,
That till the earth's last closing scene
She should unchanged remain.

Within a deep sarcophagus
Of alabaster sheen,
With sculptured lid of roses white,
She slumbered in unbroken night,
By mortal eyes unseen.

Above her marble couch was reared
A monumental shrine,
Where cloistered sisters, gathering round,
Made night and morn the aisle resound
With choristry divine.

The abbess died: and in her pride
Her parting mandate said,
They should her final rest provide,
The alabaster couch beside,
Where slept the sainted dead.

* Vuolsi così colà dove si puote
Ciò che si vuole, e più non domandare.
DANTE.

The abbess came of princely race:
    The nuns might not gainsay:
And sadly passed the timid band,
To execute the high command
    They dared not disobey.

The monument was opened then:
    It gave to general sight
The alabaster couch alone:
But all its lucid substance shone
    With præternatural light.

They laid the corpse within the shrine:
    They closed its doors again:
But nameless terror seemed to fall,
Throughout the live-long night, on all
    Who formed the funeral train.

Lo! on the morrow morn, still closed
    The monument was found:
But in its robes funereal drest,
The corpse they had consigned to rest
    Lay on the stony ground.

Fear and amazement seized on all:
    They called on Mary's aid:
And in the tomb, unclosed again,
With choral hymn and funeral train,
    The corpse again was laid.

But with the incorruptible
    Corruption might not rest:
The lonely chapel's stone-paved floor
Received the ejected corpse once more,
    In robes funereal drest.

So was it found when morning beamed:
    In solemn suppliant strain
The nuns implored all saints in heaven,
That rest might to the corpse be given,
    Which they entombed again.

On the third night a watch was kept
    By many a friar and nun:
Trembling, all knelt in fervent prayer,
Till on the dreary midnight air
    Rolled the deep bell-toll, 'One!'

The saint within the opening tomb
    Like marble statue stood:
All fell to earth in deep dismay:
And through their ranks she passed away,
    In calm unchanging mood.

No answering sound her footsteps raised
  Along the stony floor:
Silent as death, severe as fate,
She glided through the chapel gate,
  And none beheld her more.

The alabaster couch was gone:
  The tomb was void and bare:
For the last time, with hasty rite,
Even 'mid the terror of the night,
  They laid the abbess there.

'Tis said, the abbess rests not well
  In that sepulchral pile:
But yearly, when the night comes round,
As dies of 'One' the bell's deep sound
  She flits along the aisle.

But whither passed the virgin saint,
  To slumber far away,
Destined by Mary to endure,
Unaltered in her semblance pure,
  Until the judgement day?

None knew, and none may ever know:
  Angels the secret keep:
Impenetrable ramparts bound,
Eternal silence dwells around,
  The chamber of her sleep.

CHAPTER XXXV

Σοὶ δὲ θεοὶ τόσα δοῖεν, ὅσα φρεσὶ σῇσι μενοινᾷς,
Ἄνδρα τε, καὶ οἶκον, καὶ ὁμοφροσύνην ὀπάσειαν
Ἐσθλήν· οὐ μὲν γὰρ τοῦ γε κρεῖσσον καὶ ἄρειον,
Ἢ ὅθ' ὁμοφρονέοντε νοήμασιν οἶκον ἔχητον
Ἀνηρ ἠδὲ γυνή.

May the gods grant what your best hopes pursue,
A husband, and a home, with concord true:
No greater boon from Jove's ethereal dome
Descends, than concord in the nuptial home.

ULYSSES *to* NAUSICAA, *in the sixth book of the Odyssey.*

WHAT passed between Algernon and Morgana, when the twenty-eighth morning brought his probation to a close, it is unnecessary to relate. The gentleman being predetermined to propose, and the lady to accept, there was little to be said, but that little was conclusive.

Mr. Gryll was delighted. His niece could not have made a choice more thoroughly to his mind.

'My dear Morgana,' he said, 'all's well that ends well. Your fastidiousness in choice has arrived at a happy termination. And now you will perhaps tell me why you rejected so many suitors, to whom you had in turn accorded a hearing. In the first place, what was your objection to the Honourable Escor A'Cass?* He was a fine, handsome, dashing fellow. He was the first in the field, and you seemed to like him.'

MISS GRYLL

He was too dashing, uncle: he gambled. I did like him, till I discovered his evil propensity.

MR. GRYLL

To Sir Alley Capel?

* Ἐς κόρακας: *To-the-Crows:* the Athenian equivalent for our *To-the-Devil:* a gambler's journey: not often a long one.

MISS GRYLL

He speculated; which is only another name for gambling. He never knew from day to day whether he was a rich man or a beggar. He lived in a perpetual fever, and I wish to live in tranquillity.

MR. GRYLL

To Mr. Ballot?

MISS GRYLL

He thought of nothing but politics: he had no feeling of poetry. There was never a more complete negation of sympathy, than between him and me.

MR. GRYLL

To Sir John Pachyderm?

MISS GRYLL

He was a mere man of the world, with no feeling of any kind: tolerable in company, but tiresome beyond description in a tête-à-tête. I did not choose that he should bestow all his tediousness on me.

MR. GRYLL

To Mr. Enavant?

MISS GRYLL

He was what is called a fast man, and was always talking of slow coaches. I had no fancy for living in an express train. I like to go quietly through life, and to see all that lies in my way.

MR. GRYLL

To Mr. Geront?

MISS GRYLL

He had only one fault, but that one was unpardonable. He was too old. To do him justice, he did not begin as a lover. Seeing that I took pleasure in his society, he was led by degrees into fancying that I might accept him as a husband. I liked his temper, his acquirements, his conversation, his love of music and poetry, his devotion to domestic life. But age and youth cannot harmonize in marriage.

MR. GRYLL

To Mr. Long Owen?

MISS GRYLL

He was in debt, and kept it secret from me. I thought he only wanted my fortune: but be that as it might, the concealment destroyed my esteem.

MR. GRYLL

To Mr. Larval?

MISS GRYLL

He was too ugly. Expression may make plain features agreeable, and I tried if daily intercourse would reconcile me to his. But no. His ugliness was unredeemed.

MR. GRYLL

None of these objections applied to Lord Curryfin.

MISS GRYLL

No, uncle; but he came too late. And besides, he soon found what suited him better.

MR. GRYLL

There were others. Did any of the same objections apply to them all?

MISS GRYLL

Indeed, uncle, the most of them were nothing; or at best, mere suits of good clothes; men made, as it were, to pattern by the dozen; selfish, frivolous, without any earnest pursuit, or desire to have one; ornamental drawing-room furniture, no more distinguishable in memory than a set of chairs.

MR. GRYLL

Well, my dear Morgana, for mere negations there is no remedy; but for positive errors, even for gambling, it strikes me they are curable.

MISS GRYLL

No, uncle. Even my limited observation has shown me, that men are easily cured of unfashionable virtues, but never of fashionable vices.

Miss Gryll and Miss Niphet arranged that their respective marriages and those of the seven sisters, should be celebrated at the same time and place. In the course of their castle-building before marriage, Miss Niphet said to her intended:

'When I am your wife, I shall release you from your promise of not trying experiments with horses, carriages, boats, and so forth; but with this proviso, that if ever you do try a dangerous experiment, it shall be in my company.'

'No, dear Alice,' he answered; 'you will make my life too dear to me, to risk it in any experiment. You shall be my guiding star, and the only question I shall ask respecting my conduct in life, will be, Whether it pleases you?'

> Some natural tears they shed, but wiped them soon;

might have been applied to the sisters, when they stepped, on their bridal morning, into the carriages which were to convey them to the Grange.

It was the dissipation of a dream too much above mortal frailty, too much above the contingencies of chance and change, to be permanently realized. But the damsels had consented, and the suitors rejoiced; and if ever there was a man on earth with 'his saul abune the moon,' it was Harry Hedgerow, on the bright February morning that gave him the hand of his Dorothy.

There was a grand *déjeuner* at Gryll Grange. There were the nine brides, and the nine bridegrooms; a beautiful array of bridemaids; a few friends of Mr. Gryll, Mr. Niphet, Lord Curryfin, and Mr. Falconer; and a large party at the lower end of the hall, composed of fathers, mothers, and sisters of the bridegrooms of the seven Vestals. None of the bridegrooms had brothers, and Harry had neither mother nor sister; but his father was there in rustic portliness, looking, as Harry had anticipated, as if he were all but made young again.

Among the most conspicuous of the party were the Reverend Doctor Opimian and his lady, who had on this occasion stepped out of her domestic seclusion. In due course the Reverend Doctor stood up and made a speech, which may be received as the epilogue of our comedy.

THE REVEREND DOCTOR OPIMIAN

We are here to do honour to the nuptials; first, of the niece of our excellent host, a young lady whom to name is to show her title to the love and respect of all present; with a young gentleman, of whom to say that he is in every way worthy of her, is to say all that can be said of him in the highest order of praise: secondly, of a young lord and lady, to whom those who had the pleasure of being here last Christmas are indebted for the large share of enjoyment which their rare and diversified accomplishments, and their readiness to contribute in every way to social entertainment, bestowed on the assembled party; and who, both in contrast and congeniality,—for both these elements enter into perfect fitness of companionship—may be considered to have been expressly formed for each other: thirdly, of seven other young couples, on many accounts most interesting to us all, who enter on the duties of married life, with as fair expectation of happiness as can reasonably be entertained in this diurnal sphere. An old Greek poet says:—'Four things are good for man in this world; first, health; second, personal beauty; third, riches not dishonourably acquired; fourth, to pass life among friends.'* But thereon says the comic poet Anaxandrides: 'Health is rightly placed first; but riches should have been second; for what is beauty ragged and starving?'† Be this as it may, we here see them all four; health in its brightest bloom; riches in two instances; more than competence in the other seven; beauty in the brides, good looks, as far as young men need them, in the bridegrooms, and as bright a prospect of passing life among friends as ever shone on any. Most earnestly do I hope that the promise of their marriage morning may be fulfilled in its noon and in its sunset; and when I add, may they all be as happy in their partners as I have been, I say what all who know the excellent person beside me will feel to be the best good wish in my power to bestow. And now, to the health of the brides and bridegrooms, in bumpers of champagne. Let all the attendants stand by, each with a fresh bottle, with only one uncut string. Let all the corks, when I give the signal, be discharged simultaneously; and we will receive it as a peal of Bacchic ordnance, in honour of the Power of Joyful Event‡ whom we may assume to be presiding on this auspicious occasion.

* *Ὑγιαίνειν μὲν ἄριστον ἀνδρὶ θνατῷ·*
Δεύτερον δὲ, φυὰν καλὸν γενέσθαι·
Τρίτον δὲ, πλουτεῖν αδόλως·
Καὶ τὸ τέταρτον, ἡβᾷν μετὰ τῶν φίλων.

SIMONIDES.

† ATHENÆUS: l. xv. p. 694.

‡ This was a Roman deity. *Invocato hilaro atque prospero Eventu.*—APULEIUS: *Metamorph.* l. iv.

THE END.

APPENDIX A

PREFACE

TO THE

VOLUME OF 'STANDARD NOVELS'

CONTAINING

'HEADLONG HALL,' 'NIGHTMARE ABBEY,' 'MAID MARIAN,' AND 'CROTCHET CASTLE'

ALL these little publications appeared originally without prefaces. I left them to speak for themselves; and I thought I might very fitly preserve my own impersonality, having never intruded on the personality of others, nor taken any liberties but with public conduct and public opinions. But an old friend assures me, that to publish a book without a preface is like entering a drawing-room without making a bow. In deference to this opinion, though I am not quite clear of its soundness, I make my prefatory bow at this eleventh hour.

'Headlong Hall' was written in 1815; 'Nightmare Abbey', in 1817; 'Maid Marian', with the exception of the last three chapters, in 1818; 'Crotchet Castle', in 1830. I am desirous to note the intervals, because, at each of those periods, things were true, in great matters and in small, which are true no longer. 'Headlong Hall' begins with the Holyhead Mail, and 'Crotchet Castle' ends with a rotten borough. The Holyhead mail no longer keeps the same hours, nor stops at the Capel Cerig Inn, which the progress of improvement has thrown out of the road; and the rotten boroughs of 1830 have ceased to exist, though there are some very pretty pocket properties, which are their worthy successors. But the classes of tastes, feelings, and opinions, which were successively brought into play in these little tales, remain substantially the same. Perfectibilians, deteriorationists, statu-quo-ites, phrenologists, transcendentalists, political economists, theorists in all sciences, projectors in all arts, morbid visionaries, romantic enthusiasts, lovers of music, lovers of the picturesque, and lovers of good dinners, march, and will march for ever, *pari passu* with the march of mechanics, which some facetiously call the march of intellect. The fastidious in old wine are a race that does not decay. Literary violators of the confidences of private life still gain a disreputable livelihood and an unenviable notoriety. Match-makers from interest, and the disappointed in love and in friendship, are varieties of which specimens are extant. The great principle of the Right of Might is as flourishing now as in the days of Maid Marian: the array of false pretensions, moral, political, and literary, is as imposing as ever: the rulers of the world still feel things in their effects, and never foresee them in their causes; and political mountebanks continue, and will continue, to puff nostrums and practise legerdemain under the eyes of the multitude; following, like the 'learned friend' of Crotchet Castle, a course as tortuous as that of a river, but in a reverse process; beginning by being dark and deep, and ending by being transparent.

THE AUTHOR OF 'HEADLONG HALL'

March 4 1837

APPENDIX B

Textual Variants in *Headlong Hall*

BELOW is a list of substantive variants in the first three editions, which adds to those in Halliford, vol. i, and in Joukovsky's edition (see Note on the Texts). A few variants in punctuation are given where they affect the sense, and P.'s earlier spellings of Mac Laurel's dialect and of Welsh place names (*1815* and *1816*) are also listed. Obvious errors in earlier editions are not included.

Manuscript drafts of Cranium's lecture (ch. 12) and Escot's final speech (ch. 15) are preserved in BM Add. MS 36816, ff. 45v–37v. The drafts are difficult to decipher, as Joukovsky noted, owing to P.'s habit of abbreviation as well as to general illegibility and the density of corrections. The draft of Escot's speech appears to correspond closely to the printed text, but that of Cranium's lecture has a number of false starts and the paragraphs are ordered somewhat differently (e.g. the skulls of the beaver and the bullfinch—the first two in the published text—appear after those of the tiger, the fox, the peacock, the robber, and the conqueror). We have selected ten variant readings from these manuscript fragments which can be read with some certainty and are significant enough to be of interest to the reader. They are designated *MS*.

Line numbers are calculated with reference to all the text on the page, including speech-headings, epigraphs, and P.'s own footnotes.

Contents page: *Not in* 1815, 1816.

p. 1, l. 3: THE MAIL *All chapter headings and running headings were added in* 1822.

p. 1, ll. 5–12: dispelled . . . traveller.] enabled each of the four passengers,—who had dozed through the first seventy miles of the road, with as much comfort as the jolting of the vehicle, and an occasional admonition to remember the coachman, (thundered through the open door, accompanied by the gentle breath of Boreas, into the ears of the sleeping traveller,) could admit—to observe the companions of their journey. 1815

p. 1, l. 29: by one account,] *Not in* 1815, 1816.

p. 1, l. 31: a tradition having] for, a tradition has 1815

p. 2, l. 22: a man] man 1815

p. 4, l. 16: character] nature 1815, 1816

p. 5, ll. 6–8: the alacrity ... ankle, and he] he sprang out with so much alacrity as to sprain his ancle, and 1815, 1816

p. 7, ll. 18–20: disastrous. I admit, that in some respects the use of animal food retards] disastrous: but I admit, that the use of animal food retards in some measure 1815

p. 8, l. 28: them ... they] it ... it 1815

p. 8, l. 30–p. 9, l. 5: If we can ... mind] *Not in* 1815.

p. 8, l. 38: See ... Auricula] *Not in* 1815.

p. 10, ll. 23–38: Mr. Knight ... can confer] *Not in* 1815.

p. 11, l. 26: his] the 1816

p. 11, l. 28: the] his 1815, 1816

p. 12, l. 30: delivered; nor] delivered, who had ever since continued implacable; nor 1815

p. 12, ll. 36–9: 'Il est ... liv. 5] See the fourth volume of Rousseau's Emile 1815

p. 13, l. 23: Panscope] Panoscope 1815, 1816 *throughout*

p. 13, l. 28: well.] well; that is, not at all. 1815, 1816

p. 14, l. 16: the pedestal] a pedestal 1815, 1816

p. 14, l. 21: seven] three 1815

p. 14, l. 22: fit him for propping] make him prop up 1815, 1816, 1822

p. 16, l. 23: preserved] saved 1815

p. 19, ll. 5–7: nactar ... descovered ... tarrestrial] neectar ... deescovered ... tarreestrial 1815, 1816

p. 19, l. 21: to] from 1815, 1816

p. 21, l. 23: review.] Review? 1815, 1816, 1822

p. 22, ll. 4–7: readily ... wullinly ... encoorage ... leetle ... predilaction] reedily ... weelinly ... encourage ... little ... predileection 1815, 1816

p. 22, l. 15: general?] general. 1815, 1816, 1822

p. 22, ll. 19–20: SQUIRE HEADLONG Buz!] *Speech and prefix not in* 1815; Fill. 1816; Buz. 1822

p. 23, l. 2: leeve] live 1815, 1816

p. 23, ll. 7–10: Every ... leeves ... defference ... respact] Eevery ... lives ... deeference ... respect 1815, 1816

p. 23, ll. 14–23: admetted ... pheelosophers ... sic thing ... desenterestedness ... every ... beautifu' ... every ... fetness ... grund ... atween ... ae] admeeted ... philosophers ... sic a thing ... diseenterestedness ... eevery ... beautiful ... eevery ... fitness ... groond ... between ... ane 1815, 1816

p. 24, ll. 8–16: desenterested ... pheelosophical ... semply ... aunly ... paisant ... thrapple ... pheelosopher delevers ... preson ... menester ... fetness] diseenterested ... philosophical ... seemply ... anely ... peesant ... throat ... philosopher deleevers ... preeson ... meenester ... fitness 1815, 1816

p. 24, ll. 23–4: every ... endaivours ... every] eevery ... endeevours ... every 1815, 1816

p. 25, ll. 1–16: descreptions ... pairfectly ... fetness ... exceetability ... flingin' ... menestry ... place or a pansion ... o'ye ... send ... plenipotentiary] descreeptions ... parefectly ... fitness ... excitability ... throwing ... meenestry ... peension ... of ye ... seend ... pleenipotentiary 1815, 1816

p. 25, ll. 18–19: SQUIRE HEADLONG Off with your heeltaps.] *Not in* 1815.

p. 26, l. 22: had tasted] tasted 1815

p. 29, l. 31: subvert] evert 1815, 1816

p. 32, ll. 28–9: Cephalis, ... it] Cephalis; and it 1815

p. 32, l. 31: his rival] Mr. Escot 1815

p. 32, l. 32–p. 33, l. 1: The stimulus ... Mr. Escot] Accordingly, after due deliberation, he resolved on *cutting him out* 1815

p. 36, l. 11: rugged] ragged 1815

p. 36, l. 29: mossy] massy 1815

p. 37, l. 22: Egregious, by Jupiter!] Exquisite, upon my soul! 1815, 1816

p. 38, l. 21: Oh!] O! 1815, 1816, 1822

p. 39, l. 1: Oh!] O! 1815, 1816, 1822

p. 39, l. 35: from ... *Occasione*] from a passage in the *Occasione* of Machiavelli 1815, 1816

p. 41, l. 7: Explosion] Explosions 1816

p. 41, l. 13: Bedd] Beth 1815, 1816

p. 41, l. 19: Meirionnydd] Merioneth 1815, 1816

p. 41, l. 26: an accurate] a more accurate 1815

p. 48, l. 3: been expected to teach] taught 1815, 1816

p. 49, ll. 9–12: quantity . . . store] quantity . . . quantity 1815; sufficiency . . . quantity 1816

p. 50, l. 6: having] having now 1815, 1816

p. 50, l. 32: bondage] *vinculum* 1815, 1816

p. 52, l. 25: feels] will feel 1815, 1816

p. 52, l. 35: it is rather to be feared] I rather fear 1815, 1816

p. 52, l. 37: the three] those three 1815, 1816

p. 53, l. 17: from whence] whence 1815, 1816, 1822

p. 53, l. 20: addressed him:—] addressed him with: 1815, 1816

p. 54, l. 21: tid n't] did n't 1822

p. 54, l. 22: the tevil] he 1815, 1816

p. 59, l. 10: Gaster, turning down an empty egg-shell;] Gaster; 1815

p. 59, l. 31: had occurred] occurred 1815, 1816

p. 60, l. 12: Kernioggau] Kerniogge 1815, 1816

p. 60, l. 21: August] July 1815

p. 61, ll. 25–7: osteo . . . medullary,] osseocarnisanguineoviscericar-tilaginomedullary 1815

p. 62, l. 29: THE LECTURE] A LECTURE ON SKULLS. 1815, 1816

p. 63, ll. 17–21: 'Again . . . waiting.] *Not in* 1815, 1816.

p. 63, ll. 30–1: of faculties] of the faculties 1815, 1816, 1822

p. 64, l. 19: constructiveness] building *MS*

p. 64, l. 23: carnage] destruction *MS*

p. 64, l. 24: plunder] theft *MS*

p. 64, l. 33–p. 65, l. 4: a round dozen . . . enlargement] epics. You observe in this skull all the same organs of destruction and theft in this skull (*sic*) as in the highway man. The superior magnitude *MS*

p. 65, l. 3: separately] just 1815, 1816

p. 65, l. 9: differences] difference 1815, 1816, 1822

p. 65, l. 18: turned out of doors] left *corrected to* turned out of doors *MS*

p. 66, l. 3: show] trace 1815, 1816

p. 66, l. 8: similitude] resemblance 1815

p. 66, l. 15: as much] so much 1815

p. 66, l. 17: feeling the] feeling all the 1815, 1816

p. 66, l. 33: Llugwy] Conwy 1815, 1816

p. 67, l. 1: Edeirnion] Llwyd 1815, 1816

p. 67, l. 17: Gaster] *Not in* 1815, 1816.

p. 69, l. 5: should] *Not in* 1815, 1816.

p. 71, l. 23: method] methods 1815, 1816

p. 73, l. 6: outlines] outline 1815, 1816, 1822

p. 73, l. 12: Brindle-mew] Grimalkin 1815

p. 73, l. 14: the countenance] countenance 1815, 1816, 1822

p. 74, l. 5: field] stream 1815

p. 74, l. 25: the best] best 1815

p. 74, l. 30–p. 75. l. 10: aicho ... feddle ... indefference ... tessue ... pooerfu' ... lug] eecho ... feedle ... indeeference ... teesue ... pooerful ... ear 1815, 1816

p. 76, ll. 1–4: In English ... tongue:]

He asked her for bread, for with toil he was worn:
He asked for a bed to repose him till morn:
In English he spoke, and none knew what he said,
But her oatcake and milk on the table she spread:

For she guessed at his wants, and she pitied his care,
And she hastened a bed for his rest to prepare;
Then he sate to his supper, and blithely he sung,
And she knew the dear sounds of her own native tongue:

1815, 1816 [*In the second line 1816 has* He asked her for bed]

p. 77, l. 20: re-adjourned] returned 1815; went back 1816

p. 78, l. 3: pleased] well pleased 1815, 1816, 1822

p. 79, l. 1: Brindle-mew] Grimalkin 1815

p. 79, l. 3 'Consulted!'] 'Consulted,' 1815, 1816, 1822

p. 82, l. 24: does not] does 1837

p. 82, ll. 38–9: *one is*] *one's* 1815

p. 84, ll. 17–18: obviate. The ... drunk.] obviate: the ... drunk! 1815, 1816

p. 84, l. 30: peaceful] peaceable 1815, 1816, 1822

p. 85, ll. 8–9: which ... cordially:] *Not in* 1815, 1816.

p. 85, l. 23: not by legal . . . but] *Not in MS.*

p. 85, l. 27: present system] present perverted and (?) abominable systems *MS*

p. 85, l. 27–34: So far . . . general good] *Not in MS.*

p. 85, l. 31: for] to 1815

p. 86, ll. 11–13: that merciless . . . may] the slavish spirit of a merciless world will *MS*

p. 86, ll. 18–22: whom that . . . absurdities!] *Not in MS.*

APPENDIX C

Textual Variants in *Gryll Grange*

BELOW is a list of the substantive variants found in the *Fraser's* text (*F.*) of the novel (together with two punctuation/spelling variants affecting sense) which adds to those in Halliford, v. 380–4.

The list also includes a number of manuscript variants. The manuscript of *G.G.* survives only in fragments, all in the Carl H. Pforzheimer Library of the Carl and Lily Pforzheimer Foundation, New York, except for the manuscript of the poem 'A New Order of Chivalry' (ch. 18), which is in the British Library (BM Add. MS 47225), written on paper watermarked 1830. Also in the British Library is the manuscript (Add. MS 36815, ff. 183–90) of an untitled and unfinished story, first printed in Halliford (viii. 391–3) under the heading 'A Story of a Mansion Among the Chiltern Hills', from which a few sentences about St Catharine's martyrdom and about the three chapels (pp. 141 and 142 above) were taken for *G.G.*, ch. 9. In the Pforzheimer are the following fragments (TLP 112): last paragraph of ch. 1; beginning of ch. 7; and scraps of ch. 19 (p. 215 above), ch. 33 (p. 305, above). Also present is a manuscript list of chapter titles for the novel, twenty-four of them differing from the titles as printed; only four of these (recorded below) differ substantially, however, the others being merely abbreviated versions of the printed title, e.g. 'The Forest' (ch. 4) instead of 'The Forest—A Soliloquy on Hair—The Vestals'. Also in the Pforzheimer (TLP 126a and b) is the manuscript of the beginning of an untitled story about 'the Knights of St. Katharine' from which a few sentences about the disposition of the saint's body (p. 141, above) were incorporated into *G.G.*, ch. 9. Variants found in these manuscript fragments are listed below, designated *MS*, and are quoted by permission of the Carl H. Pforzheimer Library. (The fragments from chs. 7 and 33 have 'said the Doctor', 'said his wife', 'said Harry', and so on, instead of the capitalized speech-headings of 1861, but these variants have not been included below.)

British Library Add. MS 36815 also contains (f. 190[r]), besides the unfinished story referred to above, the following fragment of what must clearly have been an early draft for part of ch. 4 or possibly chs. 7 or 8—Opimian is either soliloquizing about Falconer and his Tower or reporting them to his wife or the Grylls:

inhabitant in a relic of antiquity. He answered me very good humouredly that he was doing nothing to disfigure that its external appearance would remain unchanged but that he was fitting it up internally to live in it. I said I did not see how he could make it a comfortable habitation without the addition of domestic offices which would be at best an incongruous appendage. ⟨which would⟩ He said whatever addition he might make to it would be ⟨at the⟩ in the back-ground and masked by the trees

Line numbers are calculated with reference to all of the text on the page, including speech-headings, epigraphs and P.'s own footnotes.

Title page: Epigraph not in *F*.

91: *Prefatory Note* In the following pages] In this little work *F*.

93.24–5: ELECTRICAL SCIENCE . . . CONVALESCENT] Poetical Faith *MS*

93.27–8: THE FOREST DELL . . . MARRIAGE] Love and Marriage *MS*

94.2–3: LORD CURRYFIN . . . MONOTONY] A Dinner Party *MS*

95.29: THE CONQUEST OF THEBES] The Triumph of Perseverance *MS*

99.30: grayling,] *Not in F*.

99.35: returns, compelled into a circle.] returns into its own circle. *F*.

100.12: mode] modes *F*.

104.24–8: This inscription . . . nihil.] *Not in F*.

106.38: godfather] godfather as well as the uncle *MS*

106.39: had had] had *MS*

107.3: beautiful] happy *MS*.

107.3–4: and exercised . . . men] *Not in MS*.

108.15–16: agreeably situated] comfortable *F*.

112.21–2: a velleity towards German; but I never had any.] any velleity towards German, which I never had. *F*.

113.35: sixteen and seventeen] fifteen and sixteen *F*.

118.4: 'Hair the only grace of form,'] *Solum formae decus capilli F*.

130.9: Agapêtus and Agapêtae] 'Αγαπητὸς καὶ 'Αγαπηταὶ *MS*

130.10: the next morning at breakfast] *Not in MS*.

130.10: sense of the words:] sense of the words, in the sense in which the word 'αγαπητὴ was used by St. Paul *MS*

131.1: Agapêtê] Αγαπητὴ *MS*

131.5: where there are none] where there are neither *MS*

131.12: these damsels] female domestics *MS*

131.14–15: Their respectful deference . . . cannot be mistaken] There may be affection in both cases but there is a respectful deference in one case which is wanting in the other. I cannot be mistaken in the symptom. *MS*

131.19: THE REVEREND DOCTOR OPIMIAN I am sure I am not] I am sure I am not said the Doctor / The good wife quietly allowed the Doctor to have the last word *MS*

132.15: not to] never to *F.*

141.12–13: He then brought her to the stake, and] He doomed her to die by fire. *MS*

141.14: then ordered] ordered *MS*

141.15: permitted, and] permitted: but *MS*

141.17: at the place] on the place *MS*

141.17: Intense] Preternatural *MS*

141.19–20: the summit of the loftiest mountain] a fold on the highest mountain *MS*

141.22: Here it] Here, *MS*

141.23–4: Till, in the fulness of time . . . the shrine] till a holy man, guided by a vision, discovered it, and built a shrine by its side; and the rock-sarcophagus closed above it. *MS*

142.5: a mitred Abbot as] *Not in F.*

142.19–20: She did not again seek to obtain the ring.] *Not in F.*

142.27: in the Catholic days of England. Three] In the Catholic days of England, three *MS*

142.28: Catharine . . . built them] Katharine . . . built three chapels *MS*

142.30: these chapels] these towers *MS*

142.31–3: The sisters thought . . . ruins] They were reverenced as memorials of piety and sisterly affection till the days of the Reformation, which demonstrated by the holy text of pike and gun that both the piety and the affection were Pagan and idolatrous. Of the chapels of Saint Katharine and Saint Martha there are still some graceful ruins. *MS*

142.33–4: the chapel of St. Anne] that of Saint Anne *MS*

143.24–6: the palm . . . to her; and] the wheel, the fire, the sword, the crown, the glory, and *F.*

147.1: sure that] sure *F.*

147.2: who would] that would *F.*

148.11: occurrences] occurrence *F.*

160.2: rehearsals] production *F.*

162.7: who] of whom the components *F.*

162.28–9: Marcus Oppius and Quintus Cicero] Marius and Cicero *F.* (*P. made this correction during serial publication in a note at the end of a later instalment [vol. 62, p. 62] giving an explanatory note on this passage*:

Caius Marius the younger, in the proscription of Sulla, and Marius Cicero the younger, in the proscription of the second triumvirate, were as faithful to their fathers as the circumstances they were placed in permitted; but the most conspicuous examples of filial piety to the proscribed were those above substituted. It is also more fitting to take both examples from the proscription specially referred to by Paterculus.)

162.34–6: A compendious ... cxx: 77–80.] *Not in F.*

165.7: on the younger side of thirty] about thirty years of age *F.*

170.25: not dream] not then dream *F.*

172.38: Anthologia Palatina: Appendix: 72.] *Not in F.*

176.5: figure] figures *F.*

186.18: reform in] reform of *F.*

191.6: Dryden] Pope *F.*

192.22: had begun to apprehend] apprehended *F.*

195.28: suspended] being suspended *F.*

196.19: snatched] she snatched *F.*

196.30–1: had received] received *F.*

199.12–13: of the enchanted] the enchanted *F.*

202.12: yet] *Not in F.*

204.24: I] Part I *MS*

204.25: Sir Jamramajee] and Sir Jamsetjee *MS*

205.16: *MS adds the following stanza at the end of Part I*:

Now think, when these words in their ears have been poured,
And each hero struts off with his cross-handed sword,

Should Satan lack funds Heaven's Prince to dethrone,
These are just the three worthies to raise him a loan.

205.17: II] Part II *MS*

205.18: Now] Or *MS*

206.9: we settle it thus] though some make a fuss, *MS*

206.10: par, is the Hero] par is the question *MS*

211.33: other hand] other *F.*

214.32: carried him just] just carried him *F.*

215.7–21: We have had ... exotic belonging to education] You must consider that on the other side you have Political Economy throughout the whole period and latterly you have had a mighty mass on the Purification of the Thames. To be sure Competitive Examination belongs to Education. / I think that turns the scalc allowing the full weight of the two former ingredients. *MS*

220.14–15: still many, and among them] now *F.*

220.15: bring] give *F.*

221.24: without a box to intercept the view,] *Not in F.*

228.14: the petrifaction of an antediluvian] a petrified *F.*

238.16: the game] this game *F.*

241.12: a lady] of a lady *F.*

242.6: contented] content *F.*

257.1: sure] sure that *F.*

258.24: decidedly] so decidedly *F.*

261.30: loved] had loved *F.*

262.2–3: I regret that I gave in to the punctilio: but] Yet, *F.*

262.3: the idea] this idea *F.*

263.17–19: seats; which, however ... stalls.] seats. *F.*

267.40: See Chapter XV., page 185.] *Not in F.*

273.33–4: would have] *Not in F.*

276.14: *richesses*] *richesse F.*

276.33: deceitfully allured.] deceived. *F.*

276.36: alluringly] *Not in F.*

278.16–17: This is ... fiction.] *Not in F.*

281.35–282.1: sorry, for your sake, and his,] sorry *F.*

282.36: exactly] *Not in F.*

282.36–283.1: implied it. You have] *Not in F.*

283.22: seemed] seemed to be *F.*

288.1–5: *Epigraph to ch. 31 heads ch. 32 in F. and vice versa.*

289.38–290.1: the ghost-trade . . . becoming tangible, and] ghosts *F.*

290.6: boilers] engines *F.*

296.20: Morgana, but] Morgana. But *F.*

297.2[illegible]–33: merry . . . sorry . . . merry] mery . . . sory . . . mery *F.*, *which follows the Shakespeare Society text referred to by P. in his footnote.*

300.1: host, who] host, the latter, who *F.*

302.19–21: game. They were . . . and the Doctor] game; and when they were presented, after the usual music in the drawing room, the Doctor *F.*

304.33: is not] isn't *F.*

305.16–17: We'll settle . . . lawyers say] *Not in MS.*

305.19–20: You are a thoroughly good fellow . . . choice; but that] That *MS*

305.25: arduous] difficult *MS*

307.19: knew] know *F.*

311.30: Generally,] *Not in F.*

314.10: rare] choice *F.*

315.24: mortal] living *F.*

321.23: Mr. Niphet,] *Not in F.*

321.27: None of the bridegrooms had brothers, and] *Not in F.*

322.29–31: Let all attendants . . . string.] *Not in F.*

EXPLANATORY NOTES

THE following notes are intended to explain all but the most obvious allusions and references in *Headlong Hall* and *Gryll Grange*. We owe much information in the *Headlong Hall* notes to N. A. Joukovsky's unpublished thesis (see Note on the Texts). 'J.' indicates places where we are particularly indebted to this work. We have also drawn on David Garnett's notes to his edn. of P.'s novels. Unless otherwise stated, all translations from classical authors are taken from editions in the Loeb Classical Library ('Loeb'). 'Halliford' refers to the 10-vol. Halliford edition of P.'s works.

HEADLONG HALL

xli [*Title page*] *nature to submit*: Swift's *Cadenus and Vanessa*, ll. 722–5, with 'All philosophers' for 'Or, as philosophers'.

1 *Holyhead mail*: the Holyhead Mail coach ran from London to Holyhead in Anglesey along a route close to what is now the A5, the principal route of communication with North Wales and Ireland. In 1815 the journey to Capel Curig took about 36 hours, the coach leaving London at 7 p.m and arriving early on the second morning (see John Carey's *New Itinerary*, 6th edn., 1815) (J.). The coachman and horses were changed regularly *en route*, and at each stage the passengers would be expected to tip the retiring coachman ('remember the coachman').

Boreas: the north wind.

Llanberris: Headlong Hall is imagined by P. as located somewhere near the modern town of Llanberis, under the northern face of the Snowdon range, in the county of Gwynedd. The landscape in the vale of Llanberis, by the shores of Llyn Peris and Llyn Padarn, is in fact far too mountainous to be susceptible to 'polishing and trimming' in the manner Mr Milestone proposes in ch. 3 (see note to p. 11, below).

Cadwallader: the last of the ancient kings of Wales; died at Rome in AD 664 while on a pilgrimage. It was prophesied that the Welsh would regain the sovereignty of Britain when Cadwallader's bones should be brought back to Wales. (*The Chronicle of the Kings of Britain*, tr. P. Roberts (1811), pp. 188–9.) See ch. 9 of *H.H.*

deluge . . . Snowdon: Welsh folk-history records the flooding of the land of Gwaelod by the sea in what is now Cardigan Bay. The survivors are said to have landed on the mountains of Snowdon. P. possibly found this information in *The Mythology and Rites of the British Druids* (1809) by Edward Davies, who believed the story to be a myth comparable to that of the flood in Genesis (pp. 242–4). P. used the story of the inundation in *The Misfortunes of Elphin*. (J.)

2 *bagsmen . . . riders*: commercial travellers.

. . . as 't was said to me: Scott's *The Lay of the Last Minstrel*, II. xxii. 15–16.

Headlong: David Garnett identified Headlong with Thomas Johnes (1748–1816), Lord Lieutenant of Cardiganshire, cousin of Richard Payne Knight and friend of Uvedale Price (see notes to pp. 10 and 13 below). Johnes' Gothic mansion at Hafod in Cardiganshire (now Dyfed), built in 1786–8, was being rebuilt after its destruction by fire when P. visited the area in 1811. Johnes was noted for his impulsive character as well as his extravagant support for the arts and an enthusiasm for wild scenery (J). But Headlong's taste in scenery (which he shares with his prototype Humphrey Hippy in P.'s farce *The Three Doctors*) is for the 'improved' landscape of the Brown/Repton school rather than for the wildness of the Knight/Price School. See note to p. 10 below. Richard Holmes's assumption (*Shelley: The Pursuit* (1976), p. 174) that Headlong is based on William Madocks (for whom see note to p. 41, below) appears to have no foundation.

Menander: the quoted phrase, meaning 'but also something more' (suggesting, in the context, sexual adventures), comes from a fragment of the lost play *Kybernetai* (Pilots) by Menander (*c*.343–*c*.290 BC), the best known of the writers of Greek New Comedy. (Loeb, pp. 396–7.)

beating up: i.e. searching. A term taken from hunting, and thence recruiting for the armed forces.

3 [*P.'s note*] *Foster . . . Escot . . . Jenkison*: Φωστηρ means 'star' or 'splendour'. 'Foster' has, of course, a quite different, Germanic, etymology, Ες σκοτον and *in tenebras* both mean 'into darkness'; *intuens* 'considering' or 'watching'. P.'s interpretation, 'looking into the dark side of the picture', is jocular. αιεν εξ ισων and *semper ex aequalibus* means 'always from the same thing': P.'s explanation is far-fetched. P.'s delight in

facetious etymology is evident in the character-names in his other works, especially *Nightmare Abbey* and *Melincourt*, as well as in his neologisms such as 'cataballitive' (see note to p. 81 below) and in the introductory part of Cranium's lecture (ch. 11).

[*P.'s note*] *et praeterea nihil*: i.e. 'a stomach, and nothing else'. A parody of the proverbial phrase 'Vox et praeterea nihil' ('a voice and nothing else'), first recorded in Plutarch's *Moralia* (first century AD). There is a 'Messere Gaster' in Rabelais's *Gargantua and Pantagruel* (bk. iv, chs. 57–62).

illuminati: 'persons affecting or claiming to possess special knowledge or enlightenment' (*OED*). P. further satirizes such pretension in *Nightmare Abbey*, ch. 2, where he refers more specifically to exclusive or secret societies, such as Spartacus Weishaupt's group, founded in 1776 at Ingolstadt (the setting of Mary Shelley's *Frankenstein*), who called themselves 'Perfectibilists' (*Perfektibilisten*) (see next note).

unlimited perfection: the doctrine of the indefinite perfectibility of mankind is found in William Godwin's *Enquiry Concerning Political Justice* (bk. i, ch. 5 in 2nd edn, 1796) and in Antoine-Nicolas de Condorcet's *Esquisse d'un Tableau Historique des Progrès de l'Esprit Humain* (1795; translated into English in the same year), chapter entitled 'Dixième époque'. It was enthusiastically embraced by, among others, the young Shelley (see note on 'coincide with the equator' to p. 40 below). Both philosophers foresaw an era in which human affairs would be governed by universal enlightenment through reason, but perfectibility was an unending process: Godwin writes of 'perpetual improvement' and explains: 'The term perfectible ... not only does not imply the capacity of being brought to perfection, but ... stands in express opposition to it. If we could arrive at perfection, there would be an end of our improvement.' (J.) And Condorcet specifies that perfectibility is 'indefinite in relation to us, if we cannot fix the limit it always approaches without ever reaching' (tr. J. Barraclough (1955), p. 200). Foster's oxymoron 'unlimited perfection' may be a deliberate joke by P., though Condorcet uses the phrase 'indefinite perfection' (p. 143). The argument for perfectibility was rebutted by, among others, Thomas Malthus in *An Essay on the Principle of Population, as it affects the future improvement of Society: with Remarks on the Speculations of Mr. Godwin, M.*

Condorcet, and Other Writers (1798): many of Escot's remarks in this chapter are along Malthusian lines. See also introduction.

4 *imbecility and vileness*: this passage derives from several in *Antient Metaphysics* (6 vols., 1779–99) by the Scottish judge and man of letters James Burnet, Lord Monboddo (1714–99). See vol. v, bk. 4 and vol. iii, bk. 2, *passim*. A relevant passage is quoted by P. in a footnote to *Melincourt*, ch. 25: 'The necessary consequence of men living in so unnatural a way, with respect both to houses, clothes, and diet, and continuing to live so for many generations, each generation adding to the vices, diseases, and weaknesses, produced by the unnatural life of the preceding, is that they must gradually decline in strength, health, and longevity, till at last the race dies out . . .'

toto coelo: i.e. completely.

5 *myrmidon*: i.e. assistant. The word is used of Achilles's followers, who accompanied him to the Trojan war (*Iliad*).

Silenus: the attendant of Bacchus. He is generally representated as 'a fat and jolly old man . . . crowned with flowers, and always intoxicated' (Lemprière's *Classical Dictionary*).

Achates: i.e. a loyal follower. Achates was the friend of Aeneas, and the phrase *fidus Achates* ('loyal Achates') became proverbial. See, e.g., *Aeneid*, i. 312–13.

7 *Prometheus*: Escot's interpretation of the myth of Prometheus (who brought fire to mankind, and was chained perpetually to a mountain, where vultures fed on his liver, for witholding a secret from Jupiter) was first expounded by Shelley's friend J. F. Newton in *The Return to Nature, or A Defence of the Vegetable Regimen* (1811) and taken up by Shelley in his *A Vindication of Natural Diet* (1813), part of which became a note to his poem *Queen Mab*: 'Prometheus (who represents the human race) . . . applied fire to culinary purposes . . . From this moment his vitals were devoured by the vulture of disease' (Oxford Standard Authors edn., pp. 826–7) (J.).

and lamentably less: the phrase derives from Matthew Prior's poem *Henry and Emma* (1709): 'Fine by degrees, and beautifully less' (l. 323). The version P. quotes here was a popular catch-phrase in the 19th century (see *Notes and Queries*, Series I, iii. 105 and 154).

the face of the earth: P. is again making use of Monboddo's *Antient Metaphysics*, vol. ii, bk. 2, *passim*. Monboddo further believed (vol. v) that the human race is destined either to die a lingering death or to be consumed in the cataclysm described in Revelations. Passages from his chapters on the diminution of the human race are quoted by P. in footnotes to *Melincourt*, ch. 19. See note to p. 4 above and Introduction.

Aeschylus and Virgil: Aeschylus, *Prometheus Bound*, ll. 108–10; 254–6 (Loeb, i. 227; 229). Virgil, *Georgics*, i. 84–93; 129–35.

8 *ends of their creation*: the notion that one species is expressly made to be eaten by another is ridiculed in J. Ritson's *An Essay on Abstinence from Animal Food as a Moral Duty* (1802): 'The Lamb is no more intended to be devour'd by the wolf, than the man by the tyger . . .' (p. 232); and in Shelley's *A Refutation of Deism* (1814) (J.). For Ritson see note to p. 173 below. Voltaire used the same argument in his *Dictionnaire Philosophique* (1764): 'No doubt sheep were not absolutely made to be eaten, since several nations abstain from this horror' (tr. T. Besterman (1972), p. 206).

final causes: i.e. ultimate purposes.

Lotophagi: i.e. lotus-eaters (*Odyssey*, ix), whose diet induced them to disdain action and the sense of purpose.

[*P.'s note*] *Emmerton on the Auricula*: in his *A Plain and Practical Treatise on the Culture and Management of the Auricula* (1815) Isaac Emmerton notes that the auricula (a species of primula also known as the 'bear's-ear') requires a rich soil, and he gives several recipes for appropriate composts, including goose-dung, bullock's blood, sugar-baker's scum, night-soil, urine, as well as sand, loam, and cow-dung (pp. 56 ff.).

9 *Hindoos and the ancient Greeks*: Monboddo noted with disappointment (*Antient Metaphysics*, v. 237) that the physical stature of Hindus did not support his theory that vegetarianism leads to increased height.

frugivorous animals: a close paraphrase of Rousseau's remark in *Discours sur l'origine de l'inégalité parmi les hommes* (1755), Note E, translated in Ritson's *An Essay on Abstinence from Animal Food* (1802): 'It seems, therefore, that, the teeth and intestines of man being like those of frugivorous animals, he should, naturally, be rang'd in this class . . .' (p. 41). Cf. also Shelley's *Vindication of Natural Diet* (*Works*, vi. 7–8) (J.). The debate about vegetarianism is part of the larger discussion of man's origin: Rousseau, Newton, Shelley, and Monboddo all point out that the orang-outan, regarded as an archetype of man in the 'natural' state, is a herbivore. See Introduction and note to this page, below.

where doctors disagree: cf. Pope, *Moral Essays*, iii, l. 1: 'Who shall decide, when Doctors disagree . . .?'

Jehu: noted for his furious driving of a chariot (see 1 Kings 19: 20).

wild man of the woods: i.e. the orang-outan; so defined in Rees's *Cyclopedia*, iii (1788), which records fanciful reports of civilized behaviour in orang-outans (e.g. using a napkin at the dinner-table) disseminated by Buffon in his *Histoire Naturelle* (1752). See P.'s *Melincourt*, *passim*, and the sections of Monboddo's *Antient Metaphysics* referred to above (notes to pp. 4 and 7).

10 *confusion thrice confounded*: a play on Milton's phrase 'confusion worse confounded' (*Paradise Lost*, ii. 996), describing the fall of Satan and his confederates.

peautiful tamsel: P.'s representation of Welsh accents is perhaps modelled on Fluellen in Shakespeare's *Henry V*.

[*P.'s note*] *can confer*: the text P. quotes in this footnote (added in the 2nd edn. of *H.H.*) is from the preface to the second edn. (1795) of Richard Payne Knight's (1750–1824) 'didactic poem' *The Landscape*; and the 'celebrated *improver*' is Humphry Repton (1752–1818) who, as a follower of Lancelot ('Capability') Brown (1715–83), had established himself as a fashionable improver of estates and invented for himself the term 'landscape gardener'. In his design for the remodelling of Tatton Park in Cheshire Repton had suggested that the family coat of arms might be placed on surrounding buildings and objects as a way of showing the extent of the property. Knight ridiculed this in the first edn. of *The Landscape* (1794), a poem that satirized the kind of smoothing and taming of landscape advocated by Brown and Repton. Repton replied to Knight's attack in *Sketches and Hints on Landscape Gardening* (1794). It is this reply that Knight (and subsequently P.) quotes. Milestone, whose prototype appears under the same name in P.'s farce *The Three Doctors*, is clearly intended to represent Repton, the leading exponent of the more formal, eighteenth-century school of landscape gardening. Nevertheless, in the discussion of picturesque and beautiful scenery (ch. 4) he represents not Repton, but a part of the controversy between Knight and Price which P. followed with enthusiasm and amusement when Price's works were reprinted in 1810. For Price see note to p. 13 below.

11 *pleasure-grounds*: a phrase that occurs several times in Repton's *Sketches and Hints*, e.g. ch. 7, p. 103: '... pleasure-grounds which I am frequently called upon to decorate'. Knight seized upon it in *The Landscape*: 'Till tir'd, I ask "Why this eternal round?"/And the pert gard'ner says " 'Tis pleasure-ground" ' (ii. 227–8).

Putney and Kew: in the early nineteenth century Putney was a small village situated between the Thames and an open heath. Kew Gardens, formed in 1802 out of three private domains (one of them the garden of Richmond Palace, a favourite Royal retreat) had been 'improved' in the eighteenth century, partly by Sir William Chambers and partly by 'Capability' Brown. It was much altered in the mid-nineteenth century.

capabilities: Lancelot Brown was given the nickname 'Capability' from his habit of discussing the 'capabilities' of his clients' grounds for improvement. See note to p. 10 above.

shaving and polishing: cf. 'polishing and trimming' (this page) and 'clumping and levelling' (p. 13). All these terms were used derisively of the Brown/Repton School by both Knight in *The Landscape* and *An Analytical Enquiry into the Principles of Taste* (1805), and Uvedale Price in *An Essay on the Picturesque* (1794–8).

12 *wild men, not less than ten feet high*: cf. Monboddo, *Antient Metaphysics*, iv. 51: 'it appears to me, that in all countries there have been, in very antient times, a race of wild men of extraordinary stature' (J.). Cf. iii. 146 ff. (the latter is quoted by P. in notes to ch. 19 of *Melincourt*).

Cephalis Cranium: the name derives from Greek *kephalis* ('little head') and Latin *cranium* ('the bones encasing the brain'). It refers, of course, to her father's study of craniology, the study of the configurations of heads. The subject became popular in 1814–15 largely owing to the work of Franz Joseph Gall and Johann Kaspar Spurzheim. Spurzheim lectured in London in 1814 and published a lengthy treatise, *The Physiognomical System of Drs. Gall and Spurzheim* (1815) (J.). In the same year P.'s close friend, the naturalist and astronomer, Thomas Forster (1789–1860), who had studied Gall and accompanied Spurzheim to Edinburgh, published a 'Sketch' of the 'new Anatomy and Physiology of the Brain' in *The Pamphleteer* (no. 5) introducing the term 'phrenology' by which this new 'science' became generally known. Opinion on the subject was dramatically divided. The sober scientific *Philosophical Magazine* reported Spurzheim's lectures extensively and without critical comment (xliv. 71, 215f., 305 ff., 370 ff.), whereas the *Edinburgh Review* a year later denounced the book as 'a piece of thorough quackery from beginning to end' (xxv. 227 ff.) and noted that 'Great Britain is a field for quacks to fatten in; they flock to it from all quarters of the world'. In 1817 *The Edinburgh*'s editor Francis Jeffrey co-wrote and published in Spurzheim's name an amusing poetic satire, *The Craniad*. Spurzheim called his subject 'physiognomy', not 'craniology', but there is little doubt that Mr Cranium is partly based on him. Cranium uses the phrase 'physiognomical empiricism' (p. 61) and follows Spurzheim's practice of illustrating his lectures with a large number of skulls and of coining new words ending in '-ive' and '-iveness' (see *H.H.*, ch. 5 and *Physiognomical System* (2nd edn., 1815), pp. ix–xi, for Spurzheim's apology for the need to do this). On the other hand, Cranium's necessitarian views (see ch. 14) do not derive from Spurzheim, who specifically argues that, since our mental faculties are not equally active at any particular moment, we are not completely at the mercy of our cranial 'organs' (2nd edn., pp. 495 ff.). Cranium's view is distinctly Godwinian, and it is therefore clear that P. is using him for more than one satiric purpose. Nevertheless, craniology was popularly supposed to imply a determinist philosophy, and such a philosophy is ascribed to Spurzheim in *The Craniad*, pt. ii, pp. 76 ff.

[*P.'s note*] ROUSSEAU, *Emile, liv. 5*: 'It is certain that they kiss each other more affectionately and caress each other more gracefully in the presence of men, for they are proud to be able to rouse their envy without danger to themselves by the sight of favours which they know will arouse that envy' (tr. B. Foxley (1910), p. 339).

13 *insides*: passengers travelling inside the vehicle. A post-chaise normally had two seats, with a third that could be folded out when necessary.

generalissimo: attempts to identify these four characters with actual writers in the *Edinburgh Review* and *Quarterly Review* have not been successful. (They are summarized by J., pp. 196 ff.) Broadly, P. is satirizing the partisan spirit of reviewing rather than any particular journal or reviewer (compare Coleridge's criticism of reviewing, *Biographia Literaria*, chs. 2, 3, and 21). Nevertheless, Gall's remark on 'unexpectedness' in ch. 3 derives, as P. notes, from an essay in the *Edinburgh*, and his name, Geoffrey, probably suggests Francis Jeffrey, the *Edinburgh*'s editor. Mac Laurel's name suggests both the *Edinburgh*, owing to its Scottishness, and the *Quarterly* because the Laureate, Robert Southey, wrote for it and, as J. remarks (p. 198), his defence of political inconsistency is similar to that of Mr Feathernest, P's caricature of Southey in *Melincourt*. Saintsbury noted that Mac Laurel's cast of mind is more *Quarterly* than *Edinburgh* (quoted by J., p. 198). See also note to p. 38 below.

Chromatic: in Chromatic P. satirizes the fashion for intricacy and ornament in music (see *H.H.*, ch. 13). P. used the name for a similar figure in *The Dilettanti*; the word 'chromatic' probably implied 'discord': see, e.g., Theodore Hook's popular farce *Music-Mad* (1808), pp. 30–3, where the hero, Sir Christopher Crotchet, complains 'I find a disturbance likely to take place—discord and chromatics'. P.'s dislike of gratuitous virtuosity is evident also in *G.G.*, ch.15, and in his essay on Bellini (Halliford, ix).

O'Prism: usually identified with Uvedale (later Sir Uvedale) Price (1747–1829), who, like Knight, opposed Repton's ideas and taste in gardening. J. argues that in most cases O'Prism voices the words of Knight, not Price, but many of O'Prism's speeches echo both writers, who used similar terminology in their strictures upon Repton. See note to p. 10, above. O'Prism is developed from Phelim O'Fir, a character in *The Three Doctors*.

Philomela Poppyseed: P. probably refers to Amelia Opie (1769–1853), best known as a writer of moralistic and sentimental fiction, such as *Father and Daughter* (1801), *Adeline Mowbray* (1804), and *Temper* (1812). They are much concerned with the dangers of imprudent marriage. See also note to p. 34 below.

Panscope: Panscope ('Panoscope' in the 1st and 2nd edns.) has been thought (e.g. by Richard Garnett in his edition of P.'s novels) to represent Coleridge, perhaps because *Melincourt*, *Nightmare Abbey*, and *Crotchet Castle* each have a distinct caricature of him in the figures Moley Mystic, Mr Flosky, and Mr

Skionar, respectively. But the similarity between Coleridge and Panscope is slight.

galvanistical: concerned with the study of electric currents, especially as used in the study of living organisms. Named after Luigi Galvani, who first observed electro-chemical action in 1792.

clumping and levelling: see note on 'shaving and polishing' to p. 11, above.

Almanach des Gourmands: written by A. B. L. Grimaud de la Reynière and published in eight parts, 1803–12. A compendium of recipes, anecdotes, essays on table manners and on the 'morale et metaphysique gourmandes'.

Rees's Cyclopaedia: Abraham Rees's *Cyclopaedia; or Universal Dictionary of Arts, Sciences and Literature* occupies 6 massive folio volumes in the edn. dated 1819 (published in 39 parts, 1802–20).

14 *Hercules*: the fragmentary statue is probably a joke on the phrase *ex pede Herculem*, meaning that you can measure the height of Hercules if you know the size of his foot (i.e. from a small sample you can judge the whole). Pythagoras is said to have made this calculation on the basis of the unusual size of Hercules's stadium at Olympia. The descriptions of the other statues are clearly jokes too: Neptune is out of his element, and Bacchus is immersed in water (cf. Headlong's horror of water, *H.H.*, ch. 8); Atlas is usually represented as bearing the heavens on his shoulders.

the rocks shall be blown up: Marilyn Butler (*Peacock Displayed*, p. 34) finds in Price's *Letter to Humphry Repton* P.'s source for Milestone's explosive activities. Price relates how Lord Clive, owner of Powys Castle, was advised by a landscape gardener to blow up a picturesque rocky promontory in front of the castle 'in order to make the whole ground smooth, and gently falling from the castle'.

Pagodas: the fashion for oriental elements in landscape gardening, recurrent in the eighteenth century, was given impetus by the publication of Sir W. Chambers's *Dissertation on Oriental Gardening* (1772). Repton notes that pagodas are suitable for large public gardens but not (as Milestone seems to think) for smaller private estates such as Knight's at Downton in Herefordshire: he talks of things being 'as incongruous and out of character as a Chinese temple from Vauxhall transported into the Vale of Downton' (*Sketches and Hints*, p. 103).

15 *physiognomy of the universe*: this speech is taken almost verbatim from *The Three Doctors*, II. i, as are many of Headlong's remarks in the preceding paragraph (Halliford, vii. 386–405).

[*P.'s note*] *Price on the picturesque*: P. refers to Price's *Essays on the Picturesque* (1810), where many of O'Prism's phrases occur, e.g.: 'A painter's eye . . . looks with indifference, if not with disgust, at the clumps, the belts . . . the eternal smoothness and sameness [of the "improved" landscape]' (i. 14); '. . . as intricacy in the disposition, and variety in the forms, the tints, and the lights and shadows of objects, are the great characteristics of picturesque scenery; so monotony and baldness, are the great defects of improved places' (i. 22-3); 'every thick unbroken mass of black [i.e. trees] . . . is a *blot*; and has the same effect on the horizon in nature, as if a dab of ink were thrown upon that of a Claude' (i. 278). But some of O'Prism's phrases are more reminiscent of Knight, e.g.: 'T' improve, adorn, and polish they profess,/But shave the goddess whom they come to dress' (*The Landscape* (1794), i. 265–6); 'the only quality in visible objects which is at all analogous to smoothness in tangible bodies is the even monotony of a billiard-table or a bowling green' (*Analytical Enquiry*, pt. i, ch. 5, sect. 14).

Tints variously broken and blended: O'Prism's phrase echoes the contention between Knight and Price over definitions of the picturesque and the beautiful. See next note.

[*P.'s note*] *Edinburgh Review*, No. XIV: 'There is a refined degree of novelty, which acts in a lively manner on the mind, and often, by sympathy, on the nerves; for which we shall venture to coin the name of unexpectedness' (*Edinburgh*, vii (1805), 310). The passage occurs in a review by Henry Hallam of Knight's *Analytical Inquiry*, and the context is a defence of Price's distinction between the picturesque and the beautiful (*An Essay on the Picturesque*) which Knight attacked as meaningless. Milestone's riposte perhaps reflects Repton's remarks (*Sketches and Hints*, Appendix, sect. 9) on 'conceits and whims, which lose their novelty after the first surprise'. In fact, Knight, as well as the *Edinburgh*, valued unexpectedness—'The best approach to ev'ry beauteous scene/Is where it's least expected or foreseen' (*The Landscape*, ii. 177–8)—whereas Repton tended to think of *contrast*, provided it was not too violent (*Sketches and Hints*, App., sect. 10).

16 *the Cyanean Symplegades*: the Symplegades ('clashing together') were two floating rock islands colliding with and rebounding off each other at the entrance to the Bosporus which joins the Mediterranean to the Black Sea; they were also known as the Cyaneae or Blue Rocks. Jason and his Argonauts in their quest for the Golden Fleece succeeded in navigating them after which the rocks became fixed, Fate having decreed that they could move no more once a ship had passed safely through them.

battle of Salamis: the naval battle (480 BC) in which the Greeks, as a result of a stratagem of Themistocles, defeated a much larger Persian fleet commanded by Xerxes. It was the decisive victory in the Graeco-Persian wars.

English seventy-four: a warship with seventy-four guns.

17 *increases in virtue*: a Godwinian view: see *Political Justice* (1793), bk. iv, ch. 4 and its appendix: 'Of the Connexion between Understanding and Virtue' (J.). Cf. also Condorcet, who writes of the moral principles 'whose flowering waits only upon the favourable influence of enlightenment and freedom' (*Esquisse*, tr. Barraclough, 1955, p. 192). See also Introduction and note on 'unlimited perfection' to p. 3 above.

ΕΡΙΗΡΕΣ ΕΤΑΙΡΟΙ: i.e. 'dear companions', a common phrase in Homer.

I will give up the field: cf. *Melincourt*, ch. 36: '. . . a modern man of science knows more than Pythagoras knew: but consider them only in relation to mental power, and what comparison remains between them?' (Halliford, ii. 381). Foster's reference to Newton possibly derives from Condorcet, *Esquisse* (tr. Barraclough, 1955), p. 196 (J.).

[*P.'s note*] *sources of evil*: Virgil, *Georgics*, iii. 527.

19 *heeltaps*: 'the liquor left in the bottom of a glass after drinking' (*OED*).

as to skylight, liberty-hall: 'skylight' or, more usually, 'daylight' means, with regard to drinking, 'a clear, visible space . . . between the rim of a wine-glass and the surface of the liquor, which must be filled up when a bumper is drunk' (*OED*). (Squire Headlong means that his guests may fill their glasses as full as they wish.)

[*P.'s note*] *Monboddo's Ancient Metaphysics*: on Monboddo see note to p. 4 above. The passage referred to here appears in iii.

82–3: 'Thus, I think, I have proved, that the Life of Man in his natural state, when he was guided by instinct and not by opinion, was in the open air ... The first step out of the air, which, I think Hamlet says rightly [II. ii], is *into the grave* (if not immediately, sooner, at least, than we should otherwise go), is that which the New Hollanders [i.e. Australian aborigines] have made,—into the hollows of trees.' ('New Holland' was the name given to Australia by Dutch explorers in the seventeenth century.)

20 *the authority of Moses*: Adam is described as speaking (naming the animals) just after his creation in ch. 2 of the Book of Genesis, supposedly written by Moses.

profound cosmogonist: J. compares P.'s note on Richard Kirwan's *Geological Essays* (1799) in his *Sir Proteus* (1814; Halliford, vi. 286): 'First, we discover him ... laboring to build a geological system, in all respects conformable to the very scientific narrative of that most enlightened astronomer and profound cosmogonist, Moses.' J. suggests that P.'s seeing Moses as an astronomer derived from a book he refers to elsewhere, Sir William Drummond's *The Oedipus Judaicus* (1811) in which certain passages from the Mosaic books of the Old Testament (i.e. the first five books) were interpreted as astronomical allegories. For Drummond see note to p. 25 below.

labyrinth of mind: P. had already used this phrase in connection with Locke in his poem *The Genius of the Thames* (1810); see Halliford, vi. 368.

Lavoisier: Antoine Laurent Lavoisier (1743–94) French scientist, regarded as the father of modern chemistry.

sacrificed to the few: J. compares Rousseau's *Emile* (1762), bk. iv: 'Toujours la multitude sera sacrifieé au petit nombre ...'.

21 *and uncharitableness*: cf. the Litany in The Book of Common Prayer: 'from envy, hatred, and malice, and all uncharitableness, Good Lord, deliver us.'

22 *the first creetics an' scholars o' the age*: J. notes that a Tory journal, *The British Critic*, in attacking Payne Knight's book, *The Progress of Civil Society* in 1796, warned him that amongst the *Critic*'s contributors were 'the ablest scholars or critics of the age'. Knight counter-attacked in his *Analytical Inquiry* (1805), making great play with the phrase 'the first critics and scholars of the age'.

stricken deer: a phrase that Cowper applies to himself in *The Task* (iii. 108).

Buz!: Empty the bottle!

23 *the pheelosophers of Edinbroo'*: this would seem to be a reference to some political economist, perhaps writing in the *Edinburgh Review* and maybe a distant follower of David Hume or Adam Smith. Mac Laurel's repeated phrase about the 'moral an' poleetical fetness o' things' suggests that P. has a very particular target in mind here, but we have not been able to identify it.

24 *Like those of the Roman republic*: in *Melincourt*, chs. 15 and 26. P. represents Mr Forester as dividing up his estate into small farms, 'specimens of that simple and natural life which approaches as nearly as the present state of things will admit to my ideas of the habits and manners of . . . the fathers of the Roman republic' (Halliford, ii. 281).

thrapple: 'The windpipe of any animal. They still retain it in the Scottish dialect' (Johnson's *Dictionary*, 1755).

a poet . . . against the people: alluding to Southey who had been Poet Laureate since 1813, having been an ardent republican in his younger days. Lampooned as 'Mr Sackbut' in *Nightmare Abbey* and as 'Mr Feathernest' in *Melincourt*.

25 *Off with your heeltaps*: Drink up!

organ of benevolence: see note on Cephalis Cranium to p. 12, above.

[*P.'s note*] *Drummond's Academical Questions*: Sir William Drummond (?1770–1828), scholar and diplomat; published a philosophical treatise, *Academical Questions* (1805). The following passage occurs on pp. 226–7: 'Let his interests change, and man will take as many colours as the cameleon; cruel or gentle, insolent or meek, mean or generous, and only constant to the love of self, and, therefore, only consistent in his aim to deceive, he is always guided by the hopes of individual advantage, or by the dread of personal punishment.'

26 *featherless biped*: Plato's definition of man. See his *Definitiones* (*Horoi*) in vol. v (415a) of *Platonis Opera*, ed. J. Burnet: 'man is a wingless, two-footed, flat-nailed creature'.

27 *little skeletons of silver*: as at Trimalchio's feast in *The Satyricon*. See note to p. 288, below.

[*P.'s note*] *Homer is proved to have been a lover of wine . . . bestows on it*: Horace, *Epistles*, I. xix, 6.

28 [*P.'s note*] *A cup of wine at hand . . . prompts*: *Odyssey*, viii. 70.

Un morne silence: a gloomy silence.

THREE TIMES THREE: formal toasts were drunk standing with three cheers thrice repeated.

29 *Aristotle, Plato . . . Thomas a-Kempis*: Mr Panscope's formidable catalogue of authorities, with its bizarre juxtapositions of philosophers, scholars, saints, monarchs, poets, sages, and miscellaneous writers from both the ancient and the modern world is presented mainly for comic effect but, as J. notes, P. does include in it a number of people who had written refutations of Rousseau's 'deteriorationist' *Discours sur les sciences et les arts* (1750), viz. Frederick the Great of Prussia (1712–86), Stanislaus I of Poland (1677–1766), Joseph Gautier (?1714–76), 'professeur de mathématiques et d'histoire', and Jean le Rond D'Alembert (1717–83), mathematician and fellow-editor of Diderot's on the *Encyclopédie*. P. delights in slipping in some fairly obscure names among the famous ones, for example, Gronovius and Hemsterhusius. The former was the name of two eminent German classical scholars, Johann (1611–71) and his son Jakob (1645–1716); and Hemsterhusius was the Dutch Tiberius Hemsterhuys (1685–1766) who revived the study of Greek in his homeland. Zimmermann was the Swiss writer, Johann Georg von Zimmermann (1728–95) whose *Uber die Einsamkeit* (1756; rev. edn. 1784/5) first appeared in English (as *Solitude*) in 1791.

those who value an authority more than a reason: J. notes that this phrase comes from Drummond's *Academical Questions*, p. 392: 'Sir Isaac Newton has indeed said . . . and I cite him for the advantage of those (and they are many) who value an authority more than a reason.'

the Encyclopedia Britannica: originally published in 3 vols. (1768–71), the latest (4th) edn. (1801–10) had been in 20.

the Monthly Review: founded in 1749, this was the oldest literary review current in 1815, thus the one with the longest run of back numbers.

the Variorum Classics: a generic term, not the title of any particular series; meaning an edition of a classical text with 'notes variorum', i.e. with notes by various scholars and critics.

Memoirs of the Academy of Inscriptions: the French Académie des Inscriptions et Belles-Lettres published 50 vols. of *Mémoires de littérature* between 1717 and 1809.

30 *disadvantage of being unintelligible*: J. compares a passage from Drummond's *Academical Questions* (p. 73): '. . . Lord Monboddo . . . thought this the best definition of motion, which has ever been given. It is a pity, then, that it should be unintelligible.'

as Dr. Johnson observed: see Boswell's *Life of Johnson*, ed. G. Birkbeck Hill, iv. 313: 'Johnson having argued for some time with a pertinacious gentleman; his opponent . . . happened to say, "I don't understand you, Sir." Upon which Johnson observed, "Sir, I have found you an argument; but I am not obliged to find you an understanding."'

31 *Music has charms to bend the knotted oak*: Congreve, *The Mourning Bride*, I. i: 'Music has charms to sooth a savage Breast/To soften Rocks, or bend a knotted Oak.'

heeltap . . . skylight . . . twilight: for 'heeltap' and 'skylight' see notes to p. 19 above. 'Twilight' is obscure in meaning as used here. J. cites a correspondent in *Notes and Queries* (1908) as offering the most plausible explanation: 'I suggest that "no skylight" means no light at the top of the glass, i.e. fill to the brim; and that "no twilight" means no half-light in the glass, i.e. drink to the dregs.'

32 ΟΜΑΔΟΣ ΚΑΙ ΔΟΥΠΟΣ ΟΡΩΡΕΙ!: 'a din and hubbub arose' (*Iliad*, ix. 573).

33 *the novels of . . . Miss Philomela Poppyseed*: see note to p. 13 above.

34 *puffed into an extensive reputation*: Mrs Opie's works were very favourably reviewed by the *Edinburgh Review*. In a review of her *Simple Tales* (July 1806) the writer commends her earlier novel *Adelina Mowbray* as 'perhaps the most pathetic, and the most natural in its pathos, of any fictitious narrative in the language'.

35 *Miss Philomela's opiate*: no doubt, as J. observes, a pun on the name Opie.

plan for Lord Littlebrain's park: P. is alluding here to Repton's famous 'red books', portfolios of views of an estate for which he was submitting landscaping proposals; movable flaps enabled the reader to see the view before and after 'improvement'. Often these drawings would contain figures standing for the owners of the estate.

en règle: in order, according to rule.

36 *Base, common, and popular*: *Henry V*, IV. i.

[*P.'s note*] *see Knight on Taste*: in his *Analytical Inquiry into the Principles of Taste* (pp. 375–7) Payne Knight, discussing attempts 'to introduce these charming delights of danger, pain, terror and astonishment, into the art of landscape gardening', gives the following example: 'Amidst some very grand scenery of woods, rocks and mountains, was a spacious and picturesque cave; which, as some improver of the school naturally conceived, only wanted a little terror to render it truly sublime. This, he easily supplied, by prevailing on the then proprietor to place a monstrous figure of a giant or cyclops over the entrance to it, with a huge stone suspended in his hand, and ready to fall upon the head of any person who should presume to enter. Not, however, calculating correctly the exact distance or degree of danger necessary to produce the desired effect, the stone actually did fall; and, coming nearer to the head of one of the spectators, than the laws of the system allow, it has brought the scheme into such disrepute among the ignorant mechanics and barbarous country gentlemen of the neighbourhood, that there is some danger of the benefit of the example being lost to the public.'

38 *Nightshade . . . Grub Street . . . determined opposition*: J. plausibly suggests that P. here alludes to Scott's 1808 break with the *Edinburgh Review* (after Jeffrey's unfavourable review of his *Marmion*) and his leading role in founding the rival *Quarterly Review* the following year.

mastigophoric: from the Greek μαστιγοφόρος, 'scourge-bearing'.

39 [*P.'s note*] *Capitolo dell' Occasione*: Machiavelli (1469–1527), best known as the author of *Il Principe* (1513), later wrote his *Capitoli*, four poetical descriptions of (respectively) Fortune, Ingratitude, Ambition, and Opportunity. The last, by far the shortest, is closely modelled on the Latin of Ausonius (fourth century AD), Epigram No. 33 (Loeb Ausonius, i. 175–6).

the morning: i.e. the whole of the day up to dinner-time (no regular meal was served between breakfast, about 10 a.m, and dinner, which would be served any time between 4 and 6 p.m).

les prémices des dépouilles: the first-fruits of the spoils.

their walk to Tremadoc: P.'s philosophers must have been truly heroic pedestrians as the journey from the Vale of Llanberis to Tremadoc and back (which they accomplished in daylight) covers a distance of upwards of 40 miles.

40 [*P.'s note*] *Fragments of a demolished world*: the Latin phrase comes from Thomas Burnet's *Telluris Theoria Sacra* (1681–9), i. 90 (he is describing the Alps) and is quoted, inaccurately, by J. Evans in his *Tour through Part of North Wales* (1800), p. 144, describing the road from Tan-y-bwlch to Beddgelert: 'The mountains, almost bare, consist of huge projecting rocks . . . and forcibly reminded us of Burnet's observations on Caernarvonshire, "That it was the fragment of a demolished world".'

convulsion . . . which destroyed the perpendicularity of the poles: until the late eighteenth century it was generally believed that changes in the earth's surface were the result of 'convulsions' or catastrophic events. One such convulsion was thought to have tilted the earth's axis to its present angle of 23½° from perpendicular to its plane of orbit around the sun. (Earlier Christian cosmologies related the convulsion to the Fall of Man: see, e.g. *Paradise Lost*, x. 668–71.) The idea that geological change happens slowly, without 'convulsion' was first argued by James Hutton (1726–97) of Edinburgh in a paper in *Transactions of the Royal Society of Edinburgh*, 1785, and in *Theory of the Earth* (2 vols., 1795): the view was popularized in John Playfair's *Illustrations of the Huttonian Theory* (1802).

coincide with the equator: Foster's view that the earth's return to 'perpendicularity' would accompany the moral and intellectual perfection of humankind was shared, as David Garnett noted, by Shelley in 1813: 'It is exceedingly probable, that this obliquity will gradually diminish, until the equator coincides with the ecliptic [the plane of the earth's path around the sun]: the nights and days will then become equal on the earth throughout the year, and probably the seasons also. There is no great extravagance in presuming that the progress of the perpendicularity of the poles may be as rapid as the progress of the intellect; and that there should be a perfect identity between the moral and physical improvement of the human species. Astronomy teaches us . . . that the poles are every year becoming more and more perpendicular to the ecliptic . . . [there is] a strong presumption that this progress is not merely an oscillation, as has been surmised by some late astronomers.' (*Queen Mab*, note to vi. 45–6). In a footnote Shelley names

Laplace as one of the 'late astronomers' whom he opposes. See next note.

41 *secular equation of a very long period*: this passage is taken from Shelley's *A Refutation of Deism* (1814): 'The learned Laplace has shown, that the approach of the Moon to the Earth and the Earth to the Sun is only a secular equation of a very long period, which has its maximum and its minimum' (J.). 'Secular' means 'proceeding so slowly that its speed cannot be measured'; and it is the relative positions of the moon, the earth and the sun, as Laplace discovered, that causes the precession of the equinoxes (The precession is cyclical with a period of about 26,000 years.) Shelley clearly changed his mind on the subject between the writing of the two pamphlets, and P. used them to represent opposing views (see Introduction). Pierre Simon Laplace (1749–1827), mathematician and astronomer, called 'the Newton of France', published his *Exposition du Système du Monde*, the work Shelley refers to, in 1796.

the embankment: across the mouth of the Traeth Mawr estuary, constructed 1808–11 by the ebullient William Alexander Madocks (1773–1828), industrialist, philanthropist, and Radical MP (see E. Beazley, *Madocks and the Wonder of Wales*, 1967). Meirionnydd and Caernarvon now form part of the modern county of Gwynedd. Shelley rented Madocks's house at Tan-yr-allt during 1812–13 and tried to raise money for repairing damage caused to the embankment by a storm but became rapidly disenchanted with Madocks and all his projects.

the triple summit of Moëlwyn: J. notices that Moëlwyn has, in fact, only two summits but P. may have considered the nearby summit of I Onicht as part of the same mountain.

Wyddfa: the highest peak of Snowdon.

42 *little town of Tremadoc*: i.e. 'Madocks's Town', the building of which (1805–11) was financed by Madocks, who also planned it in detail. The Manufactory there was the first water-powered woollen mill to be built in Wales. See Beazley, op. cit., pp. 102–7.

an adventurous fiddler: Sir John Rhys in his *Celtic Folklore: Welsh and Manx* (1901), i. 201–3, quotes a late seventeenth-century account of various North Welsh legends concerning musicians venturing deep into caves and getting lost, among them one about a fiddler called Ned Pugh or Iolo ap Huw.

43 *Rousseau ... in ... the Alps*: see Rousseau's *Rêveries du promeneur solitaire* (1782), 'Septième promenade': What Rousseau actually writes is as follows (our translation): 'My first emotion was a feeling of joy at finding myself among human beings where I had believed myself to be totally alone; but this emotion ... soon yielded to a more lasting feeling of distress at not being able, even in the caverns of the Alps, to escape from the cruel hands of those men furiously bent on persecuting me ... I made haste to suppress this gloomy notion [i.e. that the factory-workers were in a plot against him] and ended by laughing to myself over my puerile vanity and the comical manner in which I had been punished' (Rousseau, *Les Confessions. Les Rêveries*, ed. Louis Martin-Chauffier, Paris, Librairie Gallimard, (1951), p. 725).

44 FUNERE MERSIT ACERBO!: *Aeneid*, vi. 426–9: 'Immediately cries were heard. These were the loud wailing of infant souls weeping at the very entrance-way; never had they had their share of life's sweetness, for the dark day had stolen them from their mother's breasts and plunged them to a death before their time' (tr. W. F. Jackson Knight, Penguin Classics, 1956).

45 *le droit du plus fort*: the phrase comes from Rousseau's *Du Contrat Social* (1762), bk. i, ch. 3.

[*P.'s note*] *Tooke's Diversions of Purley*: John Horne Tooke (1736–1812). Radical politician and philologist. His *Diversions of Purley*, a series of dialogues on grammatical and philological subjects appeared in two parts (1786, with 2nd edn. 1798, and 1805). In Part Two (p. 408) Tooke derives the word 'Truth' from the indicative 3rd person singular of the verb 'To Trow', 'Troweth' (this is incorrect, however; 'truth' derives from Old English *treowe* 'faithful', and 'trow' from the verb, derived from adjective, *treowan*, 'to have *treow* or faith in' E. Partridge, *Origins. A Short Etymological Dictionary of Modern English*). Tooke writes (pp. 403–4): 'TRUE ... means simply and merely—that which is TROWED ... TRUTH supposes mankind: *for whom* and *by whom* alone the word is formed, and *to whom* only it is applicable. If no man, no TRUTH. There is therefore no such thing as eternal, immutable, everlasting TRUTH; unless mankind *such as they are at present*, be also eternal, immutable, and everlasting. Two persons may contradict each other, and yet both speak TRUTH: for the TRUTH of one person may be opposite to the TRUTH of another.'

47 *turn awry the current of enterprise*: echoing *Hamlet*, III. i: 'And thus the native hue of resolution/ Is sicklied o'er with the pale cast of thought,/ And enterprises of great pith and moment/ With this regard their currents turn awry/ And lose the name of action.'

[*P.'s note*] *tredici cani, &c*: 'As it was written in certain of his journals he had slaughtered with his own hands an infinite number of animals: five thousand and fifteen pheasants, six thousand hares, eighty-three boars and, by accident, also thirteen dogs'—from 'L'Arcivescovo di Praga' in the *Novelle Galanti* (1790) of Giovanni Battista Casti (1721–1803).

48 *dies nefastus*: an unlucky or inauspicious day.

abstract, and brief chronicle: *Hamlet*, II. ii: 'they [the players] are the abstracts and brief chronicles of the time.'

50 *tower on a . . . point of rock*: J. believes P. to be thinking here of a ruined tower, all that remains of Dolbardan Castle, standing at the north end of Lyn Peris. He notes that P. 'makes the descent on the lake side more or less perpendicular—it is in fact a fairly steep slope—so that Mr. Cranium can fall from the tower into the lake.'

elastic: i.e. propulsive.

51 *a view-holla*: 'shout given by a huntsman on seeing a fox break cover' (*OED*).

52 *against Plato and Lucretius*: see the conclusion of Plato's *Republic*, v, in which Socrates distinguishes between lovers of 'tones and beautiful colours and the like' and true philosophers who love only wisdom; and the famous conclusion of bk. iv of Lucretius's *De Rerum Natura* (ll. 1058–end), often called his 'tirade against love' but which, as F. O. Copley points out, in a note to his translation (New York, 1977), 'is rather an attempt to analyze and explain, in Epicurean terms, the phenomenon of sex in all its aspects . . . and above all the dangers that sex poses for the Epicurean ideal of pleasure . . .'.

like Achilles or Orlando: at the opening of bk. xxiv of *The Iliad* Achilles is described as unable to sleep because of his grief for Patroclus: 'Thinking thereon he would shed big tears, lying now upon his side, now upon his back, and now upon his face.' In *Orlando Furioso* (for which see note to p. 154, below) Orlando, having learned of his beloved Angelica's marriage to

Medor 'sighed and moaned and made great circular sweeps of the bed with his arms; it felt harder than rock; it stung worse than a bed of nettles' (tr. G. Waldman, 1974).

uni-multiplex: P.'s coinage.

Ladurlad in the curse of Kehama: see Southey's *The Curse of Kehama* (1810), sect. ii, st. 14, ll. 19–22: 'Thou shalt live in thy pain,/ While Kehama shall reign,/ With a fire in thy heart/ And a fire in thy brain.' Ladurlad, a peasant, is thus cursed after killing the son of Kehama, the Great Raja of the World, to protect his own daughter.

like the south over a bank of violets: *Twelfth Night*, I. i: 'O, it came o'er my ears like the sweet sound,/ That breathes upon a bank of violets,/ Stealing and giving odour!' Pope's emendation of 'south' (i.e. the south wind) for 'sound' was still generally accepted in the early nineteenth century.

'forgotten what the inside of a church was made of': *1 Henry IV*, III, iii.

featherless biped: see note to p. 26, above.

53 *'arrayed in her bridal apparel of white'*: from M. G. Lewis's 'Alonzo the Brave and the Fair Imogine' (stanza 16) in *The Monk* (1796), iii, ch. 2. Imogine is claimed by her dead lover Alonzo immediately after her marriage to a rich rival.

[P.'s note] and ten thousand times miserable!: Aristophanes, *Plutus*, ll. 850–2.

pensa que fut un diableteau: from the title of ch. 67 of 'The Fourth Book of the Noble Pantagruel' by Rabelais: 'Comment Panurge, par male paour, se conchia et du grand chat Rodilardus pensoit que feust un Diableteau'—'Panurge shits himself out of utter fear, and of the large cat Rodilardus, which he took for a Devil.'

54 *Hugh Llwyd's pulpit*: '[Hugh Lloyd's Pulpit] consists of an insulated rock of the grey shaly sort, standing erect in the centre of the river [Cynfael], forming a quadrangular pier. This by tradition is called Pwlpit Huw Lloyd, or the Pulpit of Hugh Llwyd; a singular character and poet, who, about the year 1600, lived at a house, yet standing, on the banks of this river. It is reported of him that, at different times, when particularly warmed with poetic fervour, he would ascend the rostrum to arrange his ideas, and commit them in proper order to paper; which, having done, he would recite them with great pathos and effect; and the sound of his voice was heard to

vibrate at a distance, along the hollow aquaeous amphitheatre. None ever attempted or dared to approach him, on account of the difficulty of descending. He was also believed to be skilled in magic' ([Edwin Pugh,] *Cambria Depicta: a Tour through North Wales illustrated with Picturesque Views by a Native Artist* (1816), p. 166).

'man and boy, forty years': *Hamlet*, v. i: 'I have been sexton here, man and boy, thirty years.'

55 *the bone-house*: a charnel-house, built in churchyards as a place to deposit bones which might be thrown up when digging graves.

Cadwallader: see note to p. 1, above.

the sons of little men: this phrase comes from 'Barrathon: A Poem' by Ossian. See note to p. 57 below.

NUNC TANDEM: 'Now at last.' Mr Escot's successor, Mr Forester in *Melincourt*, also believes that 'the human species is gradually decreasing in size and strength'. Having dug up an unusually fine skeleton, he has the skull polished and set in silver as a drinking vessel.

56 *the farthing rushlight of the rascal's life*: P. gave this expression to an earlier Irish character, Phelim O'Fir, the picturesque tourist, in *The Three Doctors* (Halliford, vii. 408).

57 [*P.'s note*] *long since dead*: *Iliad*, vii. 89; xxiii. 331.

Ossian: semi-legendary Gaelic poet of the third century AD. A young Scotsman, James Macpherson (1736–96), claimed to have found manuscripts (which he would never produce for scholarly scrutiny) containing fragments of Ossian's work in the Highlands and published what he asserted to be straight translations of them in 1761, followed by a complete epic *Fingal* (1765). Dr Johnson denounced these productions as 'impudent forgeries', but their fervid descriptions of great passions and heroic deeds in the setting of Northern mists and mountains made them immensely popular and influential, especially in Europe (Goethe ranked Ossian with Hamlet, for example) and Macpherson may be regarded as one of the founding fathers of the Romantic Movement. Ossian is a major influence in P.'s first published volume *Palmyra, and Other Poems* (1806)—see Halliford, I. xxxiv. P.'s quotation here comes from 'Barrathon: A Poem' (*The Poems of Ossian*, new edn., 1773, ii. 196 and 207).

these degenerate days: *The Iliad of Homer*, tr. Pope, v. 372; xii. 540.

58 *I am inclined . . . deplorably bad one*: P. may be indebted in this paragraph to Malthus's rebuttal of Condorcet's 'perfectibilianism': 'On the contrary, a candid investigation of these subjects . . . may have a tendency to convince them [perfectibilians] that in forming improbable and unfounded hypotheses, so far from enlarging the bounds of human science, they are contracting it; so far from promoting the improvement of the human mind, they are obstructing it . . .' (*Essay on the Principle of Population*, 2nd edn, 1803, p. 364).

59 *vexation of spirit*: Ecclesiastes, 1: 2, 14; 2: 17; 7: 8.

dies albâ cretâ notandos: days marked with a white stone, i.e. happy or fortunate days—an allusion to the ancient custom of marking such days on the calendar with chalk.

60 *whiskey*: 'A kind of light two-wheeled one-horse carriage, used in England and America in the late 18th and early 19th c.' (*OED*).

the Menai: the famous suspension bridge, designed by Telford, carrying the Holyhead to Bangor road over the Menai Straits was not opened until 1826.

[*P.'s note*] *Rabelais*: bk. v, ch. 21 (p. 650 of *Gargantua and Pantagruel*, tr. J. M. Cohen, Penguin, 1955). Pantagruel and his companions behold all kinds of miracle-workers in the Kingdom of the Quintessence, including one who 'by a magnificent contrivance threw houses out of windows, and thus purged them of pestilent air'.

Shenkin: Rice ap Shinken in Thomas D'Urfey's comedy *The Richmond Heiress* (1693) sings a song, 'Of Noble Race was Shenkin', accompanying himself on a harp and using a traditional Welsh melody.

Hoel and Cyveilioc: the warrior-bards, Hywel ab Owain Gwynedd (died 1170), Prince of Gwynedd, and Owain Cyfeiliog (*c.*1130–97), Prince of Powys.

Voulez vous danser, Mademoiselle?: a popular dance tune included in Thames Wilson's *A Companion to the Ball Room* (1816).

Comus and Momus: Milton represents Comus as the son of Bacchus and Circe. Momus was one of the children of Night, who features in Lucian's satires as a lampooner of the gods; he is represented as either a young or an old man wearing a mask and carrying a fool's bauble. J. quotes a letter to T. J. Hogg of 7 Sept. 1818, in which P. calls Rabelais 'the court fool of Olympus, the chief jester of Jupiter, a genuine incarnation of Momus'.

tipsy dance and jollity: Milton, *Comus*, l. 104.

negus: a flavoured and sweetened mixture of wine and hot water.

61 *inexcussibly*: not to be dislodged.

osteo ... muelos; osseo ... medullary: P. compounds these words from, respectively, the Greek and Latin terms for bone, flesh, blood, internal organs, cartilage, nerve, and marrow.

desiderative ... congeries of ideas: Mr Cranium is here wallowing in phrenological jargon. See note on Cephalis Cranium to p. 12, above.

63 *featherless biped*: Plato's definition—see note to p. 26, above.

an animal which forms opinions: J. suggests P. derives this definition from Drummond's preface to his *Academical Questions* (see note to p. 25, above). 'Polybius has defined man to be an animal that forms opinions ...'

again, man has been defined to be an animal that carries a stick ... and lords in waiting: this sentence was added in the 1822 edn. of *H.H.* It alludes to Monboddo's argument that the orang-outan's use of a stick was one evidence of the creature's humanity (*Of the Origin and Progress of Language* (2nd edn., Edinburgh, 1774), i. 290; for Monboddo see note to p. 4, above). P. had quoted the passage in *Melincourt* (1817), in which novel he introduces the splendid Sir Oran Haut-Ton, a civilized orang-outan who, though unable to speak, is by far the noblest character in the book (Halliford, ii. 202)).

64 *osseous compages of the occiput and sinciput*: i.e. the bony container of the front and back of the head, the skull.

In all animals but man, the same organ is equally developed in every individual of the species: J. notes this passage is based on pp. 19–20 of *Dr. F.J. Gall's System of the Functions of the Brain, Extracted from Charles Augustus Blode's Account of Dr. Gall's Lectures ... at Dresden* (*c.*1807): 'We are taught by natural history, that certain faculties and inclinations are so permanently inherent in every species of animals, that they are found in every individual of the species ... The rapacity and cruelty of the lion and the tiger, the skill of the beaver, the address of the elephant are found again in every individual lion, tiger, beaver and elephant, only sometimes with some little alteration owing to fortuitous circumstances.'

preparations of hemp: the rope used for hangings was made of hemp.

sung to a villanous tune: see Massinger's *The Bondman* (1634), v. iii: 'GRACCULO: You may see/ We are prepared for hanging, and confess/ We have deserved it: our most humble suit is,/ We may not be twice executed./ TIMOLEN: Twice!/ How mean'st thou?/ GRACCULO: At the gallows first, and after in a ballad/ Sung to some villainous tune.'

65 *I would advise every parent . . . the skull of his son*: J. compares the following passage from P.'s friend Thomas Forster's 'Essay on the Application of the Organology of the Brain to Education' in *The Pamphleteer* for May 1815 (v. 483): 'With regard to the intellectual faculties, this science will be of the highest utility. A father may discover by the prominences of his son's head his particular fort; and may chalk out for him an appropriate profession.'

66 *rescission*: a cutting off.

and vile reproach: *Henry V*, III. vi.

Meirion, the pre-eminent in loveliness: J. cites a note by P. to his poem *The Philosophy of Melancholy* (1812; Halliford, vi. 232): 'The Welch have a very pleasing ballad, *Morwynnion glan Meirionydd* ('The Fair Maids of Merionethshire'), which assigns with strict poetical justice, the palm of female loveliness to the young ladies of that most picturesque and beautiful county.'

Caractacus: famous king of the Silures, inhabitants of south-east Wales, taken as a prisoner to Rome, 51 AD.

Llugwy: 1st and 2nd edns. read 'Conwy'.

67 *vale of Edeirnion*: 1st and 2nd edns. have 'vale of Llwyd', probably an error for 'Clwyd'. A celebrated vale in Denbighshire, now part of the county of Clwyd.

68 *elixion and assation*: boiling or stewing and roasting or baking.

slakes his thirst in the mountain-stream: J. notes the echo here of Shelley's *Vindication of Natural Diet* (1813): 'How many thousands have become murderers and robbers . . . from the use of fermented liquors; who, had they slaked their thirst only at the mountain stream, would have lived but to diffuse the happiness of their own unperverted feelings.'

συμμισγεται τῃ επιτυχουσῃ: 'mates with the first comer'. No source for this Greek has been traced. P. may simply be using what Gibbon called 'the decent obscurity of a learned language' to express the notion of sexual promiscuity.

statue of Condillac: Etienne Bonnot de Condillac (1715–80), French philosopher. In his best-known work, *Traité des Sensations* (1754), he illustrates his theory of knowledge by supposing the human mind can be likened to a statue, devoid of innate ideas and even of the faculties of 'reflection' and 'inner perception' proposed by Locke in *An Essay Concerning Human Understanding* (1690), but endowed with the capacity for sensation of all kinds. Like the British philosopher David Hartley (1704–57), whose *Observations on Man* (1749) so influenced Coleridge and Wordsworth, Condillac argued that all knowledge is derived from sensation.

70 *'with twenty mortal murders on his crown'*: *Macbeth*, III, iv.

blue devils: figurative expression which became current in the late eighteenth century for fits of depression or despondency.

71 *'the dreary intercourse of daily life'*: Wordsworth, 'Lines Composed a Few Miles Above Tintern Abbey' (1798), l. 131.

'small talk': the first record of this phrase in *OED* is dated 1751 (in Chesterfield's Letters to his Son).

72 [*P's note*] *Rousseau, Discours sur les Sciences*: see Rousseau's *Discours sur les Sciences et les Arts* (1750), ed. G. R. Havens (1946), pp. 102–3. P. somewhat adapts the passage, which may be literally translated as follows: 'You civilized people, cultivate them [the arts and sciences]: happy slaves, you owe to them that fine and delicate style on which you pride yourselves; that sweetness of character and that urbanity of manner which makes intercourse among you so pliant and easy; in a word, the appearance of all the virtues without having any of them.'

Miss Brindle-mew Grimalkin Phoebe Tabitha Ap-Headlong: 'Brindle-mew' was doubtless suggested by *Macbeth*, IV, i: 'Thrice the brinded cat hath mew'd'. 'Grimalkin' was a nickname for an old she-cat; Phoebe is Diana, goddess of chastity; and 'Tabitha' recalls 'tabby', a slang term for an old maid.

73 *'Now, when they had eaten and were satisfied'*: a frequently recurring line in Homer (*Iliad*, i, 469 and *passim; Odyssey*, i, 150 and *passim*).

74 *[P's note] imitated from . . . Dante*: see *Purgatorio*, Canto 8, ll. 1–6.

soond is . . . an aicho to sense: cf. Pope's *Essay on Criticism*, l. 365: 'The Sound must seem an Echo to the Sense.'

75 *the lug*: the ear (Scottish).

[BALLAD]: J. notes that this ballad is founded on a local legend about Huw Llwyd (see note to p. 54, above) recorded in Edward Jones's *Musical and Poetical Relicks of the Welsh Bards* (1794), p. 78, and William Bingley's *North Wales* (1804), i. 460–2.

76 [CHORUS]: 'Pistyll' means a waterspout.

78 *Llewelyn Ap Yorwerth*: Llewellyn the Great (1173–1240), Prince of Gwynedd, who established an ascendancy over the other Welsh states.

Nidd-y-Gygfraen: i.e. 'The Raven's Nest'. J. notes that one of P.'s haunts near Maentwrog was called Llyn-y-Gygfraen, 'the Raven's Pool'.

79 *'Music has charms'*: see note to p. 31, above.

a penchant: cf. P.'s *The Dilettanti*, II. v (Halliford, vii. 381): 'METAPHOR: . . . I believe Miss Comfit has a little penchant for me. O'PROMPT: (*aside*) A *penchant*! I suppose that is the dilet-tantte word for *sneaking kindness*.'

80 *to throw off*: to begin hunting; (*figuratively*) to make a beginning to anything (*OED*).

whipper-in: 'A huntsman's assistant who keeps the hounds from straying by driving them back with the whip into the main body of the pack' (*OED*).

81 *compages*: shell, i.e. body.

caballitive: P.'s coinage from the Greek καταβαλλω, 'to cast down'.

82 *Quod victus fateare necesse est*: 'Since you are defeated you must confess', Lucretius, *De Rerum Naturae*, i., 623–4 (omitting a phrase between 'quod' and 'victus').

the eternal fitness of things: this phrase was, according to the *OED*, 'extensively used in the 18th c. with reference to the ethical theory of [Samuel Clark] [1675–1729] in which the quality of moral righteousness is defined as consisting in a "fitness" to the relations inherent in the nature of things' (J.).

83 [*P.'s note*] *Jeremy Taylor*: (1613–67), chaplain to Charles I and later Bishop of Down and Connor. The quotation is from his *The Rule and Exercises of Holy Dying* (1651), ch. 1, sect. 5: 'He that is no fool, but can consider wisely, if he be in love with this world, we need not despair but that a witty man may reconcile him with tortures, and make him think charitably of the rack.'

84 *a post-captain*: a naval officer holding the commissioned rank of captain as distinct from one accorded the courtesy title of 'captain' through being in command of a ship.

half-seas-over: i.e. half-drunk.

spiritual metamorphosis of eight into four: alluding to the Christian doctrine that a man and a woman become 'one flesh' through the sacrament of marriage.

85 *humani nihil alienum puto*: 'I think nothing human alien to me.' See P.'s note to p. 234, above.

86 *Young men . . . are driven . . . from the company of the most amiable and modest of the opposite sex*: J. compares Shelley's note to *Queen Mab* (1813), v. 189: 'Young men, excluded by the fanatical idea of chastity from the society of modest and accomplished women, associate with these vicious and miserable beings [prostitutes], destroying thereby all those exquisite and delicate sensibilities whose existence cold-hearted worldlings have denied . . .'

Nessus or Medea: Nessus was a centaur slain by Hercules for trying to rape his wife, Deianira. The dying centaur gave her some of his blood telling her it was a strong love-charm; years later, to win back Hercules' love, she smeared this blood on a garment of his as a result of which he died in agony (see Sophocles' *Trachiniae*, ll. 531–812). In Euripides' *Medea*, Medea sends poisoned garments as wedding gifts to the daughter of Creon, king of Corinth, who is supplanting her as Jason's wife. The bride's horrible death is described in ll. 1168–1203.

87 [*P.'s note*] *goes to the devil*: *Iliad*, xxii. 212–13.

89 [*Title-page*] *Butler*: The first two lines are adapted from ll. 1–4 of Butler's poem 'Opinion' (*Satires and Miscellanies*, ed. R. Lamar, 1928): 'Who does not know with what fierce Rage/ Opinions Tru, or False ingage?/ And 'cause they Govern all Mankind,/Like the Blindes leading of the Blinde ...'. The other four lines are from *Hudibras*, pt. iii, canto 2, ll. 815–18.

91 *as if it were still unenclosed*: the New Forest, an area of 92,365 acres, lying mainly in the county of Hampshire, was formed as a Royal Hunting Forest by William the Conqueror in 1079. In 1851 the so-called Deer Removal Act empowered the enclosure by the Crown of a total of 10,000 acres at any given time in exchange for the sovereign's giving up of the right to keep deer in the Forest. Six thousand acres were already subject to enclosure as a result of earlier Acts of Parliament. The provisions of the 1851 Act gave rise 'to so much bitterness locally, and to such an outburst of aesthetic feeling generally that work was stopped when about 5,000 acres had been planted up' (J. Nisbet and G. W. Lascelles in *The Victoria History of the Counties of England: Hampshire*, ii, 433).

97 [*Epigraph*] PETRONIUS ARBITER: Petronius was one of Nero's courtiers whose correct judgement on matters of taste and style earned him the title of 'Arbiter of Elegance'. He is traditionally identified with the author of *The Satyricon*, from which the quotation comes (sect. 99, ll. 1–2).

Opimian: the name of a celebrated ancient Roman wine, of the vintage of the year 121 BC when Opimius was Consul. Cf. epigraph to ch. 3 below.

of subsequent introduction: the Jerusalem artichoke was introduced into Europe from tropical America in 1617; first grown in Rome, it was called *Girasole* (or sunflower) *Articiocco* (*OED*).

me judice: in my judgement.

swindling bankers: defaulting bankers were a long-standing object of P.'s satirical invective—cf. *Melincourt*, ch. 30 and the absconding Mr Touchandgo and his clerk, Roderick Robthetill, in *Crotchet Castle* chs. 1 and 11—and an integral part of his campaign against paper money (see note on 'Restriction Act of Ninety-seven' to p. 214, below). Bank stoppages as a result of fraud or incompetence were common in the nineteenth century. J. H. Clapham notes that there were 206 bank failures between 1815 and 1830 (*An Economic History of Modern Britain ... 1820–1850* (1930), p. 266); one of the most dramatic collapses was that of the great London bank of Overend, Gurney, in 1866, shortly after *G.G.* was published.

free and independent constituency: P. had already satirized the bribery and corruption rife in Parliamentary elections before the 1832 Reform Act in *Melincourt* chs. 21 and 22 in which the noble ape, Sir Oran Haut-ton, is elected by Christopher Corporate to serve as one of the two MPs for the Borough of Onevote. In his 1856 Preface to the reissue of the novel P. noted: 'The boroughs of Onevote and Threevotes have been extinguished but there remain boroughs of Fewvotes in which Sir Oran Haut-ton might still find a free and enlightened constituency.' Agitation to extend the franchise eventually resulted in the Second Reform Bill (1867) but voting by secret ballot did not become law until 1872.

Conservative: the allusion is to Sir Robert Peel (1788–1850), whose famous 'Tamworth Manifesto' (1835) substituted the name 'Conservative' for 'Tory' as the appropriate label for the main right-wing party in British politics. By traditional squires, such as Mr Gryll is supposed to be, Peel, because of his volte-face over the Corn Laws in 1846, was seen as betraying true Tory principles.

à nil conservando: by conserving nothing.

sanitary improvements: in 1847 the Metropolitan Commission of Sewers ordered that all cesspits must be abolished and some 200,000 went out of use. 'The effect was disastrous, for now all the main sewers and underground streams discharged their new contents into the Thames which ... became a huge open sewer' (Weinreb and Hibbert, eds., *The London Encyclopedia*, (1985 edn., p. 237). See further notes to p. 98, below.

test of intellectual capacity is ... not in digestion: a quotation from Horne Tooke (for whom see note to p. 45, above) of which P. seems to have been fond—cf. his 1830 review of Moore's *Byron*: ' "The utility of reading," says Horne Tooke, "depends not on the swallow, but on the digestion" ' (Halliford, ix. 98). Here P. is turning it against his new *bête noire*, competitive examinations (see note to p. 102, below).

national education: an allusion to Sir James Kay-Shuttleworth's efforts to promote a system of national education. Kay-Shuttleworth (1804–77) was first secretary to the committee of the Privy Council on education 1839–49, co-founded the first Teachers' Training College, and devised the pupil–teacher system. The kind of curriculum he seemed to favour was often attacked, as here, for its irrelevance to the lives of those following it. Far from teaching girls how to sew, Dickens wrote to Angela Burdett-Coutts in 1856, Kay-Shuttleworth 'would have gone on to the crack of doom melting down all the thimbles in Great Britain . . . and making medals of them to be given for a knowledge of Watersheds and Pre Adamite vegetation . . . if it hadn't been for you' (*Letters of Dickens to Angela Burdett-Coutts*, ed. Johnson, p. 321). Such complaints were prominent among the testimonies offered to the Newcastle Commission on Popular Education in 1858—see P. Collins, *Dickens and Education*, pp. 150–1.

98 *a Sympathizer*: an allusion to Americans who supported, with cash and in other ways, the Irish struggle for independence under Daniel O'Connell (1755–1847) and others. Dickens mocks their violent rhetoric in 1844 in his depiction of the Watertoast Association of United Sympathisers in *Martin Chuzzlewit* (ch. 21). The Fenians were founded in New York and Ireland in 1857.

a Know-Nothing: member of an American political party also known as the American Party formed in the 1830s as a secret organization (hence its name: its members when questioned about it would say, 'I known nothing about it') dedicated to repealing the US naturalization laws and to pressing for legislation that would have excluded from political office anyone who was not a native American.

a Locofoco: a member of the extreme Radical or Equal Rights faction of the Democratic Party; so named because at a convention in the Party's New York headquarters, Tammany Hall, in 1835 when the chairman left his seat and the lights were extinguished in the hope of breaking up the tumultuous meeting, some of the Radicals re-lighted the lamps with the lucifer matches known as 'locofocos' that they had in their pockets, re-organized the meeting, and carried their motions.

Filibuster: one who makes long speeches in Congress, or any other legislative assembly, simply in order to prevent some measure from being proposed or passed. The word is a corruption of Dutch *vrijbuiter* (buccaneer or freebooter), hence P.'s reference to pirates.

friend of humanity: P. is playing with the similarity in sound between the Greek prefix 'phil' meaning 'lover or friend of' and the first syllable of 'Filibuster'.

Φιλοβωστρὴς: a word invented by P. from the Greek prefix Φιλο-('fond of') and βωστρεῖν ('to cry after').

Roaring Girl: a comedy written jointly by Thomas Middleton and Thomas Dekker, first produced in 1611.

in a parliamentary sense: i.e. meaning something very different from what the word would mean in ordinary discourse, a reference to the way in which the traditional insistence on civil language during Parliamentary debates leads to distortion of meaning.

sui generis: unique.

killed the fish in the river: 'In 1800 salmon had still been swimming up to London and beyond, but by mid-century ... no fish of any kind could survive in the river and even the swans had deserted it' (Weinreb and Hibbert, eds., *The London Encyclopedia* (1985), p. 237).

nose ... in great indignation: *The Tempest*, IV. i. The fetid odour from the poisoned Thames became particularly noisome during the very hot, dry summer of 1858, the time of 'the Great Stink', as it was called. The riverside windows of the Houses of Parliament had to be hung with sheets soaked in chloride of lime to counteract the stench. There were debates on the subject in both the Lords and the Commons, and Disraeli (who described the Thames as 'a Stygian pool reeking with ineffable and unbearable horror') introduced a bill to authorize the cleansing of the river which became law before the session ended (see *Annual Register* for 1858, ch. 8, pp. 214–17).

98–9 *The Wisdom has ordered the Science to do something*: various undated scraps of paper in the Pforzheimer Library (TLP 122) show drafts and a final fair copy in P.'s hand of a little scene in Latin verse entitled 'The Wisdom of Parliament' in which 'Sapientia' asks 'Scientia' to do 'something' about the black sewer ('nigram cloacam') of the Thames and ladles out money for the purpose (summarized by permission of the Carl H. Pforzheimer Library).

99 [*P.'s note*] DYCE: Alexander Dyce (1798–1869), editor of Shakespeare and other Elizabethan and Jacobean dramatists.

honourable friend: i.e. Member of Parliament (in the House of Commons MPs refer to members of their own party as their 'honourable friends').

[*P.'s note*] *compelled into a circle*: Virgil, *Georgics*, ii. 401. The actual lines are as follows: '. . . redit agricolis labor actus in orbem/atque in se sua per vestigia volvitur annus': 'the farmer's toil returns, moving in a circle, as the year rolls back upon itself over its own footsteps.'

halibut . . . excommunicated: Professor Saintsbury, celebrated as a gourmet as well as a littérateur, commented in his Introduction to a 1903 reprint of *G.G.*, 'The Doctor is wrong about halibut'.

100 *bream pie*: P. may have invented the detail about this pie's figuring in the indictments prepared against the monks by Henry VIII's commissioners but he himself clearly shared Opimian's taste here; recipes in his own hand survive for both bream pie and 'stew for bream' (see Halliford, ix. 451–2).

a lecturing lord: P. is primarily thinking in this passage of his recurrent butt, the former Lord Chancellor, Lord Brougham (1778–1868), and his founding of the Social Science Association in 1857. See note on 'Science of Pantopragmatics' to p. 137, below.

101 *Doctor Johnson . . . lectures*: see Boswell's *Life of Johnson*, ed. G. Birkbeck Hill (1934), ii. 7. Talking to Boswell in February 1766, Johnson observed, 'I know nothing that can be best taught by lectures, except where experiments are to be shewn. You may teach chymistry by lectures.—You might teach making of shoes by lectures!'

bienséance: decorum.

tenson . . . Gai Saber: a *tenson* was a contention in verse between two Troubadors, medieval Provençal poets. The old Provençal name for the art of poetry was *le gai saber*, rendered into English as 'the Gay Science'.

102 *Spirit-rapping*: the Scottish-born Spiritualist Daniel Home (1833–86) came to London from America in 1855 and held a number of seances (one attended by Browning who later satirized him as 'Mr Sludge the Medium' in his *Dramatis Personae*, 1864) at which spirits of the dead were said to be communicating with the living through a series of rapping

noises; other phenomena such as the tilting or turning of chairs and tables also occurred. Home was much patronized by European royalty and returned to London in 1860 when he held seances at the home of Milner Gibson, the President of the Board of Trade (his wife had become a convert), at Lord Lytton's, and elsewhere in the fashionable world.

contest . . . in the Clouds: Aristophane's *Clouds*, first produced at the Athenian festival of the Great Dionysia in 432 BC, would have been especially congenial to P. and to his satiric purpose in *G.G.* In *Clouds* Aristophanes satirizes the new forms of education and kinds of morality that were becoming popular in the Athens of his day. Socrates is made to appear as the grotesque embodiment of all this new philosophy. An old farmer who wants to wriggle out of paying his creditors brings his son to Socrates to be taught the new rhetoric which will enable him to win any argument, in court or elsewhere, even when in the wrong. Socrates arranges a debate between personifications of Right and Wrong (P.'s 'Just and Unjust') in which Right is totally outmanoeuvred and defeated.

march of mind: in the early and mid-nineteenth century a favourite slogan of those who, like Brougham, believed that it was a great age of intellectual progress thanks to the spread of education through such agencies as the Society for the Diffusion of Useful Knowledge.

competitive examination: following the Northcote-Trevelyan Report on the Civil Service, Parliament adopted the principle that entry to the Service should be by competitive examination (1855). For an interesting discussion of the possible reasons for P.'s strong hostility to this development see M. Butler's *Peacock Displayed*, pp. 248–50.

could go backwards: adaptation of Hamlet's words to Polonius in *Hamlet*, II. ii.

unknown tongues: alluding to the Irvingites, followers of Edward Irving (1792–1834), founder of the 'Catholic Apostolic Church', who attracted large audiences to hear his highly histrionic sermons which were punctuated by outbreaks of 'speaking in tongues' on the part of the congregation in imitation of the original Pentecost (Acts 2).

[*P.'s note*] *but suppose it to be true*: Juvenal, *Satires*, ii. 152–3.

104 [*Epigraph*]: the inscription is on an ancient marble at the Villa

Albana in Rome. It is quoted in C. Fea's edn. of Horace (1811), ii. 216–17, in an annotation to *Epistles*, I. xi. 22–3 ('And you—whatever God has given for your weal, take it with grateful hand, nor put off joys from year to year').

[*P.'s note to epigraph*] *senarii*: a metrical form, common in early Roman drama, in which the lines have six feet, some spondaic, and some iambic.

GREGORY GRYLL, ESQ.: according to Cole (*Biographical Notes*) P. 'used to avow that in the character of Mr Gryll he had repaired the injustice with which he had hitherto treated the landed gentry'.

Epicuri de grege porcus: Horace, *Epistles*, I. iv. 16.

[*P.'s note*] *Stoic philosophy*: Epicurus taught (*c*.300 BC) that the highest good was pleasure; his distinction between baser and higher pleasures was ignored by his critics who represented his creed as one of sensual hedonism. The Cynics were followers of Diogenes who taught (*c*.350 BC) that happiness was obtainable only through satisfying one's natural needs in the cheapest and easiest way. The aim should be to train one's body to reduce those needs to a minimum. Cynics were nicknamed 'dogs' because of their shamelessness, resulting from their belief that natural functions of the body cannot be indecent and could therefore be performed in public. Zeno of Citium (335–263 BC) was a leading exponent of the Stoic philosophy, teaching that the only real good is virtue and the only real evil moral weakness; everything else, such as poverty, pain, death, should be indifferent to the philosopher. Cleanthes (331–232 BC), Zeno's disciple, succeeded him as head of the Stoic School.

[*P.'s note*] *the greater portion has perished*: *Bruta Animalia ratione uti*, 'Beasts are Rational', one of the liveliest of Plutarch's *Moralia* (Loeb, xii. 489–533). Plutarch's dialogue is inspired by the episode in bk. x of *The Odyssey* when Odysseus makes the enchantress Circe restore to human shape those of his comrades whom she has transformed into swine. (Spenser's more austere version of Plutarch's idea appears at the very end of the canto to which P. refers his readers.)

106–7 *Morgana . . . forms of men*: enchantress in Boiardo's *Orlando Innamorato*. Panizzi, in his edn. (see notes to pp. 219 [*re* Berni's *Rifacciamento*] and 249 [*re* Berni and Domenichi], below) writes, 'This fairy is FORTUNE'. She imprisons knights by enchantment (bk. 2, canto ix); they are released by Orlando who manages to grasp her forelock—the only way to control her (2, xiii). She has 'power over . . . the forms of men' in that she changes Zinto into a dragon and back to human form (2, xiii). Her 'garden' may be her underworld (2, viii); or P. may be thinking of her fellow-sorceress Falerina's garden at Orcagna ('un giardino nobile et felice': 1, xvii). See note on Falerina, p. 154, below. Matteo Maria Boiardo (1434–94) was an Italian nobleman born at Ferrara. His poem, posthumously published in 1495, was the first of the epic poems in Italy, and won enormous popularity until it was superseded in 1516 by Ariosto's *Orlando Furioso* (see note to p. 154 below). Both poems deal with the exploits of Orlando and other followers of Charlemagne in crusades against the Saracens.

108 [*Epigraph*] ALCAEUS: Alcaeus of Mytilene in Lesbos (flourished *c.*600–570 BC) was a lyric poet whose work survives only in fragments. The fragment P. quotes here is numbered 347(a) in D. A. Campbell's *Greek Lyric*, i (Loeb). The dog star is Sirius and the 'dog days' when this star rises and sets with the sun (3 July–11 Aug.) are traditionally supposed to be the hottest days of the year. Dogs were popularly supposed often to run mad at this time.

[*Epigraph*] PETRONIUS ARBITER: *Satyricon*, sect. 34. Falernian wine, i.e. wine from the region of the Roman Campagna called Falernus Ager, was famous for its quality. For 'Opimian' see note to p. 97 above.

a good living: in the Church of England a 'living' means a parish priest's situation, with particular reference to the income connected with it.

109 *a solitary round tower*: the original of Mr Falconer's tower may well have been the inhabited one 'in two or three stories' familiar to P. from the Windsor Forest rambles of his earlier days and described in a nostalgic essay, 'The Last Day of Windsor Forest', written shortly after completing *G.G.*—see Halliford, viii. 145–54. This tower was not a round one, however: 'from its form [it] was commonly called the Clock-case'.

110 [*P.'s note*] . . . *Was first my thought*: P. is quoting Pope's translation of *The Odyssey* (pub. 1725/26), x. 165–78.

111 *Sir Calidore*: the hero of bk. vi of Spenser's *Faerie Queene*, described as the most courteous of all knights. P. began,

probably in 1816, to write a satirical novel named *Calidore*. In what we have of the novel the hero, an Arthurian knight, returns to England and is extremely puzzled by, among other things, the use of paper money (see Halliford, viii, 303 ff.).

scholar's acquisition: for Horne Tooke see note to p. 45, above. The Advertisement to ch. 7 of his *Diversions of Purley*, Part One, begins: 'I presume my readers to be acquainted with French, Latin, Italian and Greek: which are unfortunately the usual boundaries of an English scholar's acquisition.'

112 *Calderon*: Lope de Vega (1562–1635) is regarded as the founder of the national Spanish drama and Pedro Calderon de la Barca (1600–91) as one of its greatest exponents.

Porson: Richard Porson (1759–1808) who became Regius Professor of Greek at Cambridge in 1792.

réunions: social gatherings especially of people with common interests.

[*P.'s note*] *Art of Dining*: 'What use is it, gentlemen, to learn the language of a people who do not have a national cuisine?' The book P. is quoting from was written by the journalist Abraham Hayward (1801–84) in 1852.

113 *thrice great Hermes*: *Il Penseroso*, ll. 85–6.

114 *the Homeric question*: this question occurs at least half-a-dozen times in the *Odyssey*, e.g. i. 170; x. 325; xiv. 187.

115 [*Epigraph*] PERSIUS: Satire V, ll. 52–3. Persius (AD 34–62) was the Roman author of six poetic satires, most of them with a moralizing rather than an abusive tone, commenting on the degeneracy of Roman society.

Vesta: Roman goddess of the hearth whose shrine near the Forum contained a sacred flame ministered to by six women known as the Vestals who were supposed to remain virgin during their period of office (thirty years).

[*P.'s note*] *Peigniot. Paris. 1800*: 'Congreve the best comic writer of England: his most celebrated plays are . . .'. P. is mocking Peigniot's mistranslation of some of Congreve's titles: *The Mourning Bride* becomes *L'Epouse du Matin* (The Morning Bride) and in *The Way of the World* 'way' is taken to mean 'road' rather than 'fashion'.

116 *men of Gotham*: legendary fools. The story goes that the men of Gotham in Nottinghamshire once deliberately played all sorts of idiotic tricks, such as trying to rake the moon out of a pond,

in order to discourage an unwanted visit from King John and his court. A collection of tales of stupidity was published in the sixteenth century entitled *Merie Tales of the Mad Men of Gotham.* P. had earlier referred to them in the catch sung by Mr Hilary in ch. 11 of *Nightmare Abbey.*

117 *Hamadryad*: in classical legend a tree-nymph.

an inclosure act: see note to p. 91, above.

crown land: land owned by the Sovereign and administered by the Crown Commissioners.

Numa: Numa Pompilius, the second of the legendary kings of Rome (traditionally 715–673 BC), a philosopher who founded the College of the Vestal Virgins.

118 *Arbiter Elegantiarum*: Petronius. See note to p. 97, above.

Electra: daughter of Agamemnon and Clytemnestra, bitterly opposed to her mother's adulterous union with Aegisthus. Outraged by their murder of Agamemnon, she helps her brother Orestes to wreak vengeance on them. She figures in Aeschylus's *Choephoroe*, the second play in his Oresteian Trilogy, as well as being the subject of plays by both Sophocles and Euripides.

[*P.'s note*] *creavit unda.—ibid*: 'smoother than ... the round garden mushroom that is born in rain', *Satyricon*, sect. 109.

[*P.'s note*] STEEVENS: i.e. George Alexander Stevens (1710–84), comic writer. His theatrical entertainment, *A Lecture on Heads* (first authentic edn. 1785) was reprinted many times in the first two decades of the nineteenth century.

[*P.'s note*] *Artis Amatoriae*, iii, 249, 'Ugly is a bull without horns; ugly is a field without grass, a plant without leaves, or a head without hair.'

[*P.'s note*] *her own Vulcan*: in classical mythology Vulcan, the god of fire and of the forge, was the husband of Venus. Apuleius (2nd century AD) was the author of the only Latin novel to survive in its entirety, the *Metamorphoses*, better known under its English title of *The Golden Ass*.

[*P.'s note*] Περικειρομένη: 'The Girl who gets her Hair cut short.' It is, in fact, the girl's lover not her husband who crops her hair. Menander was the leading exponent of the Greek New Comedy. He made his début in 321 BC.

119 *says Aristotle*: we have not been able to discover where.

the Phormio: Terence (*c.*190–159 BC), Roman comic dramatist. *Phormio*, his sixth play, was produced in 161 BC. Young Antiphuo, in his father's absence from home, has fallen in love with, and imprudently married, a beautiful but penniless young girl whom he has seen mourning, with dishevelled hair, for her dead mother.

. . . θυμός: 'But why does my heart thus hold converse with me?' *Iliad*, xi. 407.

Vestals were: in his *Natural History*, bk. xvi, sect. 85, Pliny the Elder (AD 23–79) mentions a very ancient lotus tree in Rome 'called the Hair Tree, because the Vestal Virgins' offering of hair is brought to it'.

[*P.'s note*] *Electra, v. 449*: Electra here adjures her sister to throw away the sacrifices she is taking, at Clytemnestra's behest, to place on Agamemnon's grave: 'Fling them away and cut/ A tress of thine own locks.'

120 *Lares*: household gods.

doubled down: the source of Opimian's allusion is untraced.

Lipsius de Vestalibus: Justus Lipsius (1547–1606), humanist and classical scholar, published his *De Vesta et Vestalibus Syntagma* in Antwerp in 1603.

[*P.'s note*] *fire divine*: Ovid, *Fasti* (a poetical treatise on the Roman calendar) bk. iii, 29–30. Rhea Silvia, an early Vestal, was in legend the mother of Romulus, founder of Rome, having been impregnated by Mars. The literal translation of these lines in which Ovid makes her describe her vision is: 'I was by the fire of Ilium (i.e. Troy) when the woollen fillet slipped from my hair and fell before the sacred hearth.'

121 *vitta*: a headband or fillet.

[*P.'s note*] *the Vestal train*: *Fasti*, vi. 441.

122 [*Epigraph*] *Alcestis*: ll. 788–9.

something better: see Boswell's *Life of Johnson*, ed. Birkbeck Hill, iii. 242. What Johnson actually said was, 'Had I learnt to fiddle, I should have done nothing else.'

123 *La Morgue Aristocratique*: aristocratic haughtiness.

vinculum: bond.

The Seven Pleiads: cluster of stars in the constellation Taurus, especially the seven larger ones. In Greek myth these stars were originally the daughters of the sky-supporting Titan, Atlas. See note to p. 131, below.

St. Catharine: see note to p. 140, below.

124 δμωαὶ: slave-girls.

know the tree by its fruit: allusion to Matthew, 7: 15.

Nausicaa: a beautiful young princess who appears in bk. vi of *The Odyssey*. She succours Odysseus when he is cast up on the shores of her father's kingdom.

125 *Paradisi viola!*: apparently from some untraced 'monastic verses'; cf. P.'s description of an unnamed character in a MS fragment now in the Pforzheimer Library (TLP 112), who had 'an especial reverence for Virgin Martyrs: thought Saint Katharine the ideal perfection of the class and thought of her as what she had been called in monastic verses, the pearl of heaven and the violet of paradise' (quoted by permission of the Carl H. Pforzheimer Library).

126 [*Epigraph*]: the opening lines of 'Colin's Complaint', written about 1712 by Nicholas Rowe (1674–1781) who became Poet Laureate in 1715.

Ajax: King of Salamis, one of the leaders of the Greek host against the Trojans. A man of gigantic stature, one of his chief attributes was his massive shield.

. . . *and rivers*: *Odyssey*, x. 351–2: 'Children are they [Circe's handmaids] of the springs and groves, and of the sacred rivers that flow forth to the sea.'

as Orlando did with Morgana: for Morgana see note to pp. 106–7, above. In *Orlando Innamorato*, the enchantress Morgana can be controlled only by having her golden forelock seized. Orlando, succeeding in grasping this metaphorical key, is able to gain the literal silver key with which he can release the knights whom she has imprisoned (bk. 2, cantos ix–xiii).

[*P.'s note*] ADDISON: *Ovid. Met.* L. ii: Addison's translations appeared in *Ovid's Metamorphoses . . . Translated by the Most Eminent Hands* (1717).

127 *æs triplex of treble X*: i.e. the threefold bronze (shield) of the strongest kind of beer. The Latin phrase comes from Horace (*Odes*, I. iii. 9).

128 *conjunct tetrachords*: the Greek scale is founded on five tetrachords (series of four notes); four of them are conjunct, that is,

overlapping, and the other disjunct. The girls' names are taken from the first and second tetrachords (B, C, D, E and E, F, G, A). The key-note—the lower A—was not reckoned part of a tetrachord. See P.'s explanation in his review of Müller and Donaldson's *History of Greek Literature* (Halliford, x. 179 ff.).

129 *inherit an estate*: probably a sly allusion to Sir Walter Scott's *Rob Roy* (1817) in which the hero, Francis Osbaldistone, inherits the family estates after his six cousins, sons of his uncle Sir Hildebrand, all die violent deaths (only one of them, however, actually breaking his neck); the deaths of five of the young men are detailed in ch. 37 and the sixth, the villainous Rashleigh, is killed (by Rob Roy) in the last pages of the book.

speculum: mirror.

130 [*Epigraph*] PERSIUS: Satire v, ll. 151–3.

Agapêtus and Agapêtae: from the Greek *agape* meaning 'brotherly love'. *Agapêtae* were female celibates and widows who in the early Christian Church (third century AD) devoted their lives to the service of the clergy. Often denounced by Fathers of the Church such as Jerome, they were suppressed by the Lateran Council in 1139.

better to marry: '... it is better to marry than to burn', 1 *Corinthians* 7: 9.

131 *till one disappeared*: originally seven stars were distinguishable as the Pleiades (see note to p. 123 above) but one, Electra, known as 'the lost Pleiad', became invisible to the naked eye. According to one legend, she hid herself in shame at having married a mortal; according to another, in grief at the fall of her city of Troy.

132 *Agapêmonê*: name of a spiritual community of both sexes founded in 1859 at Spaxton in Somerset by the Revd Henry Prince (1811–99) who styled himself 'the Beloved' and claimed to be a reincarnation of the Holy Ghost. He and his followers lived in luxurious circumstances and Prince had bought the equipage of the late Queen Adelaide with four white horses. In 1860 an action was brought against him by the brother of one of his female followers. For a detailed account of Prince see Aldous Huxley's essay 'Justifications' (*The Olive Tree*, 1936).

Family of Love: sect founded by the German Anabaptist Henrick Niclaes (Henry Nicholas) who visited England about 1552. It attracted many adherents mainly in East Anglia and

Essex. A royal proclamation ordered its suppression in 1580 and Thomas Middleton's farce *The Familie of Love* appeared in 1608 but, as Opimian says, the sect continued in existence for some long time afterwards.

133 *the world has grown honest*: *Hamlet*, II, ii.

hylactic: yelping; from the Greek ὑλακτικός ('given to barking'). The only citation for this adjective in *OED* is to P.'s use of it here.

bank bubbles: cf. P.'s 1856 Preface to *Melincourt*: 'multitudinous bubbles have been blown and have burst; sometimes prostrating dupes and imposters together; sometimes leaving some colossal jobber upright in his triumphal chariot, which has crushed as many victims as the car of Juggernaut.' The collapse in early 1856 of the Tipperary Bank with a deficit cxcccding £400,000, and the suicide of its fraudulent director, Sadleir (a source for Dickens's swindler Merdle in *Little Dorrit*), would have been one of the more spectacular examples to which Opimian could have pointed.

harlequin's wand: Harlequin, whose name derives from the Arlecchino of the *commedia dell'arte*, was a traditional figure in British pantomime with his black mask and his magic wand with which he effected the 'transformation scene' that was *de rigueur* for such entertainments.

societies . . . for teaching everybody everything: the Social Science Association. See note on 'Science of Pantopragmatics' to p. 137, below.

Old Parr: Thomas Parr of Alderbury near Shrewsbury, supposed to have been born in 1483, died in 1635, and became a byword for longevity.

[*P.'s note*] *Jeremy Bentham*: untraced. Possibly P. is quoting from Bentham's table-talk during the period when he was 'extremely intimate' with him, 'dining with him tête-à-tête, once a week for years together' (Halliford, I. clxxv).

[*P.'s note*] PETRONIUS ARBITER: in Roman mythology Cerberus was the three-headed dog that guarded the entrance to the underworld. Fulgentius, a fifth-century commentator on Virgil, applies the myth of Cerberus to quarrels and litigations in the courts and quotes Petronius as saying of one Euscios, 'The barrister was the Cerberus of the courts' (Petronius, *Fragments*, no. 8).

134 *High-pressure steam boilers*: see note on 'explosions of powder-mills' to p. 208, below.

crush their occupants: Eric Robinson notes in his 'Thomas Love Peacock: Critic of Scientific Progress' (*Annals of Science* (1954), 69–77) that the *Annual Register* for 1859 details a number of instances of new buildings collapsing in London and elsewhere.

would not poison children: Robinson (op. cit.) cites cases from the *Annual Register* for 1858 and 1859 of sweets and buns containing poisonous matters being knowingly sold to children and adults; in a case in Bradford there were said to have been eighteen deaths as a result.

would not poison patients: Robinson (op. cit.) cites cases of 'Poisonings by mistake' reported in the *Annual Register* for 1856 and 1859 which resulted from the incompetence of those entrusted with the sale of poisonous ingredients.

Greek kalends: the calends were the first day of the month in the Roman calendar: 'at the Greek calends' meant 'never' since there was no such term in the Greek calendar.

ships in ordinary: i.e. not in commission.

135 [*P.'s note*] *Eurip. Suppl. 207: Herm.*: in modern texts the relevant line-number in Euripides' *Suppliants* is 199. ('Herm' is an abbreviation for the German Hellenist scholar G. Hermann, whose edition of *Suppliants* appeared in 1811.)

136 [*Epigraph*] DIPHILUS: Diphilus of Sinope (second half of third century BC) was the author of about 100 New Comedy plays, only a few of which have come down to us, in fragmentary form (several are also known from Plautus's adaptations). The first half line is from *The Little Tomb*, and the rest is from *The Boy-Lovers* (J. M. Edmonds, *The Fragments of Attic Comedy*, vol. iiiA, p. 127).

137 *Lord Facing-both-ways*: P.'s name, taken from Bunyan's *Pilgrim's Progress* (where he is named as one of By-ends's relatives in the town of Fair-speech) for Lord Brougham.

Science of Pantopragmatics: in 1857 the aged Brougham founded a 'National Association for the Promotion of Social Science' which held its first annual conference at Birmingham in October. The *Annual Register* 'Chronicle' for 1857 recorded (pp. 200–1) the objects of the Association as being 'To aid the development of the social sciences, and to guide the public mind to the best practical means of promoting the Amendment of the Law, the Advancement of Education, the Prevention and Repression of Crime, the Reformation of Criminals, the Establishment of due Sanitary Regulations, and the recognition of sound principles in all questions of Social Economy.' P. derives his satirical name for the Association from the Greek 'pan' (all) and 'pragmatic' meaning 'officious, meddlesome'; no doubt he intended the first two syllables to recall the word 'pantomime' also.

[P.'s note] Hamlet: 'Lord Michin Malicho' is P.'s name for the former and future Prime Minister, Lord John Russell (1792–1878) who was Foreign Secretary under Palmerston when P. was writing *G.G.* He had presided over the Law Amendment section of the first conference of the Social Science Association (see previous note).

138 *'words, words, words'*: *Hamlet*, II. ii.

crambe repetita: literally, cold cabbage warmed up; stale repetitions (Juvenal, *Satires*, vii. 154).

[P.'s note] The Coke's Prologue: 'Jack of Dover' means a fool (see above, p. 177) and it may have been in Chaucer's day also the name for some kind of pie or fish which the Host in *The Canterbury Tales* accuses the Cook of warming up again in order to sell it as just cooked.

[P.'s note] Huddleston: actually Huddesford, the Revd George (1749–1809), political satirist, whose *Bubble and Squeak* was published in 1799.

140 *Saint Catharine*: the cult of St Catharine of Alexandria originated in the ninth century at Mount Sinai and was introduced into Europe by the Crusaders. During the Middle Ages it flourished exceedingly and the picturesque details of her legend, especially of her mystic marriage to Christ, appealed greatly to the imagination of artists. She seems to have become a private cult of P.'s in his later years when, apparently, he 'collected prints and engravings' of her (Halliford, I. cxix) and began a story about an Order of 'Knights of St Katharine', dedicated to protecting pilgrims to her shrine (MS in the Pforzheimer Library [TLP 126a and b], accompanied by a pen-and-ink drawing of the saint by P.).

141 *[P.'s note] Vita di Santa Catarina, L.II. Vinezia, 1541*: the author of

this life of the saint, published in Venice, was the voluminous, all-purpose Italian writer Pietro Aretino (1492–1556), chiefly celebrated for a series of lascivious sonnets. A 'fine, vellum-bound copy' of this life of St Catharine, dated 1636, was one of P.'s favourite books (Halliford, I. cxix–xx).

142 *Empress Catharine II*: Catharine the Great, Empress of Russia 1762–96.

Charles Fox: Charles James Fox (1749–1806), the great Whig champion of political liberty in successive Parliaments from 1769 onwards. In his later years he resided at St Anne's Hill near Chertsey in Surrey.

[*P.'s note*] *Illustrations of Jerusalem and Mount Sinai*: i.e. Francis Arundale, *Illustrations of Jerusalem and Mount Sinai, Beirout, from drawings by F. Arundale . . . with a descriptive account of his tour and residence in those remarkable countries* (Colburn, 1837).

143 *Guildford*: St Martha on the Hill, the parish church of Chilworth, a village about 3 miles south-east of Guildford, was rebuilt on its Norman foundations 1848–50. St Catherine's Chapel, built 1317, is still 'a picturesque ruin' on high ground beside the Godalming Road three-quarters of a mile south of Guildford. Each of these edifices could be seen from the other were it not for trees. St Anne's Hill, Chertsey, however, is about 10 miles north of Guildford and a chapel on it could never have been visible from St Martha's or St Catharine's. It is not known where P. got his story of the three sisters (which he had already set out in an unfinished story about three modern sisters, one named Catherine, printed in Halliford, viii. 391–3, under the title 'A Story of a Mansion among the Chiltern Hills'—MS in British Library [Add. MS 36815, ff. 171–81]). Possibly, it was his own adaptation and extension of the local legend recounted by Eric Parker in his *Highways and Byways in Surrey* (1908): 'St. Catherine and St. Martha, in the wonderful days of the giants, were sisters who built chapels on neighbouring hills. They had but one hammer between them, and they hurled it high over the valley to one another, St. Martha catching it from St. Catherine, driving in a nail and hurling it back again.'

church of St. Catharine: medieval church in Gothic style of St Katharine's Hospital, originally founded by King Stephen's queen, Matilda, in 1148. The church was demolished in 1825 to make way for St Katharine's Docks, opened in 1828.

the window in West Wickham church: the North Chapel of the parish church of St John the Baptist, West Wickham (now in the London Borough of Bromley), has in its windows some very early sixteenth-century stained glass including the figure of St Catharine referred to by Falconer. The figure, not particularly striking, is about 2½ feet tall and shows the saint trampling on a crowned male figure; she has her crown, wheel, book, and sword but not the palm or the fire.

[*P.'s note*] *Epod. 16, 13*: *Epodes*, bk. xvi, ll. 11–14: 'The savage conqueror shall stand, alas! upon the ashes of our city, and the horseman shall trample it with clattering hoof, and (impious to behold!) shall scatter wantonly Quirinus' bones, that now are sheltered from the wind and sun.'

144 *genius*: i.e. *genius loci*; in classical mythology a tutelary or guardian spirit of a particular locality.

Naiad . . . Oread . . . Nereids . . . Oceanitides: different varieties of nymphs, or female spirits of Nature, in classical mythology. The Naiades were nymphs of rivers, brooks and lakes, the Oreads of mountains, the Nereides of the Mediterranean Sea, and the Oceanitides of the ocean.

Martin . . . Peter and Jack: the chief characters in Swift's *A Tale of a Tub* (1704), which ridicules the Roman Catholic Church in the figure of Peter and the Calvinist Church under the name of Jack; the Church of England is represented by Martin (Luther).

Beaumont and Fletcher: Francis Beaumont (1584–1616) and John Fletcher (1579–1625) were joint authors of *A Maid's Tragedy* and many other plays.

145 [*Epigraph*] PETRONIUS ARBITER: *Satyricon*, sect. 115. For Petronius see note to p. 97, above.

147 *concurrent*: concurring.

148 [*Epigraph*] NONNUS: Nonnus of Panopolis in Egypt, believed to have lived during the fifth century AD, whose epic poem in 48 books, *Dionysiaca*, celebrates the exploits of Bacchus or Dionysus, the god of wine. The lines P. quotes and translates come from bk. xii, ll. 260–1 and 268–9. According to Sir Henry Cole (*Thomas Love Peacock: Biographical Notes*) P. 'had a great admiration for Nonnus' *Dionysiaca* . . . He delighted to ask an Oxford first-class man who Nonnus was, and to find he could get no information. He would frequently quote passages from Nonnus, which modern ideas would object to as indelicate;

and Hogg would say to him, "You think you have a privilege to be indecent in Greek".'

electric telegraph: see note to p. 209, below.

149 *Sicilian tyrant*: Phalaris, tyrant of Acragas (*c*.570/65–554/49 BC), who became legendary for his ingenious cruelty, especially for his invention of a hollow brazen bull in which his victims were roasted alive.

propitiated his genius by copious libations: 'genius' here means guardian spirit; 'libation' is the term for the pouring out of a drink-offering to the gods.

Philemon: Philemon (368/60–267/63 BC) from Syracuse, who during his long life was the author of over ninety comedies which proved very popular with Athenian audiences.

[*P.'s note*] *Florid. 16*: *Suidas* (Latin: 'fortress') is the name given to an historical and literary encyclopaedia compiled about the end of the tenth century AD. P.'s ref. here means 'the entry for Philemon'. Apuleius (see note to p. 118 above) compiled the *Florida*, a collection of excerpts from declamations on various themes. According to the *Oxford Classical Dictionary*, they 'contain much that is curious and amusing while the description of the death of the comic poet Philemon has real beauty'.

151 *'Farewell and applaud'*: 'Valete, plaudite', the traditional valediction after dramatic performances in the classical theatre.

'from the loop-holes of retreat': from *The Task* (1795) by William Cowper, iv. 88.

Gods of Epicurus: see note to p. 104 above.

152 *womanhood in its prime*: see *Headlong Hall*, ch. 7.

under the sun: Ecclesiastes, 4: 1.

received with increasing favour: an allusion to the success of the Oxford Movement set on foot in 1833 by Pusey, Newman, and Keble which sought to bring back into the service of the Church of England much of the ritual and ornaments which had been dispensed with at the time of the Reformation.

[*P.'s note*] *De Naturâ Deorum, l.i.c. 41*: 'I beg of you to realize in your imagination a vivid picture of a deity solely occupied for all eternity in reflecting "What a good time I am having! How happy I am!".'

153 [*P.'s note*] *Ecclesiastical Sonnets*, i. 21: the poem is 'The Virgin', now no. 25 in the second book of *Ecclesiastical Sonnets*. P.'s

numbering is incorrect, but he is probably referring to the 1838 edition of *The Sonnets of William Wordsworth*, in which this poem is no. 21 in the second book; it does not have this number in any other edition.

154 [*Epigraph*] PHILETAERUS: (fourth century BC): author of over twenty comedies which survive only in fragments; said by one ancient authority to have been Aristophanes's son. These lines from his play *Cynagis* (*The Huntress*) appear on pp. 22–3 of vol. ii of J. M. Edmonds's *The Fragments of Attic Comedy* (1959).

Morgana and Falerina and Dragontina: enchantresses in *Orlando Innamorato*. For Morgana, see note to pp. 106–7 above. Falerina is mistress of the garden at Orcagna, which is destroyed by Orlando (bk. 2, canto v) in revenge for her making the sword Ballisarda with which she intended to have him killed (2, iv). Dragontina imprisons Orlando magically (1, xxxviii).

Armida: enchantress in Tasso's *Gerusalemme Liberata (Jerusalem Delivered)*. Like Boiardo's and Ariosto's Angelica, Armida is sent into the Christian camp to enslave the crusading knights (bk. iv). She imprisons Tancred in her magic castle (bk. vii), then detains her lover Rinaldo in her palace, causing him to neglect his duties as the Christian champion (bk. xvi) (cf. also Alcina detaining Ruggiero in Ariosto's *Orlando Furioso*). Ultimately she meekly submits to Rinaldo's domination and they are reconciled (bk. xx). Torquato Tasso (1544–95), brought up in the court of Urbino, completed his epic in 1574; it was published in 1580–1 without his permission while he was confined in an asylum. It is an attempt to remodel the Crusader stories of Boiardo and Ariosto into a Virgilian epic, with greater unity of plot and a higher moral tone.

Alcina: enchantress in *Orlando Furioso*. Her beauty deceives first Astolfo (whom she turns into a myrtle when tired of him), then Ruggiero, whom she detains in voluptuous indolence, until he is given Angelica's magic ring which enables him to see that Alcina's beauty is a delusion: in reality she is a withered hag, an 'ancient harlot' (canto 7). Ludovico Ariosto (1474–1533), Italian poet and playwright, is known almost solely for *Orlando Furioso* (published 1516; extended version 1532), the greatest of Italian epic poems, and a model for, among other poems, Spenser's *Faerie Queene*.

156 *in which Dante met Virgil*: in canto i of the *Inferno*.

Angelica fled: Angelica is a central figure in *Orlando Innamorato* and *Orlando Furioso*. In both, she is in constant flight from Charlemagne's court to her home in Cathay, drawing after her many knights who are enslaved by her beauty, and involving them in various adventures on her behalf—notably Rinaldo and Orlando. She is specifically described as fleeing through a wood in *Orlando Furioso* canto I, stanza 3 ff.

the semblance of Armida: see note on Armida above. Armida attempts to ensnare Rinaldo by causing him to dream, as he sleeps by a stream in an enchanted wood, that he sees her bathing there: 'This, though her form a syren's charms display'd/ Was but a semblance and a delusive shade' (*Gerusalemme Liberata*, bk. xiv, Hoole's translation).

the wounded deer: *As You Like It*, II. i.

Matilda: name of the beautiful lady whom Dante sees, from the other side of a stream, singing and picking flowers in the Garden of Eden (*Purgatorio*, canto 28).

Rosalind and Maid Marian: Rosalind is the heroine of *As You Like It*; banished from her usurper-uncle's court she takes refuge, disguised as a boy, in the Forest of Arden. Maid Marian, beloved of Robin Hood, dwelt with him and his band in Sherwood Forest. P.'s novel *Maid Marian* was published in 1822.

[*P.'s note*] PETRARCA: *Sonetto* 240: actually Sonnet 281 and so numbered in Robert M. Durling's edn. and translation of *Petrarch's Lyric Poems* (Harvard, 1976) where his translation reads: 'Now in the form of a nymph or other goddess who comes forth from the deepest bed of Sorgue and sits on the bank.' The Sorgue is a river in Provence, near Avignon.

157 *Odes of Pindar*: Pindar (518–438 BC), the great poet of Thebes, celebrated various victors in the athletic meetings of ancient Greece with a series of magnificent odes, classified under the names of the particular Games at which the victory was achieved, for example, the Olympic, the Pythian (the latter Games were held at Delphi in the third year of each Olympiad).

second Olympic: celebrating the victory in the chariot-race at the Olympia of Theron, tyrant of Acragas, Sicily, in 476 BC. This ode contains a memorable description of the life of the blessed after death. For Charles Fox see note to p. 142, above.

the ninth Pythian: celebrating the victory in the race in full armour won by Telesicrates of Cyrene at the Pythian Games in 474 BC. According to Greek myth, the great city of Cyrene (North Africa) was founded in honour of the beautiful Cyrene, daughter of the King of the Lapithae. Pindar described how Apollo saw her guarding her father's cattle and 'wrestling alone with a monstrous lion' and loved her. He transported her to Libya 'where he made her queen of a land rich in flocks and fruits'.

[*P.'s note*] . . . φιλοτάτων: the opening words of the wise old centaur Chiron's reply to Apollo when the latter questions him about whether Cyrene is worthy of the love of a god; translated (Loeb) as follows: 'Secret, O Phoebus! are the keys of wise Persuasion that unlock the shrine of love'.

158 *Semele*: daughter of Cadmus, the legendary founder of Thebes. She was beloved by Jupiter which roused the jealousy of Juno who persuaded her to trick Jupiter into appearing to her in his own shape whereupon the fire of his thunderbolts killed her but made her son (Dionysus) immortal.

Epicurean: see note to p. 104, above.

triple brass: see note to p. 127 above.

Chimaera: in classical mythology a triple-bodied monster described by Homer (*Iliad*, vi. 181) as 'in the fore part a lion, in the hinder a serpent, and in the midst a goat'.

[*P.'s note*] *Antigone*: l. 781.

[*P.'s note*] *triste bidental*: in his translation of Persius's satires (1874) J. Conington renders l. 27 of Satire II ('triste jaces lucis evitandumque bidental') as follows: '[or because] you are [not] this moment lying in that forest, a sad trophy of [Jupiter's] vengeance for men to turn from'. For Persius see note to p. 115, above.

[*P.'s note*] THE BROTHERS: ll. 121–2.

vain to make schemes of life: cf. P.'s words about Shelley in his 'Memoirs' of the poet: 'He had many schemes of life . . .' (Halliford, viii. 75).

159 *. . . never did run smooth*: *A Midsummer Night's Dream*, I. i.

[*P.'s note*] *Principles of Taste*: *An Analytical Enquiry into the Principles of Taste* (1805) by Richard Payne Knight. See also *Headlong Hall*, Introduction and chs. 2 and 4. The first of

Knight's mottos comes from Horace (*Epistles*, II. i. 101) and translates: 'what likes and dislikes are there, that you would not think easily changed?' The second is from the *Odyssey*, xviii. 136–7: 'for the spirit of men upon the earth is even such as the day which the father of gods and men brings upon them'.

[*P.'s note*] THEODECTES: *apud Stobaeum*: Theodectes (*c*.375–34 BC), student of Aristotle's, was an orator and composer of popular verse riddles. *Apud Stobaeum* means 'in Stobaeus', Stobaeus being the author of an anthology of excerpts from poets and prose writers compiled probably early in the fifth century AD. See edn. by Wachsmuth and Hense (1848), iv. 524.

[*P.'s note*] JUV. *Sat. x. 352–3*: 'We ask for wife and offspring; but the gods know of what sort the sons, of what sort the wife, will be.'

161 *Homer's view of the conjugal state*: for Jupiter and Juno see *Iliad*, i. 536–end; for Venus and Vulcan see *Odyssey*, viii. 266–365.

Agamemnon . . . congratulates Ulysses on . . . an excellent wife: in the *Odyssey* (xi. 440 ff.). The ghost of Agamemnon, after lamenting his wife's treachery, says to Ulysses: 'Wherefore in thine own case be thou never gentle even to thy wife. Declare not to her all the thoughts of thy heart . . . Yet not upon thee, Odysseus, shall come death from thy wife, for very prudent and of an understanding heart is the daughter of Icarius, wise Penelope.'

Xantippe: second wife (probably) of Socrates, a woman of legendary ill-temper.

Euripides . . . a woman-hater: the legend that Euripides became a misogynist as a result of the infidelity of two successive wives seems to have no foundation in fact.

Cicero, who was divorced: Cicero divorced his wife of thirty years, Terentia, in 47/46 BC to marry a younger woman, Publilia, but almost immediately divorced her too.

Marcus Aurelius: 'Faustina, the . . . wife of Marcus, has been as much celebrated for her gallantries as for her beauty. The grave simplicity of the philosopher was ill calculated to engage her wanton levity . . . Marcus was the only man in the empire who seemed ignorant or insensible of the irregularities of Faustina . . .' (Gibbon, *The Decline and Fall of the Roman Empire*, ch. 4).

Dante . . . left his wife behind him: in fact, Dante was absent from Florence on a diplomatic mission when his political opponents triumphed (1302) and condemned him to death. For the rest of his life he was an exile but always hoping and scheming to return to his beloved city. His three sons followed him into exile and, according to some authorities, so also did his wife, Gemma Donati, and daughter.

Milton, whose first wife ran away from him: at the age of 35 Milton suddenly married, in 1643, Mary Powell aged 17, from an Oxfordshire Royalist family. Finding life in Milton's home somewhat gloomy, Mary went on a visit to her family after a month or so and apparently refused Milton's order to return home; they were reconciled two years later, however, and she bore him four children before dying in 1652.

Shakespeare . . . scarcely shines in the light of a happy husband: the fact that Shakespeare's wife is mentioned in his will only once, in an interlineated afterthought which bequeathes her his 'second-best bed', was the foundation for the tradition in Shakespearian biography, from Malone onwards, that he was unhappy in his marriage. See S. Schoenbaum, *Shakespeare's Lives* (1970), pp. 174, 247, 325, 343 ('Shakespeare was not a very happy married man', J. P. Collier, *Life of William Shakespeare*, 1844).

162 . . . *semita vitae*: Horace, *Epistles*. i, 18, 103.

[*P.'s note*] PATERCULUS: *l. ii. c. 67*: Velleius Paterculus (*c.*19 BC–after AD 30), soldier and chronicler whose *Historiae Romanae* is a compendium of Roman history which becomes much more detailed the nearer it gets to the author's own day.

[*P.'s note*] *'Freinshemius: Supplementa Liviana', cxx. 77–80*: Johann Freinsheim (1608–60) was a German scholar chiefly known for his supplements to the Roman historians Curtius and Livy containing the missing books of those authors, supplied by himself. The Supplements to Livy were first published in 1679. P. owned no less than three eighteenth-century edns of Livy with Freinsheim's Supplements. The Proscription referred to was that of Mark Antony, Lepidus, and Octavian in 43–42 BC. It was a device to rid themselves of enemies and political opponents; over 2,000 individuals were listed as 'proscribed', which meant that they were to be hunted down and executed throughout Italy. The co-operation of their families and slaves was sought by means of bribes and threats. M. Oppius and his aged father were both proscribed. The father being too weak to flee unaided, the son carried him on his shoulders and brought him safely to refuge in Sicily. Quintus Cicero, a soldier and administrator, was the younger brother of the great orator; he and his son were both betrayed by their slaves in the Proscription. After his capture the son refused to reveal his father's hiding-place, even under torture. P. substituted these examples of filial loyalty for the two specified in the *Fraser's* text of *G.G.* (see above p. 334).

164 [*Epigraph*]: Horace, *Odes*, iii. 29, ll. 41–8. The translation is Dryden's (ll. 65–72) with 'be storm or calm' for 'be fair or foul'.

Miss Ilex: the ilex is the holm-oak, an evergreen.

165 '*over sharp wits whose edges are very soon turned*': Ascham's *The Scholemaster*: 'Quicke wittes are . . . like ouer sharpe tooles, whose edges be verie soone turned' (*OED* s.v. 'turn', iii. 9,b).

[*P.'s note*] HUDIBRAS: Part I, canto i, ll. 45–50.

167 *orchestra*: the semi-circular space in front of the stage in a Greek theatre where the chorus danced and sang.

discussed: here used in the old semi-humorous sense of to investigate, try the quality of, food or drink, i.e. to consume (*OED* sense 8).

168 *Siberian dinner*: sarcastic reference to the fashion already established by the mid nineteenth century for 'Russian dinners'—see *OED Supp.* s.v. 'Russian', sense d, and the following citation from *London at Table* (1851): 'We have already alluded to a Russian dinner, which is the best and most economical. It is always served hot from the kitchen, and . . . the joints served at the sideboard by an experienced artist, are more palatable and tempting than when carved on the table.' The 'Russian screen' referred to by Opimian was presumably one placed between the sideboard and the dining-table to conceal the operations of the 'experienced artist' from the eyes of the guests. Macborrowdale's 'Siberian' alludes no doubt to the temperature of the meat by the time it actually reaches the guests under this system. P. had already protested against this system of serving dinners in an article entitled 'Gastronomy and Civilisation', written jointly with his daughter, Mary Meredith, and published in *Fraser's*, Dec. 1851; see Halliford, ix. 397–8.

turkey-poult: i.e. a young turkey, a pullet.

[*P.'s note*] *Tatler, No. 148*: in this paper (21 Mar. 1710) Addison examines 'the diet of this great city' and complains, as Opimian says, about the fashion for French cookery. Immediately before the passage quoted comes Addison's description of his pleasure at finding there was at least some roast pig on the table but, he says, he could not eat any of it after being told that it had been 'whipped to death'. (The punctuation of the original differs from that in the passage as quoted by P. and 'I thought' should read 'Methought'.)

and hoists the enemy's: Sir Robert Peel again. See note on 'Conservative', p. 97, above.

170 *infinite variety*: the phrase that Enobarbus applies to Cleopatra in *Antony and Cleopatra*, II. ii.

171 [*Epigraph*] ANACREON: lyric poet, born about 570 BC at Teos who later lived on Samos and eventually in Athens. His poetry, according to the *Oxford Classical Dictionary*, 'is concerned mostly with pleasure'. The fragment P. quotes here is numbered 116 in J. M. Edmonds's *Greek Lyric*, ii. (Loeb).

Condiments of sea-water and turpentine: cf. C. T. Seltman, *Wine in the Ancient World* (1957), p. 91: writing of the Greek custom of mixing water with wine, Seltman says, 'According to a late authority [which he does not specify] a favoured proportion was one part of wine to three parts of water.' Athenaeus (*Deipnosophistae*, Loeb, i. 143; see note to p. 172, below) says that wines 'carefully treated with sea water do not cause headache' and instances the Myndian and the Halicarnassian, noting that the satirist Menippus calls Myndus 'salt-water drinker' and adding, 'The wine of Cos also is very highly treated with salt water'. Pine resin was added to strengthen and preserve wine (cf. modern *retsina*).

Christophero Sly: the drunken tinker in the Induction to *The Taming of the Shrew* who, duped into believing that he is actually a great lord, calls for 'a pot o' the smallest ale'.

172 *Rhianus*: Cretan author of epic poems and epigrams, born around 275 BC. This particular epigram does not appear in the Palatine Anthology Appendix included in Loeb (for Loeb edn. of the Anthology see note to p. 180). P. must have been citing another appendix to the Anthology compiled by a modern scholar. The epigram is quoted by Athenaeus (for whom see note to p. 172, below)—Loeb, v. 231—where it is translated as follows: 'This flagon, Archinus, contains exactly one-half resin from pine cones, one half wine. And I know not the flesh of a

leaner kid than this; yet Hippocrates who sent them is worthy to be praised on all accounts.'

Alcaeus: see note to p. 108, above.

Nonnus: see note to p. 148, above.

Athenaeus: (*c.* AD 200) Athenaeus of Naucratis in Egypt, author of *The Deipnosophists* (*The Sophists at Dinner*), an account of a banquet, extending over several days, attended by poets, philosophers, doctors, lawyers and others. These guests discuss all aspects of banqueting and conviviality and Athenaeus introduces into the discussion many relevant quotations and excerpts from a very wide range of ancient writers.

Pliny: Pliny the Elder (see note to p. 119, above).

[*P.'s note*] *Trelawny's Recollections*: *Recollections of the Last Days of Shelley and Byron* (1858) by Edward John Trelawny. Fletcher's strictures on Greek food and wine appear in ch. 18.

173 *coarseness of ours*: the Greek enharmonic scale made regular use of quarter intervals, whereas the smallest interval in classical Western music is the semitone.

Ritson: Joseph Ritson (1752–1803), antiquary and collector of ancient regional poetry; fiercely polemical, he attacked the authenticity of Percy's *Reliques* and exposed various forgeries. The words 'I will not go so far . . . melody' occur almost *verbatim* in P.'s essay on Bellini (Halliford, ix) whom P. praised for his melodies while deploring his harmony. But P. also praised the 'torrent of sound' of Beethoven's harmony in *Fidelio* (ibid.) (*cf.* Mr Falconer's 'Niagara of sound').

'*the Campbells are coming*': Scottish folk-song originating in 1715 when John Campbell, 2nd Duke of Argyll, successfully led the Government troops against the Jacobite rebels.

. . . *swelled the note of praise*: stanza 10 of Gray's *Elegy written in a Country Church-Yard*. The verb is actually 'swells'.

175 *relievo*: relief.

. . . *call for different dishes*: Horace, *Epistles*, II. ii. 61–2.

176 *minutiae and kaleidoscopical colours*: P. is no doubt here alluding primarily to the group of young painters (Holman Hunt, Millais, Rossetti) who styled themselves 'The Pre-Raphaelite Brotherhood' and whose work with its emphasis on minute fidelity to Nature created such a sensation in the artistic world in the early 1850s.

buz: to finish (a bottle) to the past drop.

177 *competitive examination man*: see note to p. 102, above.

pragmatical: meddlesome.

to the top of his bent: *Hamlet*, II. ii.

[*P.'s note*] EUTRAPEL, *p. 28*. 'Eutrapel' was the pen-name of Noel du Fail, Seigneur de la Herissaye (died about 1585). David Garnett implies that P.'s quotation comes from a work called *Oeuvres Facetieuses* a copy of which we have not been able to locate.

178 *Heautontimorumenos*: title of one of the comedies of the Roman dramatist Terence (second century BC).

A fox that has lost his tail: from *Aesop's Fables*.

pour avoir de la considération: the last words of Voltaire's *Lettre d'un Turc sur les Fakirs et sur son ami Bababec* (1750). Bababec is, in fact, the name of the fakir who is persuaded to leave his chair of nails, in which he sits naked with a great heavy chain to his neck, to wash and dress and live a good life. But women no longer come to him for spiritual guidance when he does this, so after two weeks he returns to his chain and nails to regain their respect. Cf. P.'s earlier use of this allusion (Halliford, ix. 293).

a distinguished meddler with everything: i.e. Lord Brougham—see note on 'Science of Pantopragmatics' to p. 137, above. The *Annual Register* for 1857 in its report of Brougham's founding of the Social Science Association called him 'a veteran philanthropist, whose life, now protracted into a vigorous age, has been devoted to the great purposes of the social, physical and intellectual advancement of mankind' ('Chronicle', pp. 200–1).

179 *La Dive Bouteille*: see bk. v, ch. 45 of Rabelais's *Histories of Gargantua and Pantagruel* in which Panurge solemnly consults the oracle of 'the Divine Bottle' and hears in response the one word 'trink'.

180 [*Epigraph*] *Anthologia Palatina*: *v. 72*: i.e. Epigram no. 72 in bk. v of The Palatine Anthology (so called because known only from a unique manuscript in the Palatine Library at Heidelberg), a vast collection of Greek epigrams made about AD 980 by an unknown Byzantine scholar. It appears in Loeb as *The Greek Anthology* (5 vols., tr. W. R. Paton, 1916). This epigram appears on p. 163 of vol. I where it is attributed to Palladas of

Alexandria (fourth century AD) and is translated as follows: 'This is life, and nothing else is; life is delight; away, dull care! Brief are the years of man. Today wine is ours, and the dance, and flowery wreaths, and women. Today let me live well; none knows what may be tomorrow.'

181 *... spiegasti l'ali*: the final aria (actually 'Tu che a *Dio* spiegasti ...') in Donizetti's *Lucia di Lammermoor* (1835), in which the hero (Edgar, the Master of Ravenswood) laments the death of Lucy. The opera is based on Scott's *The Bride of Lammermoor*. Rubini (1793–1854) was a celebrated Italian tenor whose career in England was at its height in the 1830s and early 1840s. P.'s admiration of him is evident in his essays on music (Halliford, ix), particularly in the essay 'Bellini' in which he also defended 'redundancies of ornament' as expressions of feeling.

'My mother bids me bind my hair': poem by Anne Hunter (1742–1821), set to music by Haydn and included in his *Six Original Canzonettas* (1794).

182 [*P.'s note*] *Braham ... in 1832*: John Braham (1774–1856), the great operatic tenor, said in answer to a question as to whether English ballads were less admired as a 'result of the developing taste for Italian music': 'there is always a beauty and an appeal to the heart in ballads which will never be lost except to those who pretend to be fashionable, and to despise the voice of nature' (*Report from the Select Committee appointed to enquire into the Laws affecting Dramatic Literature*. Parliamentary Papers 1831–2 [679], vii. 94).

[*P.'s note*] *Barbazan*: Etienne Barbazan (1696–1770) published his *Fabliaus et contes des poetes francois des XI, XII, XIII, XIV & XVes siecles tirés des meilleurs auteurs* in 1756. The poem, 'Du Vair Palefroy', by Huon le Roy, was added in the 1808 edn. by Meon (i. 164–208), a copy of which was in P.'s library.

185 *Brindley*: James Brindley (1716–73), the son of a cottier, or small farmer, in Derbyshire, remained illiterate all his life but had a prosperous career as a repairer of old machinery and as a civil engineer. He constructed over 365 miles of canal including the Bridgewater (Manchester and Liverpool) and Grand Trunk (Trent and Mersey).

[*P.'s note*] *Donaldson History of Greek Literature, vol. iii. p. 156*: i.e. K. P. Müller's *History of the Literature of Ancient Greece*, translated and completed by John William Donaldson, 3 vols., 1858. P.

reviewed Donaldson at length in *Fraser's Magazine*, March 1859 (Halliford, x. 163–229). Hermogenes (second century AD) achieved celebrity as an orator at the age of 15 when he attracted the attention of the Emperor Marcus Aurelius; his first work on rhetoric was published when he was 17 and became a standard manual of instruction.

[*P.'s note*] PONT-Y-PRID: town in Glamorgan, Wales, which takes its name from the celebrated stone bridge spanning the river Taff with just one arch, built in 1755 by William Edwards (1719–89), a self-taught mason.

Haydn ... good morning: the incident occurred during Haydn's residence in London, and the 'pupil' was an English nobleman. What Haydn actually said was, 'I see, my lord ... that it is you who are so good as to give lessons to me, and I am obliged to confess, that I do not merit the honour of having such a master' (*The Lives of Hadyn and Mozart*, tr. from the French of L. A. C. Bombet [i.e. Stendhal], 1818, pp. 187–8.

186 *sold it to Pope Clement the Sixth*: Joanna I of Naples (*c.*1327–82) was suspected of complicity in the murder of her first husband, Andrew of Hungary. When his brother, the King of Hungary, invaded Naples she took refuge in Provence, then part of the Neapolitan Kingdom. In return for the Pope's protection against her rebellious nobles Joanna sold him the city of Avignon (19 June 1348).

[*P.'s note*] PLATO: *Alcibiades i*: PERSIUS: '*Sat*'. iv: the *First Alcibiades* of Plato is a dialogue between Socrates and the 18-year-old Alcibiades who is about to embark on a political career. Socrates demonstrates to him that he is ignorant of the most important things that should be known by anyone dealing with public affairs. Persius's Fourth Satire deals in passing with the question of personal qualifications for taking part in affairs of state (ll. 1–16), using Alcibiades as an example. Persius is not, of course, concerned with competitive examinations: the overall purpose of the poem is to counsel self-knowledge.

187 *Chimborazo*: a mountain in Ecuador.

188 [*Epigraph*] TASSO: *Aminta*: the final lines of Act I of Tasso's pastoral drama *Aminta*, first performed in 1573.

Niphet: from the Greek νιφετός, 'a snow shower'.

fountain-fire: *Thalaba*, bk. x, st. 22.

Cimarosa's Orazii: *Gli Orazi e I Curiazi*, one of the last operas composed by Domenico Cimarosa (1749–1801), based on the classical legend of the two sets of brothers, the Roman Horatii and the Alban Curiatii who fought each other to decide the war between their two cities. Horatia, sister of the Horatii, is married to one of the Curiatii and vainly seeks to prevent the battle. When the only combatant to survive proves to be one of her brothers she curses him in a frenzy of grief for her dead husband; her brother rebukes her for her lack of patriotism whereupon she invokes the gods' curses on him and he strikes her down with his sword. In calling her Camilla P. was doubtless thinking of Cimaroso's source, Corneille's *Horace*, where the heroine is named Camille.

189 *Horace ascribes to all singers*: the sentence is a close translation of Horace, *Satires*, II. iii. ll. 1–3.

190 *Protean*: versatile; from the name Proteus, a minor sea-god of classical mythology who had the power to change himself into all manner of shapes.

191 . . . *all mankind's epitome*: *Absalom and Achitophel*, ll. 545–6.

Melpomene: one of the nine Muses, the Muse of Tragedy.

192 *L'amour vient sans qu'on y pense*: 'one falls in love without realizing it'.

console even Orlando for the loss of Angelica: Orlando pursues Angelica throughout *Orlando Furioso* but she eventually marries Medor, a young Saracen soldier (canto 19). When Orlando discovers this he goes mad with grief. He is never consoled for the loss of Angelica.

letting . . . wait upon I would': *Macbeth*, I. vii.

193 *Triton*: the merman of classical mythology.

'*dank and dripping weeds*': the phrase is from Milton's translation of the Fifth Ode of Horace, *Lib.* 1, where the actual words are 'dank and dropping weeds'. P. may intend an ironic effect here since Horace's poem, which compares falling in love to a shipwreck, describes the ex-lover hanging up his wet sea-clothes in Neptune's temple as a thank-offering to the god for preserving him.

195 [*Epigraph*] ARIOSTO: canto 25 of *Orlando Furioso*.

the great horse-tamer: J. S. Rarey, from Ohio, achieved great fame in Britain in 1858/9. His treatise *The Taming of Wild*

Horses went through six editions in two years, and he gave demonstrations to groups of subscribers and lessons to wealthy individuals (at £25 each). He became 'the Lion of the London Season', earning, it is estimated, £20,000 in one year (see *A New Illustrated Edition of J. S. Rarey's Art of Taming Horses*, 1858).

Absalom: favourite son of King David; accidently hanged when his hair caught in the boughs of an oak tree which he was riding under (2 *Samuel* 18: 9).

[*P.'s note*] *Hartleap Well*: ll. 39–40. P. is quoting from an early version of the poem, which was revised in 1820.

196 *Atalanta*: in Greek mythology a faster runner and a more intrepid hunter than any man.

[*P.'s note*] *wondering minds*: *Aeneid*, vii. 814.

198 *Antaeus*: son of Poseidon and Earth, who challenged all comers at wrestling and was invincible as long as he remained in contact with the earth. Hercules killed him by lifting him into the air and crushing him.

Pygmalion: the story of Pygmalion appears in bk. x of Ovid's *Metamorphoses* (ll. 243–97). He falls in love with a beautiful ivory statue of a woman that he has made and Venus brings it to life for him. Ovid describes the ivory as *niveum* (snowy), so the reference is a very appropriate one in relation to Miss Niphet.

199 *accidents by flood and field*: *Othello*, I. iii. It is in listening to Othello's account of his adventures that Desdemona falls in love with him.

200 *balloons*: ascent in balloons had become more common in the 1820s and 1830s when coal gas replaced hydrogen as a much cheaper means of inflation. By the late 1850s the English Channel and the Alps had been crossed and a new distance record was set in 1858 with a flight from St Louis to Henderson, New York. Ballooning was nevertheless still dangerous because the gas was inflammable and there was no means of propulsion (other than air currents), and therefore no efficient way of steering.

Vitruvius: Roman architect, involved in the rebuilding of Rome in the reign of Augustus. His book *On Architecture*, written probably before 27 BC, is the chief written record of classical architecture and was immensely influential in the Renaissance.

In ch. 5 he describes the construction of theatres as it relates to musical harmony ('the ancients contrived to strengthen the voice by the harmonic construction of the theatres'), and gives specific directions for the placing of the *echeia* (brass echoing vases) under seats in the auditorium. P. is not quite correct in stating that Vitruvius says the vases 'gave an accordant resonance in the fourth, the fifth and the octave'. In fact Vitruvius prescribes thirteen vases (for a small theatre) resonating in the second, the fifth, the octave, the ninth, the eleventh, the twelfth, and the double octave. See W. Newton's translation of Vitruvius (1791) i. 102 ff and plates 38 and 39. P. was fully conversant with the principles of Greek harmony: see note to p. 128 above. Curryfin's efforts are probably misguided because (*a*) Vitruvius was describing open amphitheatres made of stone, and (*b*) the effect of the vases depended on the player's special skill in directing his voice at a particular vase accordant with the pitch of his voice at that moment.

201 [*P.'s note*] GARDINER's *Music of Nature*: the composer William Gardiner (1770–1853) was best known as a church musician who used melodies and motifs from Mozart and others in his own arrangements and compilations. *The Music of Nature*, published in 1832, is, according to the title-page, 'an attempt to prove that what is passionate and pleasing in the art of singing, speaking and performing upon musical instruments is derived from the sounds of 'The Animated World'. The passage P. quotes is on p. 249. A corno di bassetto is a basset-horn or tenor clarinet.

diatessaron ... *diapente* ... *diapason*: the musical intervals of, respectively, a fourth, a fifth, and an octave. See note on Vitruvius, above.

202 *Reform*: Lord John Russell (see note to p. 137, above) was a prime mover in the introduction of the first Reform Bill (1831) and attempted to introduce another in 1853–4. It came to nothing, chiefly because the Crimean War intervened.

Gracchus: the brothers Gracchus—Gaius Sempronius (153–121 BC) and Tiberius Sempronius (163–133 BC)—were Tribunes at Rome, famous for their reforming legislation, furthering the cause of the people against the patricians.

Sisyphus: condemned to roll a stone uphill for ever in the Greek underworld, having betrayed a secret of Zeus.

202–3 *Limbo of Vanity*: the abode of 'all things vain and transitory' in *Paradise Lost*, iii. 440 ff. Milton's source is the Limbo of Vanity on the moon in *Orlando Furioso*, canto 34.

203 *Knowledge-is-Power Club* . . . THE POWER OF PUBLIC OPINION: MacBorrowdale is perhaps referring to popular agitation immediately before the passing of the Reform Bill. 'Knowledge is Power' was the motto of the 'unstamped' (i.e. unlicensed) radical newspaper *The Poor Man's Guardian*, first published in 1830 by Henry Hetherington, who in 1837 helped to found the Chartist movement. The phrase (attributed to Francis Bacon: *nam et ipsa scientia est potestas*, [*Meditationes Sacrae*]) was also in vogue shortly before P. wrote *G.G.* Brougham used it in his address at the first annual meeting of the National Association for the Promotion of Social Science in 1857. (See note on 'Science of Pantopragmatics' to p. 137, above.)

Hudibras: see, for example, Butler's account of the fight in Part I, canto ii.

204 '*L'Ordène de Chevalerie*': a thirteenth-century poem (included in Barbazan's collection—see note to p. 182 above) telling how the author won his freedom from Saladin by explaining the trappings and the significance of knighthood. It was translated by F. S. Ellis as *The Ordering of Chivalry* (1892). In P.'s poem the speech of the Sovereign is a close paraphrase of Hue de Tabaret's instructions to the Sultan.

Sir Moses . . . Aaron . . . Jamramajee: P. possibly has in mind Sir Moses Montefiore (knighted 1837) and Sir Jamsetjee Jejeeboy (knighted 1842 and made a baronet in 1857); 'Sir Aaron' is a fictional generic figure: nobody named Aaron had been knighted when P. wrote (see W. Shaw, *The Knighthood of England*, 1906). Montefiore was briefly a member of the stock exchange in the 1820s; Jejeeboy a merchant in Bombay. Both used their wealth philanthropically. In 'Gastronomy and Civilisation' P. objects to foreign names for English dishes 'on the same principle that we object to a Sir Moses, and a "Sir Jamsetjee"' (Halliford, ix. 386–7). In the MS version of the poem (see Appendix C) P. uses 'Jamsetjee' rather than 'Jamramajee'.

shroffing: an Anglo-Indian word: to 'shroff' is to 'examine coinage in order to separate the genuine from the base' (*OED*). As a noun, 'shroff' means, more broadly, a money-changer.

Balls Three: the traditional sign of a pawnbroker's shop.

205 *Ponsard*: François Ponsard (1814–67), writer of farces and historical plays and a noted opponent of Romantic literature. *La Bourse* was performed and published in Paris in 1856. There is no English translation. The conversation P. gives (with much left out) takes place during the Napoleonic Wars and can be rendered as follows:

ALFRED: You say the stock market has fallen now that we've conquered in war? Would it rise if we were beaten?

DELATOUR: We feared that such a success, so glorious for France, would defer hope for the peace we dreamed of.

ALFRED [who has lost all his money]: By God! The exchange has no feelings. Words are wasted on hungry men—for honour's sake! ... I'm not going to speculate any more— at least not until I'm married—but I look forward to making some money out of Waterloo.

206 *scrip above par*: share certificates sold above their face value (stock-exchange jargon).

207 [*Epigraph*] RABELAIS: *l.v.c. 45*: 'Trinq is a panomphean word: that is, a word understood and celebrated in all nations, and it means drink ... And here we maintain that it is not laughing but drinking that is the distinguishing mark of mankind. I don't mean drinking purely and simply, because even animals drink; I mean drinking good cool wine' (bk. v, ch. 46 in Penguin translation).

[*P.'s note*] JUV. *Sat. xiv. 41–43*: 'You will find a Catiline among any nation or in any climate; nowhere will there be a Brutus or a man like Brutus's uncle [i.e. Cato].' Catiline led a conspiracy against Rome and died in battle in 62 BC. Cato, who opposed Catiline, won a reputation for virtue and integrity in public life. Brutus was admired for his respectability and gravity in spite of his part in the conspiracy against Julius Caesar. Shakespeare's representation of him is probably flattering.

208 *Explosions of powder-mills ... factories*: Eric Robinson in his 'Thomas Love Peacock: Critic of Scientific Progress' (*Annals of Science* [1954], p. 71) notes: 'A very high proportion of the accidents reported in the *Annual Register* for the years 1856–60 was caused by the explosion of steam-boilers in collieries, ironworks, dockyards, factories, ships and locomotives.' Many of these resulted from the use of high-pressure boilers instead of the much safer but more expensive low-pressure ones that Opimian, and P. himself, approved of (see p. 134, above). The *Annual Register* for 1858 and 1859 records a total of 65 deaths

resulting from boiler explosions in factories, steam-boats, etc. (including 6 people killed on Brunel's *Great Eastern*); 236 died as a result of colliery explosions in 1857 and a further 133 in 1858; powder-mill explosions accounted for 12 deaths during 1859; and so on.

209 *hand-book*: i.e. guidebook. Handbooks (e.g. those published by Murray and by Baedeker) became extremely popular in the 1830s and 1840s.

Mauvaise auberge!: poor lodging.

electric telegraph: electric telegraphy was commercially developed in the 1840s, largely (in Britain) through the Electric Telegraph Company, founded 1845. There were attempts to lay a transatlantic cable in 1857 and 1858; the first successful link was made in 1866.

[*P.'s note*] CICERO: *Ep. ad Div. vii. i*: 'For I do not think you have any desire for Greek entertainments, especially since you like the Greeks so little that you are accustomed to avoid the *via Graeca* when you go to your country house.' Quoted with slight alteration, from Cicero's letter to his friend Marius: *Epistolae Familiares*, bk. VII, letter i (P. refers to an older title of this collection: *Epistolae ad Diversos*).

210 *Scattered their bounty*: source untraced. David Garnett compares Gray's *Elegy*, stanza 16: 'To scatter plenty o'er a smiling land.'

[*P.'s note*] *Prometheus*: l. 212 of *Prometheus Bound*, referring to Prometheus's mother, Earth.

The Red Indian: European experience of Negroes in the West Indies and of American Indians, together with the emergence of the scientific study of human origins (e.g. in Charles Darwin's *Origin of Species* (1853)), led to heated debate about racial differences. Dr Opimian's emphasis on the nobility of the Red Indians was not uncommon at the time (see, for example, *Westminster Review*, lxxiv (1860), 348–9); it owed less, perhaps, to scientific anthropology than to the mythologizing of the Red Indian as a noble 'savage' in, for example, the influential writings of Chateaubriand.

211 *Saint Domingo*: now known as Hispaniola. The part called Haiti was the first West Indian territory to gain self-government after European colonization. The aboriginal race (the 'Caraibs') had been exterminated by Spanish settlers in the sixteenth century, not by Negroes, who were imported later as

slaves. The Negroes, led by Toussaint L'Ouverture, rebelled against their French masters in 1791.

degenerates in America: this apparently strange opinion was common in some quarters of the British press (and not unknown among inhabitants of the Northern states of America) on the eve of the Civil War. See, for example, the survey of attitudes in *Edinburgh Review*, cviii (1858), 568: 'sectional conflict has delivered over the State to the management of an inferior and perpetually declining order of men . . . it is . . . too clear, that the average character of the American people has sunk far below its traditional representation.' Cf. P.'s letter to Lord Broughton, 13 Jan. 1860: 'Burke is said to have said, that the human mind degenerates in America. Is there anywhere chapter and verse for this quotation? Whoever said it, I think it is strictly true . . .' (BM Add. MS 47225, f. 104). Broughton's reply has not come to light, and the 'quotation' is untraced.

atrocities in their ships: possibly a reference to the unsanitary conditions and high mortality rate on slave ships. See *Chambers Journal*, xiii (1860), 281, for a first-hand account of a slave-voyage in which nearly 100 people died between Africa and Cuba. On non-slave merchant ships discipline was lax: 'almost every trip of a merchant vessel exhibits the existence of that tyranny which is substituted when subordination fails' (*Edinburgh Review*, cviii (1858), 573).

Courts of Justice: in the 1850s the violence of Southern representatives in Congress had become a byword. The most famous case was that in which Charles Sumner, a member for Massachusetts, was beaten to the ground by a spokesman for Southern slave owners in May 1856.

extensions of slavery: probably a reference to Kansas–Nebraska Bill of 1854, by which slavery was not legally excluded from the new territories in spite of opposition from the Abolitionists and other political groups of the North. American slavery was fiercely debated in Britain in the 1840s and 1850s: see, for example, the controversy between Carlyle and J. S. Mill in *Fraser's Magazine*, Dec. 1849 onwards. P. may also have had in mind the extension of powers to retrieve slaves absconding to free states following the Dred Scot case in 1857, which inspired the latter part of Harriet Beecher Stowe's novel *Dred*. Mrs Stowe had won enormous popularity in Britain with her earlier novel about slaves, *Uncle Tom's Cabin* (1852).

revivals of the slave-trade: by 1858 it was known to the British, who had abolished their slave trade in 1807, that France had reintroduced theirs and planned to extend it: America was not alone in its surreptitious trading, even re-enslaving freed Negroes in Liberia. The case of the ship *Regina Coeli*, intercepted in October 1857, brought the issue to the attention of the British public.

free trade: in 1846 Lord John Russell reduced the tax payable on imported sugar, allowing the non-British slave-colonies an advantage over British colonial exporters. British 'planters' vociferously pointed out that the Government was encouraging foreign slavery in the name of free trade. A large proportion of British plantations were bankrupted in the late 1840s. See W. A. Green, *British Slave Emancipation* (Oxford, 1976).

212 *Friar Bacon*: a figure based on the historical Roger Bacon (*c.*1220–*c.*1292), scientist and Franciscan monk. As legend had it, Friar Bacon magically contrived a brass wall around England to repel invaders. See Robert Greene's play *Friar Bacon and Friar Bungay* (*c.*1589).

white slavery: comparison between black slaves and white factory workers was common from the late eighteenth century onwards, and was particularly prevalent in the 1850s following Harriet Beecher Stowe's visit to England. Opimian's view that 'white slavery' was the worse of the two because it degrades 'a better race' was fairly widely held; see, for example, *Englishman's Magazine*, 1 (1852), 125–8, quoted by D. A. Lorimer in *Colour, Class and the Victorians* (Leicester, 1978).

done under the sun: see note to p. 152, above.

[*P.'s note*] CICERO *ad Atticum*: *iv. 16*: 'It is also well known that there is not a scrap of silver on the island, nor any hope of loot except in the form of slaves; and I do not think you can expect any of them to be skilled in literature or in music.'

[*P.'s note*] GIBBON: *c.i*: quoted from *The Decline and Fall of the Roman Empire* (1776–88), ch. 1, note 7. Gibbon refers to Pomponius Mela's *De Situ Orbis* (Description of the Earth), iii. 6. Mela was writing shortly after Rome had established a footing in Britain in AD 43. His treatise can be read in *Collection des Auteurs Latins*, ed. M. Nisard (Paris, 1845).

213 *'I've seen, and sure I ought to know'*: quotation untraced.

[*P.'s note*] *The anger of the Gods* . . .: Juvenal, *Satires*, xiii. 100.

[*P.'s note*] *the foot of Punishment* . . .: Horace, *Odes*, III. ii. 31–2.

214 *Plato* . . . *Xenophon* . . . *Plutarch*: Opimian refers to Plato's *Symposium* (the word means 'banquet'), a dialogue on the nature of love; to the *Symposium* of Xenophon (*c.*428–*c.*354 BC), probably written slightly later than Plato's, which it closely resembles; and to the 'Amatorius' of Plutarch, a dialogue on love which forms part of his *Quaestiones Conviviales* ('Table Talk').

Athenæus: see note to p. 172, above.

a great Indian reformer: P. doubtless has in mind Lord Macaulay (1800–59), author of *A History of England* (5 vols., 1848–61), who was responsible for the reformation of the legal system in India (his penal code became law in 1860) and took part in founding the Indian educational system. P. first met him on 31 Dec. 1851 when, Macaulay recorded in his diary, 'We had out Aristophanes, Aeschylus, Sophocles, and several other old fellows, and tried each other's quality pretty well. We are both strong enough in these matters for gentlemen . . .' (Halliford, I, clxxiv). This sounds like a highly congenial encounter for P. but he may have suffered on other occasions from what Sydney Smith called Macaulay's 'torrents of talk' (Lady Holland, *Memoir of Sydney Smith* (1855), i. 363).

Temple of Juggernaut: at Puri, in Orissa. 'Juggernaut' was the local cult-name of the Hindu god Vishnu. He was celebrated by the building of a large pyramidal temple in the late twelfth century, and the annual drawing of his chariot from the temple to his 'country house' by thousands of pilgrims, many of whom, according to legend, were crushed by the chariot.

Restriction Act of ninety-seven: in 1797 the Government, owing to a monetary crisis brought about by the war with France, ordered the Bank of England to issue paper money instead of gold coin. The order remained in force until 1821. P.'s criticism of paper money is expressed in *Melincourt*, in his 'Paper Money Lyrics' and, perhaps most amusingly, in the fragmentary novel *Calidore*, ch. 4 (see Halliford, viii).

215 *'free to confess'* . . . *'in another place'*: further examples of parliamentary euphemism. The former phrase was adopted by the younger Pitt in 1788 and became established jargon. The latter is a phrase used by members of the House of Lords in referring to debates of the House of Commons and vice versa.

Purification of the Thames: see note to p. 98, above.

Competitive Examinations: see note to p. 102 above.

216 *Mercuries are made*: cf. Apuleius, *Apologia*, c. 43: Non enim ex omni ligno . . . debetur Mercurius exculpi ('You cannot make a Mercury out of any kind of wood'). Equivalent to 'you cannot make a silk purse out of a sow's ear'.

'Mass . . . I cannot tell': *Hamlet*, v. i. Cf. P.'s letter to Thomas L'Estrange, 11 July 1861 (Halliford, viii. 253): 'In the questions which have come within my scope I have endeavoured to be impartial, and to say what could be said on both sides. If I have not done so, it is because I could find nothing to say in behalf of some specific proposition, as in *Gryll Grange*, page 171 [i.e. end of ch. 19].'

[*P.'s note*] . . . *sidere afflavit*: 'Your flatulent and formless flow of words is a modern immigrant from Asia to Athens. Its breath fell upon the mind of ambitious youth like the influence of a baleful planet.' *Satyricon*, sect. 2.

217 [*Epigraph*] LA ROCHEFOUCAULD: 'The violence we do to ourselves in order to avoid loving are often more cruel than the demands of those we love.' No. 369 of the *Maximes* (edn. of 1678).

'To observe romantic method': Butler's *Hudibras*, pt. II, canto i, l. 1 ff.

218 [*P.'s note*] BOJARDO: *l. ii. c.8. Ed. Vinegia: 1544*: *Orlando Innamorato*, bk. 2, canto viii, st. 57–9.

[*P.'s note*] *I have translated Fata, Fate*: see note on Morgana to pp. 106–7 above. Panizzi, in his edn., uses the word 'fortune'; either way, Morgana is represented as having allegorical significance.

219 *Capitolo dell' Occasione*: for Machiavelli's poem see note to p. 39, above. P.'s apparent assumption that Machiavelli's poem was written later than *Orlando Innamorato* may not be correct.

Berni's rifacciamento: Boiardo's poem was thoroughly revised by Francesco Berni (pub. 1541). Panizzi, who restored the original text in his edn. of 1830–4, criticizes Berni in the accompanying 'Essay on Romantic Narrative Poetry of the Italians' for making the poem a continuous series of cantos with no book divisions, and remarks that not one of the critics who praised the *rifacciamento* had read Boiardo from beginning to end.

220 *Livy . . . Niebuhr*: historians of radically different kinds. Livy

(*c*.59 BC–AD 17) wrote a vast *History of Rome* (142 books, of which only 35 survive intact), which established him as Rome's leading historian and exercised great influence on post-Renaissance curricula. His early books are pure mythology, and his work is distinguished throughout by moral and panegyrical qualities rather than methodological rigour. Barthold George Niebuhr's *Römische Geschichte* (1811–32; English translation 1828–42) was immediately recognized as the first sustained attempt to write the history of Rome in a scientific spirit. It was of great importance in an era which saw the rise of 'scientific' (i.e. literary-historical) criticism of the Bible.

221 *post-horses*: post horses were kept at inns. The traveller would ride from one to another, changing horses at each stage or 'post'. This form of travel was common from the sixteenth century.

magnetic rock in the Arabian Nights: in 'The Story of the Third Calender'.

'Before the starry threshold of Jove's court': Milton, *Comus*, ll. 1–4.

222 *crosshead of a side-lever steam-engine*: a crosshead is 'the bar at the end of the piston-rod of a steam-engine, which slides between straight guides, communicating motion to the connecting rod' (*OED*), i.e. it moves with an up and down motion, not swinging side to side as the connecting rod does. P. supervised trials of steamships (his 'iron chickens') for the East India Company.

[*P.'s note*] *Ode to Tranquillity*: ll. 11–16.

223 *Hebes*: Hebe was the cup-bearer of the gods in Greek mythology. She had the power of rejuvenating both gods and men.

'*... who wants nothing*': the title of ch. 3 of Johnson's *Rasselas*. The point is that Rasselas, having everything he wants, is dissatisfied because he *wants to want* something.

224 [*Epigraph*] PLAUTUS: *In Pseudolo*: these lines are not by Plautus. They form ll. 20–2 of a largely spurious prologue in Renaissance edns. of his comedy *Pseudolus*, e.g. in the 1684 Amsterdam edn. by Gronovius, a copy of which was in P.'s library.

cut a nine: quotation untraced—possibly from a parody of Wordsworth.

sextillions: a sextillion is the 6th power of a million: 1 with 36 cyphers. Presumably P. means that Curryfin executed the sequence 898 thirteen times in a row.

[*P.'s note*] ... *through the circle*: Pindar's Ninth Olympian Ode celebrates Epharmostus, the champion wrestler at the Olympic Games held in 468 BC. Lines 138–40 read: 'having vanquished ... by the cunning skill that swiftly shifts its balance but never falls, amid what loud applause did he pass round the ring.'

226 *mythological fiction*: see Aristophanes' speech in Plato's *Symposium* (189a–193d). It seems certain that Plato intends to satirize the notion that there is a 'right' person for each of us.

227 *'the cynosure of neighbouring eyes'*: Milton's *L'Allegro*, l. 80.

228 *silex*: silica (flint).

Artium ... *Et Socii*: this means 'the syndics and fellows of the society of arts'. P. doubtless chose his Latin words for the sake of the acronym ASSES. He perhaps refers to the Society of Arts (later the Royal Society of Arts) which held open examinations for the first time in 1858, chiefly for candidates who had been educated at Mechanics' Institutes. In fact, the appearance and use of the 'celt' had been described in *Philosophical Transactions of the Royal Society* in 1796.

a hole for a handle?: later celts did, in fact, have holes or depressions for handles; but the earlier were presumably tied to a handle by leather thongs. Opimian's scepticism is based on a narrow preconception.

[*P.'s note*] *in every other encyclopædia*: *Aelia Laelia Crispis* is a sepulchral inscription found near Bologna in the seventeenth century dedicated to 'neither man nor woman, nor hermaphrodite, nor girl, nor youth, nor old woman ...' P. himself wrote a short essay on this enigma (Halliford, x. 321 ff.) including an English translation, and concluded that it was a reflection on mortality, an 'epitaph' on mankind generally.

229 [*P.'s note*] *Miscellaneous Works*: Butler's poem (written *c.*1676) is a satire on the Royal Society. *Satires and Miscellaneous Poetry and Prose*, ed. R. Lamar (1928), pp. 3 ff.

230 [*Epigraph*] PERCY's *Reliques*: the poem is 'Love will find out the Way', from bk. iii of the Third Series of *Reliques of Ancient English Poetry* (1765), the enormously influential ballad collection of Thomas Percy, Bishop of Dromore (1729–1811).

the Seven against Thebes: see Aeschylus' play of that name. Polynices with six companions leads an attack on Thebes, each at one of the seven gates of the city, which is ruled by his brother, Eteocles.

231 *like a Spartan*: i.e. laconically. Spartans were renowned in ancient Greece for, among other things, their directness of speech. 'Laconic' derives from 'Lacedaemon', the name for Sparta.

232 *agistor . . . another office*: *OED* cites an 1862 source for the word with this meaning and it is still used today to denote this office.

233 *coryphæus*: the leader of the chorus in Greek tragedy.

halycon: in classical legend the halycon built its nest floating in the sea and during its breeding period had the power to charm the winds and waves, so that the weather was calm for the week before and the week after the winter solstice (22 and 23 Dec.).

'the bird of dawning singeth all night long': *Hamlet*, I. i. Shakespeare's source for the idea that the cock crows all night on Christmas Eve is unknown. There is some evidence that the idea had become, by the nineteenth century, a popular superstition. See the New Arden *Hamlet* (ed. H. Jenkins), p. 431.

234 *the gigantic sausage*: a large sausage had been part of the traditional fare at Christmas, since at least the sixteenth century. See, for example, T. Tusser: *Five Hundreth Points of Good Husbandry* (1590 edn.).

old October: beer brewed in October and kept until Christmas. October and March were regarded as the best months for brewing.

Eumaeus: swineherd in *Odyssey*, xiv. ll. 418 ff. Odysseus, returning to Ithaca disguised as an old beggar, is hospitably received (though unrecognized) by his faithful swineherd who kills and roasts 'a fatted boar of five years old' to feast him. Homer uses of him the adjective 'δῖος', meaning 'goodly', but this could also be translated 'god-like', or 'divine'.

heptachords: series of seven notes: the seven maids and the seven men.

[*P.'s note*] *De Officiis: i.9*: Chremes is a character in Terence's play *Heauton Timorumenos* (The Self-Tormentor); the famous phrase *homo sum: nihil humani a me alienum puto* ('I am human: therefore I consider nothing human is foreign to me') occurs at l. 77.

235 [*Epigraph*] ALEXIS: *Tarantini*: Fragment 219 in J. M. Edmonds's *Fragments of Attic Comedy*, ii (1959), 477–9. 'The Tarentines', like Alexis's other extant plays, exists only in fragments; it was written at about 360 BC. Alexis, author of 245 plays, was a leading writer of the so-called Middle Comedy and an associate (and perhaps teacher) of Menander.

the old-fashioned game of cards: quadrille is, like tredrille, a variant of ombre (see note to p. 238, below); all became fashionable in the early eighteenth century.

the more recently fashionable dance: quadrille, a dance normally for four couples, in which those taking part form a square. Originally a kind of country-dance, it became fashionable around 1815, chiefly, it is said, because it was a favourite with the Duke of Devonshire.

carpet-dance: i.e. an informal dance. The carpet would be taken up on more formal occasions, as at the Twelfth Night dance (ch. 31).

236 *Minuet de la Cour*: the minuet, normally a dance for two people, had first been fashionable at the French court of Louis XIV and XV, and was very antiquated by the time the polka became popular in the 1840s. P.'s phrase '*Minuet de la Cour*' probably does not refer to a particular kind of ballroom minuet: the term appears only to have been used of a theatre-dance in the late eighteenth century.

a dry, solemn, studious game: whist, the forerunner of bridge, became popular after the publication of Hoyle's *Short Treatise of Whist* in 1742—long before Mr Gryll's 'younger days'—and was always considered a more serious game than ombre and its derivatives. Compare Lamb's essay 'Mrs Battle's Opinions on Whist' (1821, in *Essays of Elia*): 'Quadrille . . . was her first love; but whist had engaged her maturer esteem. The former, she said, was showy and specious . . . a game of captivation to the young and enthusiastic. But whist was the *solider* game: that was her word. One or two rubbers . . . gave time to form rooted friendships, to cultivate steady enemies.' Whist was for some time also known as 'whisk': hence there may be some word-play on 'whisking' and 'whist' in Opimian's next speech.

237 *Late dinners*: the fashionable hour for dinners was 6 or 7 p.m., or later. In provincial society dinner was at 2 or 3 p.m., tea at 5 or 6 p.m., followed perhaps by supper at 9 or 10. Opimian's dislike of late dinners was shared by P. 'We add to the

business-imposed late hours of dining the affectation of later' (*Gastronomy and Civilization*, Halliford, ix. 393). The sentiment was common: see for example, Thomas Walker's *The Original* (2nd edn., 1838) p. 379.

Gay's ballad: 'A Ballad on Quadrille'; now known to be the work of Congreve. This ascription was first made by Scott in his edition of Swift's works (1814; xiii. 319): P. doubtless knew the poem from one of the many eighteenth-century editions of Gay's poems. The Latin line means 'you must not dance the quadrille'.

[*P.'s note*] *Boileau*: 'Every age has its own pleasures, its spirit and its customs.' From *L'Art Poétique*, iii. 374, by Nicolas Boileau (1636–1711), the leading French critic and literary theorist of the 'Classical' period.

238 *'the merry dance I dearly love'*: quotation untraced.

Rape of the Lock: in canto iii, ll. 25–100 Pope describes in detail a game of ombre. Geoffrey Tillotson, in his edn. of *The Rape of the Lock and other Poems* (1940), p. 361, points out that Miss Ilex is confusing two different games: ombre and tredrille. Ombre is usually played with a 40-card pack (the 8s, 9s, and 10s being removed), of which only 27 are in the hands of the three players. But the 13 remaining cards (the 'stock') can be drawn by the players when they discard their weak cards before play begins. Tredrille, on the other hand, is usually played with 30 cards, one of the red suits being omitted entirely along with the 8s, 9s and 10s of the other three suits. See Richard Seymour, *The Compleat Gamester* (1754), p. 45.

239 *... where Lycid lies*: Milton's *Lycidas*, ll. 142–51 (with 'deck' for 'strew').

... smooth-shaven green: Milton's *Il Penseroso*, ll. 63–6.

[*P.'s note*] *I quote from memory*: the passage is accurately quoted from *Elements of the Philosophy of the Human Mind*, ch. 5, pt. 2, sect. 4 (3rd edn., 1808, p. 310). The point of the remark is that artifice in poetry is seldom noticed because most readers are not sufficiently attentive. Dugald Stewart (1753–1828) was professor of moral philosophy at Edinburgh and an immensely influential lecturer. P. apparently did not rate him highly: see his essay 'The Epicier' (Halliford, ix. 294).

Moore: Thomas Moore (1779–1852) was, like Burns, a popular writer of lyrics which were often set to music. P.'s dislike of

Moore's work is evident also in his review of Moore's *The Epicurean*, 1827 (Halliford, ix. 1 ff.). The stanza quoted is from 'Oh breathe not his name', in Moore's *Irish Melodies*, with slight variants.

240 *Excelsior*: see Longfellow's poem 'Excelsior'. Opimian's view seems somewhat pedantic.

... burning gold: Tennyson's 'A Dream of Fair Women', ll. 125–8. Tennyson's retort is interesting: he wrote in the margin of his copy of *G.G.* 'Think on me that am with Phoebus' amorous pinches black and wrinkled deep in time' (*Antony and Cleopatra*, I. v). See J. L. Madden, *Notes and Queries*, ccxiii (1968), 416–17.

241 *Queen of Bambo*: Bambo is on the river Niger in West Africa. P. refers jokingly to 'the king of Bambo' in *Crotchet Castle*, ch. 6.

Æthiop: P. refers to Millais's illustration at the head of the poem 'A Dream of Fair Women' in the 1857 edition of Tennyson's *Poems*. P.'s epithet 'grinning' is not accurate. The illustration shows Cleopatra pointing to the 'aspick's bite' on her breast.

Ptolemy Auletes: Ptolemy XI, nicknamed 'Auletes' ('flute-player'), King of Egypt in the mid first century BC. Ptolemy I, founder of the dynasty, was the son of Philip of Macedon and half-brother of Alexander the Great. The culture of the court at Alexandria was Hellenic, but only the second Ptolemy (Philadelphus) is known to have enthusiastically fostered the arts. P.'s 'lady of Pontus' is a mystery. It is not known who Cleopatra's mother was: probably either Cleopatra Tryphaena (Ptolemy's wife and sister) or an Alexandrian courtesan. Possibly Opimian is confusing the former with Cleopatra, daughter of Mithridates VI of Pontus (died 63 BC), who married King Tigranes of Armenia.

like Theseus: a reference to the legend that Theseus, attempting to rescue Prosperina from the Underworld, was punished by being tied or (in some accounts) held by snakes to a rock known as the Stone of Forgetfulness. See Apollodorus's *Epitome*, i. 24 (ed. Sir J. G. Fraser, 1921, ii. 153) and Virgil's *Aeneid*, vi. 617–18.

[*P.'s note*] MOORE's *Epicurean, fifth note*: Thomas Moore's *The Epicurean*, note to p. 25. P.'s scathing review of Moore's prose romance (Halliford, ix. 1 ff.) alludes to this passage among others. Moore's source for his note is Joannes Cornelius De Pauw's *Philosophical Dissertations on the Egyptians and Chinese* (tr. 1795), i. 50. De Pauw is quite clear that the Ptolemies (and therefore Cleopatra) were Europeans. The phrase 'countrywomen of Cleopatra' is Moore's.

[*P.'s note*] DIO, *xlii. 34*: Dio's *Roman History* (Loeb, iv, 168–9). Cassio Dio, probably born AD 160–70 was a Roman orator and consul. His history is in eighty vols. concluding in the year AD 229.

243 [*Epigraph*] ANACREON: the quoted lines are excluded from modern editions of his poems as spurious. They can be read in, for example, the editions by J. Fischer (Leipzig, 1776), poem no. 62 'Εἰς Μυρίλλαν' ('To Myrilla'). The poem was translated by Moore (*Odes of Anacreon*, 1800, no. 66).

Atalanta: see note to p. 196, above.

244 '*poetry of motion*': a catch-phrase applied to dancing, in use at least as early as 1815. See *OED Supp*. s.v. 'poetry'.

248 [*P.'s note*] '*the key of fortune few can wisely use*': see note to p. 218, above.

[*P.'s note*] *several stanzas forward*: bk. 2, canto ix, st. 25.

249 . . . *my fee was scorn*: source as yet untraced.

[*P.'s note*] *Berni and Domenichi*: on Berni, see note to p. 219 above. Panizzi (*Essay on Romantic Narrative Poetry of the Italians*) was even more scornful of Domenichi's version (pub. 1545) than of Berni's.

[*P.'s note*] . . . *obrepere somnum*: 'It is allowable that sleep should occasionally interrupt a long work.' Horace, *Ars Poetrica*, l. 360.

[*P.'s note*] *the Milanese edition of 1539*: Panizzi's edition has 'valore' instead of 'furore' in l. 3; also 'O Dio redentore' instead of 'ahime Dio Redentore' in l. 1. But he does not always use the Milanese version: e.g. in bk. 2, canto xx, st. 14, he has 'Con una *vesta* bianca' where the Milanese text has 'testa bianca'.

251 [*Epigraph*] HOMERUS *in Odysseâ*: *Odyssey*, i. 365. After an appearance by Penelope in her husband's hall 'the wooers broke into uproar throughout the shadowy halls, and [l. 366] all prayed, each that he might lie by her side'. Entertainingly adapted by P. to introduce Harry Hedgerow and his friends wooing the 'vestals'.

252 *'dallying with the innocence of love'*: 'And dallies with the innocence of love', *Twelfth Night*, II. iv.

254 [*Epigraph*] ALCAEUS: fragment no. 335 of Alcaeus's poems, as printed by D. A. Campbell in his *Greek Lyric* vol. I (Loeb): 'We should not surrender our hearts to our troubles, for we shall make no headway by grieving, Bacccchis: the best of remedies is to bring wine and get drunk.'

Francatelli: Charles Elmé Francatelli (1805–76), *maître d'hôtel* and chief cook to Queen Victoria, was the author of *The Modern Cook* (1846) and *The Cook's Guide* (1861), copies of which were in P.'s library.

258 [*Epigraph*] SCRIBE: *La Vieille*: Eugène Scribe (1791–1861), a prolific dramatist, many of whose works were translated or adapted into English. *La Vieille* (The Old Woman) is a short comic opera, apparently performed on the English stage in P.'s lifetime: 'You must fall in love once in your life, not for the sake of the moment when you are in love—because then you experience nothing but torments, jealousy and regrets—but because little by little the torments become memories, which delight us when we are older [. . .] and when you see old people sweet, comfortable and tolerant, you can say, as Fontenelle said, "Love has passed that way".' Bernard de Fontenelle (1657–1757), poet, playwright, and author of philosophical and scientific works, is best known for his *Nouveaux Dialogues des Morts* (1683). The quotation is untraced.

259 *an ancient sorrow*: P. may be remembering *Richard III*, IV. iv: 'If ancient sorrow be most reverend . . .'

Genius: i.e. spirit.

260 [*P.'s note*] *Orlando Furioso, c.i. s. 75*: 'Then he meekly approached the damsel; he was almost human in his gesture of humility, like a dog dancing round his master who has just returned after a few days' absence. Bayard still remembered her: for she had tended him in Albracca' (tr. Guido Waldman, OUP, 1974).

universal lover: P.'s cousin, Harriet Love, wrote to his granddaughter, 'No better description can be given of him during that period [i.e. 1811–19] than that (written by himself) by Miss Ilex in *Gryll Grange* . . . in which she speaks of him as "a sort of universal lover, making half-declarations to half the young women he knew"—&c. &c.' (Halliford, I. civ).

261 ... *and dwelt apart*: Wordsworth's sonnet 'London, 1802', referring to Milton.

... *transferred it*: Miss Ilex's speech is reminiscent, in its imagery, of Shelley's veiled autobiographical poem *Epipsychidion*, ll. 345–83. P. had written his *Memoirs of Percy Bysshe Shelley* shortly before the completion of *GG* (Halliford, viii. 39 ff. and notes). In it he had defended Shelley's pursuit of an ideal companion against the prurient disapproval of T. J. Hogg's *Life of Percy Bysshe Shelley* (1858) and E. J. Trelawny's *Recollections of the Last Days of Shelley and Byron* (1858).

262 *punctilio*: i.e. trivial formality.

... *the slumber of the dead*: from *The Pleasures of Hope*, 1799, pt. ii, l. 18) by Thomas Campbell (1777–1844).

263 [*Epigraph*] PETRONIUS ARBITER: the attribution of this passage to Petronius, originally made by John of Salisbury (died 1180) in his *Liber Policraticus*, is no longer accepted. P. would have found it in his 1781 Petronius, ed. P. Burmann—one of the many edns. of this favourite author in his library.

[*Epigraph*] SHAKESPEARE: *As You Like It*, II. vii.

[*Epigraph*] CALDERON: 'In the theatre of the world all are actors.' A leading theme in Calderon's play *El Gran Teatro del Mundo*, especially in the first scene. P.'s specific 'quotation' does not occur in Calderon's dramas. For Calderon, see note p. 112, above.

[*Epigraph*] *French Proverb*: 'Not all actors are in the theatre.'

spermaceti candles: tallow candles.

argand lamps: a form of oil lamp which, because of its cylindrical wick, gives more brilliant and more constant light than earlier types; invented by Aimé Argand in 1784.

Spirit-rapping Society: see note to p. 102 above.

[*P.s note*] '*More worthy ... the tightened cord*': Horace, *Epistles*, I. x. 48.

[*P.'s note*] *the Athenian theatre ... not in use*: in fact, there was no curtain at all in the Greek theatre. It was an invention of the Romans some time after 133 BC. Ovid and Virgil both refer to a curtain rising to conceal the stage though later, it seems, the Roman theatre curtain rose and fell as in the modern theatre. See W. Beare, *The Roman Stage* (rev. edn., 1964), pp. 267–74.

264 *Erebus*: in Greek mythology, a place of darkness between earth

and Hades (the Underworld). Ulysses visits Erebus in *The Odyssey*, xi.

266 *like Socrates of yore*: in Aristophanes' *Clouds* (see note to p. 102, above) Socrates ('priest of the subtlest nonsense') summons the chorus of clouds (l. 269), who are his deities.

267 [*P.'s note*] ὀμβροφόροι: *Clouds*, ll. 275–90 and 299 ff.

267–8 '*before their renown was around them*': source as yet untraced.

268 *Stygian*: infernal (the word is derived from 'Styx', a river in the classical underworld).

269 '*... with looks profound*': Gray's 'Ode for Music' (also known as the 'Installation Ode'), l. 3.

'*... that loves the ground*': Gray's 'Ode to Adversity', l. 28.

Alastor: P. suggested the title of Shelley's poem, 'Alastor', as he recalls in *Memoirs of Percy Bysshe Shelley* (Halliford, viii. 100): 'The Greek word αλαστωρ is an evil genius, κακαδαίμων ... The poem treated the spirit of solitude as a spirit of evil.'

270 '*Men are become as birds*': source of this quotation untraced.

271 [*P.'s note*] *Homer*: *Il. xxiv*: ll. 527–33.

[*P.'s note*] *... memorare omisit*: 'No one received unmixed wine from the urn that contained blessings: he omitted to record this.' Christian Heyne (1729–1812), Professor of Eloquence at Göttingen University, was one of the great German classical scholars of the eighteenth century; he produced his edition of the *Iliad* in 1802.

[*P.'s note*] POPE: Pope's translation of the *Iliad*, xxiv. 663–70.

272 *the Ayes have it*: Parliamentary phrase, meaning that the vote is carried.

273 '*Shadows ... shadows we pursue*': from Edmund Burke's speech on withdrawing from the election at Bristol, Sept. 1780. Burke had been MP for Bristol for 6 years, but, recognizing that the electorate was against him, he conceded defeat before the votes were counted.

[*P.'s note*] *Ecclesiazusae*: at the end of Aristophanes's comedy ('Women in Assembly') the chorus invites the characters to a table laden with (in one word): 'Plate-fillet-mullet-turbot-brain-morsel-pickle-vinegar-asafoetida-honey-poured-on-the-top-of-the-ouzel-thrush-pigeon-dove-cutlet-roast-marrowbone-dipper-leveret-syrup-giblets-wings.'

274 [*Epigraph*] LEYDEN's *Scenes of Infancy*: the last four lines (with slight variants) of Part 1 of John Leyden's descriptive poem (1803). Leyden (1775–1811) was a poet, editor and linguist who assisted Scott with his *Minstrelsy of the Scottish Border* and published collections of seventeenth- and eighteenth-century Scottish poetry. The lines P. quotes embody Leyden's youthful love for 'Aurelia', remembered after many years' separation.

[*P.'s note*] *Bacco in Toscana*: ll. 147–151. Francesco Redi (1626–98), an Aretine scientist, physician, and man of letters. *Bacco in Toscana*, written between 1666 and 1685, is a dithyrambic poem celebrating the supposed arrival of Bacchus in Tuscany. It was published with copious notes and is an example of Renaissance Hellenism that appealed to P. and his circle. Leigh Hunt's translation (1825) reads: ''Tis the true old Aurum Potabile/Gilding life when it wears shabbily:/Helen's old Nepenthe 'tis,/That in the drinking swallowed thinking/And was the receipt for bliss.' Nepenthe was a drug which drove away care; Helen of Troy used it to soothe her husband Menelaus and his Greek visitors, oppressed by memories of the Trojan War (*Odyssey*, v).

Venus Calva: see the two following notes. The name is now thought to be due to the existence of a bald female statue, once regarded as a statue of Venus.

275 [*P.'s note*] JULIUS CAPITOLINUS ...: Capitolinus was a minor historian in the reign of Constantine (third to fourth century AD); his biography of the Emperor Maximinus and his son (P. refers to the latter) can be read in *Scriptores Historiae Augustae*, 376–8 (Loeb, ii). Capitolinus connects the story with the siege of Rome by the Gauls in AD 382 (see Servius's account, referred to in the next note but one), and to a similar event when Maximinus attacked the Senate in AD 235.

276 *Penelope ... Angelo*: Penelope is Odysseus's faithful wife in the *Odyssey*; Helen the cause of the Trojan War in the *Iliad*. In *Orlando Furioso* Fiordiligi is faithful to her husband Bradamant, whereas Angelica encourages several of Charlemagne's knights to pursue her (see note to p. 156, above). Sacripant faithfully pursues Angelica and is last heard of setting out again after her; Rinaldo wavers between love and hatred of Angelica (this episode is dealt with more fully in *Orlando Innamorato*). Imogen is in *Cymbeline*, Calista in Beaumont and Fletcher's *The Lover's Progress* (see note to p. 310, below) and Angelo in *Measure for Measure*.

[*P.'s note*] SERVIUS *ad Aen. i. 720*: Servius's lengthy commentary on the *Aeneid* was written early in the fifth century. The passage is taken from his note to *Aeneid*, i. 720, and is translated in Curryfin's speech. (See the edn. of G. Thil and H. Hagen (1881), pp. 200–1.) The italicized passage gives an example of the use of the verb *calvo* in the active voice.

[*P.'s note*] SALLUST: *Hist. iii*: Fragment 109 in pt. 3 of Sallust's *Historiae*, a history of the Roman Republic from 78 to 67 BC, written at about 39 BC. See *Historiarum Reliquae*, ed. B. Maurenbrecher (1966), p. 156.

[*P.'s note*] PLAUTUS *in Casinâ*: act II, scene ii, l. 3 of the play *Casina*.

[*P.'s note*] . . . Δολίη: 'Guileful Aphrodite'. A phrase common in Homer.

[*P.'s note*] *Il. xiv. 217*: 'beguilement that steals the wits even of the wise.'

[*P.'s note*] . . .δολοπλοκε: 'Wile-weaving daughter of Zeus'. Sappho, Fragment 1, l. 2. Sappho, born about 630 BC on Lesbos, is the earliest of the 'monodic' poets (i.e. writers of songs sung to the harp) whose works survive.

277 *Plato holds* . . .: see, for example, *The Symposium*, 183E and 210E–211B.

278 *perverted by fiction*: Inez died in the manner P. specifies in 1325. The story became popular in European literature. During P.'s lifetime there were several plays and operas on the subject. P. reviewed a ballet-opera at the King's Theatre, Haymarket, in 1833 and criticized it for omitting the coronation of Inez (Halliford, ix. 429–30). In fact, all these versions omit Don Pedro's devastation of the country and the coronation, and in some of them Pedro's first wife is metamorphosed into a rival lover. An earlier operatic version given at the King's Theatre in the 1790s has, indeed, a happy ending when Alfonso relents and decides not to have Inez killed.

280 [*Epigraph*] TIBULLUS: *Elegies*, I. i. 69–72 (see the Loeb *Catullus, Tibullus and the Pervigilium Veneris*, p. 196). Tibullus (*c.* 50–19 BC) is best known for his amatory poems (of which this is one) addressed to his mistress 'Delia'.

282 . . . *on with the new*: see A. Cunningham, *The Songs of Scotland* (1825), ii. 352. The poem is called 'It's gude to be merry and wise'. Cunninghan thought the lines quoted (actually 'It's best to be off . . .') existed as a proverb long before they were incorporated into the poem.

'. . . *how I love you*': source as yet untraced.

283 *to have and to hold*: a phrase from the marriage service in the Book of Common Prayer.

. . . *that melancholy smile?*: Southey's *Thalaba*, xi. st. 33 (with 'that' for 'which' and 'resist' for 'gainsay').

. . . *nulla giova*: 'It's no use being a late repentant.' The source of this quotation is untraced.

. . . *look on me*: the first two lines of a song in Sheridan's *The Duenna*, I. ii.

Mirror of Knighthood: Butler's ironic description of Hudibras: pt. I, canto i, l. 16.

284 . . . *sayings of philosophers*: ibid., pt. I, canto iii, ll. 1011–12 (with 'myself' for 'himself').

. . . *the everlasting flint*: *Romeo and Juliet*, II. vi (with 'never wear' for 'ne'er wear out').

. . . *set him free*: Southey's *Thalaba*, ix, st. 8.

287 '*masterless passion*': source as yet untraced.

288 [*Epigraph*] PETRONIUS: *c. 34*: Trimalchio is the rich vulgarian the description of whose lavish banquet forms the central episode in what survives of *The Satyricon*. Whilst his guests are admiring the luxury of the feast, he suddenly causes this elaborate *memento mori* to be produced for him to moralize over.

twelfthcake . . . The characters: on Twelfth Night a large cake was served; it was accompanied by a set of figures printed on paper or card, which were placed in two hats or basins (one for girls and one for boys). Each child drew a card from the container and was thus assigned a 'character' for the evening. (See *Notes and Queries*, 8th series, vii. (1895), 58.) There were, apparently, many characters other than the king and queen, as Thackeray implies in the 'Prelude' to *The Rose and the Ring* (1854): 'the Lover, the Lady, the Dandy, the Captain, and so on.' The custom, which died out in the latter half of the century, replaced the ancient custom of having a bean baked in the cake: the person who found the bean in his or her slice was crowned 'bean-king' or '-queen'.

[*P.'s note*] *Miscellaneous Sonnets, no. 39*: ll. 7–8 of the first of four poems entitled (in 1820) 'Personal Talk': P. refers to the 1838 edition of *The Sonnets of William Wordsworth*.

290 *Divorce Court*: the 'Court for Divorce and Matrimonial Causes'

was set up in 1857; and the Divorce Act of the same year introduced many aspects of current British divorce law. The divorce rate in the years immediately following was significantly higher than it had been.

Commercial bubbles: speculative schemes were particularly prevalent in the 1820s and the 1840s, and legislation in 1855 and 1856 made it easier for companies to obtain certificates of incorporation and limited liability, which opened the way for further abuses.

high-pressure boilers: see note on 'explosions of powder-mills' to p. 208, above.

Peace Society: P. probably has in mind the Society for the Promotion of Permanent and Universal Peace, founded in 1816.

Apollyon: in Bunyan's *Pilgrim's Progress*, a diabolic figurc who tempts Christian, on his journey to the Celestial City, with the riches of the material world.

Pantopragmatic Society . . . Lord Facing-both-ways . . . departments: see notes to p. 137, above.

291 *Peregrine Pickle*: in Smollett's *Peregrine Pickle*, ch. 48.

Niebuhr: see note to p. 220, above. The view expressed here is common in his later personal writings. See, for example, his letter to Mme. Hensler, 19.12.1830, in *Life and Letters of Barthold George Niebuhr*, 2nd edn. (London, 1852), ii. 416 ff.

Vandals on northern thrones: P. refers to Tsar Nicholas I, who, as well as notoriously persecuting intellectuals after the 1848 revolutions in Europe, also helped provoke the Crimean War; and to Frederick William IV of Prussia (and possibly his brother William who was regent during Frederick's insanity 1858–61). Nicholas died in 1855. His successor Alexander II was much more liberal, but his reforms were not under way when P. wrote.

[*P.'s note*] COLLINS: *Ode on the Manners*: ll. 3–4.

292 *southern anarchy*: P. probably refers to wars in the Italian states in the years before Italy was unified (1861). In 1859 Piedmont, in league with France, was at war with Austria over its dominion in Northern Italy; the central Italian states rose against Austria; and there was an insurrection in southern Italy and Sicily. There was much internal opposition between rival nationalist and republican groups. The British public had

become increasingly aware of political conditions in Italy after the Peace Conference in 1856 following the Crimean War.

[*P.'s note*] HOR.: *Carm.* II. *i*: 'You make your way through life among fires covered over with treacherous ashes.' Horace: *Odes*, II. i. 7–8.

293 [*P.'s note*] *La Vedova Scaltra a. 3, s.10*: 'The Artful Widow.' This dialogue is from act II, scene iii, in the translation by Frederick Davies (Penguin, 1968): 'Your servant, ma'am. Can I get you some coffee? (*She shakes her head.*) Some chocolate? (*She shakes her head.*) Some punch? (*She nods.*) She's English all right (*aside*).'

[*P.'s note*] *Pamela Fancuilla, a.* I. *s. 15*: this play, originally titled *Pamela*, is based on Richardson's novel; first performed in 1750 and translated into English in 1756. A modern translation would read: B: My friend, will you drink with us?/ C: Tea would be welcome./ A: It is a wholesome drink./ B: Would you like arrack?/ C: Arrack? Yes./ B: Allow me./

294 [*Epigraph*] *the Persae of Aeschylus*: ll. 840–2 of *Persae* ('The Persians'). After the battle of Salamis (480 BC) the ghost of Darius, King of Persia, advises his countrymen to be content with their lot and not to embark on further wars with Greece.

296 [*P.'s note*] *Il. xxii. vv. 500, 501*: 'Astyanax, that aforetime on his father's knees ate only marrow and the rich fat of sheep.'

[*P.'s note*] JUVENAL: *Sat. x. v. 346, sqq.*: 'If you ask my counsel, you will leave it to the Gods themselves to provide what is good for us, and what will be serviceable for our state; for, in place of what is pleasing, they give us what is best.'

298 [*P.'s note*] *Thalaba*: bk. ii, st. 33.

299 *burlesques*: theatrical parodies. By the mid nineteenth century burlesque had degenerated from satire (as in Fielding's *Tom Thumb* or Sheridan's *The Critic*) to extravagantly absurd and fantastic skits closely related to pantomine.

[*P.'s note*] *Equites*: l. 516 of *Knights*, Aristophanes's second surviving comedy and the first that he himself produced (424 BC).

300 . . . *the table alternately*: no source for this story has been found.

executor, administrator, or assign: a standard legal phrase in wills and other documents, meaning the person(s) who legally act(s) for one after death.

feralis cœna: funeral dinner.

301 *Ygdrasil*: P. here confuses the squirrel Ratatosk ('rat-tooth') with the serpent Nidhogger. These creatures of Norse mythology are assigned their proper roles in *Melincourt*, ch. 40 (Halliford, ii. 434): 'The ash of Yggdrasil overshadows the world: Ratatosk, the squirrel, sports in the branches: Nidhogger the serpent gnaws at the root.'

[*P.'s note*] *Pharsalia, l. i. vv. 458–462*: 'Truly the nations on whom the Pole star looks down are happily deceived [in thinking that death is only a point in the midst of a continuous life]; for they are free from that king of terrors, the fear of death. This gives the warrior his eagerness to rush upon the steel, his courage to face death and his conviction that it is cowardly to be careful of a life, which will come to him again.' P. is, perhaps, applying this passagc to the Norse myth of Valhalla, where heroes carouse after death until summoned to fight at the day of doom. Lucan's epic poem *Pharsalia* (written before AD 65) recounts the civil wars of Caesar and Pompey and Caesar's military campaigns against the 'barbarians'. The poem was a favourite of P. and Shelley.

302 *Noscitur à sociis*: i.e. you can judge someone by the company he keeps.

303 [*Epigraph*] *Prometheus*: *Prometheus Bound*, ll. 837–43.

306 [*Epigraph*] SOUTHEY: *The Grandmother's Tale*: ll. 1–5 (pub. in *English Eclogues*, 1799).

merveilleuses histoires . . . pays natal: 'stories of the marvellous told around the hearth' popular with those 'who have not left their native land'. From Chateaubriand's *Atala* (1801; pp. 77–8 in the edn. of J. M. Gautier, Geneva), a story of two Red Indian lovers partly based on fact. It was enormously popular and the speech from which these lines are taken was, along with others, set to music and sung on concert platforms throughout Europe. It is a speech of melancholy longing for one's native land, and there is clearly some irony in Miss Gryll's directing the quotation at Dr Opimian.

Patroclus: see *Iliad*, xxiii.

Darius: see note to p. 294, above.

Polydorus: in Euripides' play *Hecuba*, the ghost of Polydorus, Hecuba's son, appears in Thrace after Troy has fallen.

Bride of Corinth: in Goethe's poem (published 1798) difference of religion plays an important part. The young man, a 'heathen' from Athens, journeys to Corinth, where he enjoys the nocturnal company of his (dead) betrothed, whose family is Christian. They exchange a golden headband and a silver bowl. When the young woman's mother disturbs them, she upbraids the mother for not having given her a proper pagan burial, prophesies that the young man will die in the morning, and demands that a funeral pyre be built for them both.

307 [*P.'s note*] *Tales of Wonder, v. i. 99*: M. G. Lewis is best known as the author of the 'Gothic' novel *The Monk* (1796). *Tales of Wonder* was published in 1801. P.'s page reference is to the first of three editions published in that year.

[*P.'s note*] *Phlegon*: little is known about Phlegon except that he was a freed-man of the Emperor Hadrian (AD 117–38) and that he wrote, in addition to 'On Wonderful Things', a history of his own times, a treatise on longevity and one on the Olympiads. 'On Wonderful Things' can be read, in Greek, in A. Westerman (ed.) *Paradoxographoi* (1839) or in Latin in the edn. of J. Meursius (1620); there is no English translation.

Niceros: *Satyricon*, sects. 61–3.

308 *Virgil's Copa*: the *Copa* is a short poem describing a *copa*, a barmaid and dancer. The poem is now generally thought not to be Virgil's. Opimian follows Servius (see note to p. 276, above) in attributing the poem to Virgil (Loeb, ii. 449–51). It contains veiled references to illicit sexual activity: this may account for Opimian's view that it is 'only half panegyrical'.

demi-caractère: used of 'a dance which retains the form of the character dance but is executed with steps based on the Classical technique' (*OED* Suppl. 1st citation 1776).

309 *Genii*: guardian spirits.

little Iphis, the delight of the family: the name Iphis has disappeared from modern editions of Petronius which read *ipsimi nostri delicatus*, i.e. our master's favourite or pet. The 1781 Leipzig edn., however, reads *Iphis nostri delicatus* following a conjecture of Heinsius for a corrupt MS reading. He derives the name from the girl/boy figure in Ovid's *Metamorphoses*, ix.

Cappadocian: Cappadocia was in the east of what is now Turkey. According to Lemprière's *Classical Dictionary*, Cappadocians were regarded by the Greeks as a backward people.

[*P.'s note*] *Chian life*: only in this passage of Petronius is a Chian life synonymous with a life of luxury, but the inhabitants of Chios (the Aegean island now known as Skios) were renowned for their wine and cooking (Athenaeus, i. 25E).

310 *Lover's Progress*: Curryfin's tale is taken from Beaumont and Fletcher's tragedy *The Lover's Progress* (1623); the lines quoted are from III. i and IV. i.

. . . *tout honneur*: with the best and most honourable intentions.

311 *the Castle Spectre*: an extremely popular play written by M. G. Lewis in 1796, when tales of terror became a common subject for drama as well as novels.

Wieland: the hero, a religious enthusiast seeking direct communication with God, misguidedly assumes that a ventriloquist's words are of supernatural origin. Subsequently an inner voice commands him to kill his wife and four children. He commits suicide when he discovers his mistake. *Wieland* (1798) is the best-known novel of Charles Brockden Brown (1771–1810), the originator of 'Gothic' fiction in America.

Cock-lane ghost: a famous imposture. In 1762 in Cock Lane, London a 'ghost' rapped on a wall in the house of Thomas Kent and, when questioned, answered in coded knocks that Kent's wife, recently dead, had been poisoned by her husband. The 'ghost' turned out to be the 12-year-old daughter of Kent's neighbour, who had a grudge against him. See *Sketches of Imposture, Deception and Credulity* (anon., 1838), pp. 149 ff. Dr Johnson gives an account of the enquiry. See Boswell's *Life of Johnson*, ed. G. B. Hill, i. 409.

312 *buried treasure*: Miss Niphet's story unites two fairly common motifs in folklore, that of the ghost which cannot rest until it has atoned for some crime committed during life, and that of a ghost which directs someone to buried treasure. The closest analogue in folklore is 'The Old Lady of Little Dean' (see K. Briggs, *Dictionary of British Folk-Tales* (1970), and the sources given there); but this was not printed until 1866. There may be a more direct literary source.

313 *bogle*: a phantom, goblin, or sprite. The word has a Scottish flavour, having appeared in the works of Burns, Scott, and others.

[*P.'s note*] *Tales of Superstition and Chivalry*: by Anne Bannerman (died 1829). Pub. anonymously in 1802. The lines are from

'The Fisherman of Lapland': the shadow is cast by the (invisible) 'ghost' of a drowned fisherman.

[*P.'s note*] *The Three Brothers*, vol. iv. p. 193: Mr Gryll's quotation appears on p. 193 of this novel by Joshua Pickersgill, with 'shades' for 'shadows'. The story actually begins on p. 189 and continues for many pages of Gothic excitement.

314 '... *a bogle may fly away with it*': the tale rests on the superstition that a heartfelt wish or curse will be fulfilled when it involves demonic agency. The best-known example is Chaucer's *Friar's Tale*.

[*P.'s note*] *Tam O'Shanter*: ll. 163–6 of *Tam O'Shanter* by Burns, the story of which has much in common with MacBorrowdale's tale. 'Wawlie' means 'sturdy, buxom'.

315 *Saint Laura*: the only recorded saint of this name was an abbess at Cuteclara near Cordova in Spain. She is said to have been martyred in 864 when the Moslem authorities had her thrown into a cauldron of boiling lead. The source of P.'s story of the abbess is untraced; but the rejection from a sacred place of the corpse of an unholy person is a fairly common folk-tale motif.

[*P's note*] DANTE: *Inferno*, iii. 95–6.

318 [*Epigraph*] ... *Odyssey*: ll. 180–4. The translation is apparently P's. On Nausicaa see note to p. 124, above.

Sir Alley Capel: a speculator in money. The name derives from 'the Alley' (or Exchange Alley) and Capel Court, the place in London where dealing in stocks was carried on.

319–20

Pachyderm ... *Larval*: 'Pachyderm' means 'thick-skinned', insensitive; 'Enavant', 'forward!'; 'Geront', 'old'. 'Larval' derives, perhaps, from French 'larve' ('mask'): 'larval' is a medical term applied to 'certain diseases in which the skin of the face is disfigured as if covered by a mask' (*OED*).

321 ... *wiped them soon*: *Paradise Lost*, xii. 645, with 'shed' for 'dropped', referring to Adam and Eve on their expulsion from Paradise.

'*his saul abune the moon*': source as yet untraced.

322 [*P.'s note*] SIMONIDES: Greek lyric poet (*c.*556–468 BC) whose work survives only in fragments. The lines quoted are now attributed to Simonides or to Epicharmus, and are quoted in Athenaeus, xv. 694 (Loeb, vii. 221).

[*P.'s note*] ATHENAEUS: *l. xv. p. 694*: Anaxandrides, whom Athenaeus quotes, was a dramatist of the early and mid fourth century BC. Only fragments of his plays are now extant: the quoted lines are from a lost play called 'The Treasure'. (Loeb *Athenaeus*, vii. 223.)

[*P.'s note*] *Metamorph. l. iv*: the quotation is from bk. iv, ch. 18: 'the god of happy and successful outcomes having been called upon'.

PREFACE OF 1837

323 *out of the road*: by 1837 the old road which crossed the Llugwy into Capel Curig had been replaced by a new road, by-passing the village and running along the north bank of the river (the present route of the A5); but the inn was still recommended to tourists. See W. Bingley, *Excursions in North Wales*, 3rd edn., 1839, frontispiece (map) and p. 43. The coach journey was considerably faster than in 1815: travellers arrived at Capel Curig at about 7 p.m, roughly 24 hours after leaving London. See note to p. 1, above, and *Leigh's New Pocket Road-Book*, 6th edn., 1837.

successors: 'rotten boroughs' were urban electoral constituencies, which, owing to population decline, no longer had a substantial number of voters and could therefore easily be controlled by the crown or a wealthy landowner. They were abolished by the Reform Act of 1832. 'Pocket properties' (or 'pocket boroughs') is a more general term for a borough effectually controlled by a single landowner or family, who was said to 'carry the borough in his pocket'. See note to p. 97, above.

march of intellect: a catch phrase (along with 'march of mind') in the late 1820s and 1830s, the period which saw the foundation of the Mechanics' Institutions as well as the Society for the Diffusion of Useful Knowledge. The phrase figures prominently in *Crotchet Castle* as well as *GG*. See note to p. 102, above.

324 '*learned friend*': i.e. Lord Brougham. The phrase is Parliamentary jargon for a Member who is also a lawyer. For Brougham see notes to pp. 100 and 137, above.

THE WORLD'S CLASSICS

A Select List

SERGEI AKSAKOV: A Russian Gentleman
Translated by J. D. Duff
Edited by Edward Crankshaw

A Russian Schoolboy
Translated by J. D. Duff
With an introduction by John Bayley

Years of Childhood
Translated by J. D. Duff
With an introduction by Lord David Cecil

JANE AUSTEN: Emma
Edited by James Kinsley and David Lodge

Mansfield Park
Edited by James Kinsley and John Lucas

Northanger Abbey, Lady Susan, The Watsons, *and* Sanditon
Edited by John Davie

Persuasion
Edited by John Davie

Pride and Prejudice
Edited by James Kinsley and Frank Bradbrook

Sense and Sensibility
Edited by James Kinsley and Claire Lamont

CHARLOTTE BRONTË: Jane Eyre
Edited by Margaret Smith

Shirley
Edited by Margaret Smith and Herbert Rosengarten

FANNY BURNEY: Camilla
Edited by Edward A. Bloom and Lilian D. Bloom

Evelina
Edited by Edward A. Bloom

ANTON CHEKHOV: The Russian Master and Other Stories
Translated by Ronald Hingley

Five Plays
Translated by Ronald Hingley

WILKIE COLLINS: The Moonstone
Edited by Anthea Trodd

No Name
Edited by Virginia Blain

The Woman in White
Edited by Harvey Peter Sucksmith

JOSEPH CONRAD: Lord Jim
Edited by John Batchelor

The Nigger of the 'Narcissus'
Edited by Jacques Berthoud

Nostromo
Edited by Keith Carabine

The Secret Agent
Edited by Roger Tennant

The Shadow-Line
Edited by Jeremy Hawthorn

Typhoon & Other Tales
Edited by Cedric Watts

Under Western Eyes
Edited by Jeremy Hawthorn

Victory
Edited by John Batchelor
With an introduction by Tony Tanner

Youth, Heart of Darkness, The End of the Tether
Edited by Robert Kimbrough

DANIEL DEFOE: Moll Flanders
Edited by G. A. Starr

Robinson Crusoe
Edited by J. Donald Crowley

Roxana
Edited by Jane Jack

CHARLES DICKENS: David Copperfield
Edited by Nina Burgis

Dombey and Son
Edited by Alan Horsman

Little Dorrit
Edited by Harvey Peter Sucksmith

Martin Chuzzlewit
Edited by Margaret Cardwell

The Mystery of Edwin Drood
Edited by Margaret Cardwell

Oliver Twist
Edited by Kathleen Tillotson

Sikes and Nancy and Other Public Readings
Edited by Philip Collins

BENJAMIN DISRAELI: Coningsby
Edited by Sheila M. Smith

Sybil
Edited by Sheila M. Smith

FËDOR DOSTOEVSKY: Crime and Punishment
Translated by Jessie Coulson
With an introduction by John Jones

Memoirs from the House of the Dead
Translated by Jessie Coulson
Edited by Ronald Hingley

JOHN GALT: Annals of the Parish
Edited by James Kinsley

The Entail
Edited by Ian A. Gordon

The Provost
Edited by Ian A. Gordon

ELIZABETH GASKELL: Cousin Phillis and Other Tales
Edited by Angus Easson

Cranford
Edited by Elizabeth Porges Watson

Mary Barton
Edited by Edgar Wright

North and South
Edited by Angus Easson

Ruth
Edited by Alan Shelston

Sylvia's Lovers
Edited by Andrew Sanders

THOMAS HARDY: A Pair of Blue Eyes
Edited by Alan Manford

Jude the Obscure
Edited by Patricia Ingham

Under the Greenwood Tree
Edited by Simon Gatrell

The Well-Beloved
Edited by Tom Hetherington

The Woodlanders
Edited by Dale Kramer

ALEXANDER HERZEN: Childhood, Youth and Exile
With an introduction by Isaiah Berlin

Ends and Beginnings
Translated by Constance Garnett
Revised by Humphrey Higgens
Edited by Aileen Kelly

HOMER: The Iliad
Translated by Robert Fitzgerald
With an introduction by G. S. Kirk

The Odyssey
Translated by Walter Shewring
With an introduction by G. S. Kirk

HENRY JAMES: The Ambassadors
Edited by Christopher Butler

The Aspern Papers and Other Stories
Edited by Adrian Poole

The Awkward Age
Edited by Vivien Jones

The Bostonians
Edited by R. D. Gooder

Daisy Miller and Other Stories
Edited by Jean Gooder

The Europeans
Edited by Ian Campbell Ross

The Golden Bowl
Edited by Virginia Llewellyn Smith

The Portrait of a Lady
Edited by Nicola Bradbury
With an introduction by Graham Greene

Roderick Hudson
With an introduction by Tony Tanner

The Spoils of Poynton
Edited by Bernard Richards

Washington Square
Edited by Mark Le Fanu

What Maisie Knew
Edited by Douglas Jefferson

The Wings of the Dove
Edited by Peter Brooks

RUDYARD KIPLING: The Day's Work
Edited by Thomas Pinney

The Jungle Book (in two volumes)
Edited by W. W. Robson

Kim
Edited by Alan Sandison

Life's Handicap
Edited by A. O. J. Cockshut

The Man Who Would be King and Other Stories
Edited by Louis L. Cornell

Plain Tales From the Hills
Edited by Andrew Rutherford

Stalky & Co.
Edited by Isobel Quigly

WALTER PATER: Marius the Epicurean
Edited by Ian Small

The Renaissance
Edited by Adam Phillips

ANN RADCLIFFE: The Italian
Edited by Frederick Garber

The Mysteries of Udolpho
Edited by Bonamy Dobrée

The Romance of the Forest
Edited by Chloe Chard

SIR WALTER SCOTT: The Heart of Midlothian
Edited by Claire Lamont

Redgauntlet
Edited by Kathryn Sutherland

Waverley
Edited by Claire Lamont

TOBIAS SMOLLETT: The Expedition of Humphry Clinker
Edited by Lewis M. Knapp
Revised by Paul-Gabriel Boucé

Peregrine Pickle
Edited by James L. Clifford
Revised by Paul-Gabriel Boucé

Roderick Random
Edited by Paul-Gabriel Boucé

Travels through France and Italy
Edited by Frank Felsenstein

LAURENCE STERNE: A Sentimental Journey
Edited by Ian Jack

Tristram Shandy
Edited by Ian Campbell Ross

ROBERT LOUIS STEVENSON: The Master of Ballantrae
Edited by Emma Letley

Treasure Island
Edited by Emma Letley

LEO TOLSTOY: Anna Karenina
Translated by Louise and Aylmer Maude
With an introduction by John Bayley

The Raid and Other Stories
Translated by Louise and Aylmer Maude
With an introduction by P. N. Furbank

War and Peace (in two volumes)
Translated by Louise and Aylmer Maude
Edited by Henry Gifford

ANTHONY TROLLOPE: The American Senator
Edited by John Halperin

An Autobiography
Edited by P. D. Edwards

Ayala's Angel
Edited by Julian Thompson-Furnival

Barchester Towers
With an introduction by James R. Kincaid

The Belton Estate
Edited by John Halperin

Can You Forgive Her?
Edited by Andrew Swarbrick
With an introduction by Kate Flint

The Claverings
Edited by David Skilton

Dr. Thorne
Edited by David Skilton

Dr. Wortle's School
Edited by John Halperin

The Duke's Children
Edited by Hermione Lee

The Eustace Diamonds
Edited by W. J. McCormack

Framley Parsonage
Edited by P. D. Edwards

He Knew he was Right
Edited by John Sutherland

Is He Popenjoy?
Edited by John Sutherland

The Kelly's and the O'Kelly's
Edited by W. J. McCormack
With an introduction by William Trevor

The Last Chronicle of Barset
Edited by Stephen Gill

Orley Farm
Edited by David Skilton

Phineas Finn
Edited by Jacques Berthoud

Phineas Redux
Edited by John C. Whale
Introduction by F. S. L. Lyons

The Prime Minister
Edited by Jennifer Uglow
With an introduction by John McCormick

The Small House at Allington
Edited by James R. Kincaid

The Warden
Edited by David Skilton

The Way We Live Now
Edited by John Sutherland

VILLIERS DE L'ISLE-ADAM: Cruel Tales
Translated by Robert Baldick
Edited by A. W. Raitt

VIRGIL: The Aeneid
Translated by C. Day Lewis
Edited by Jasper Griffin

The Eclogues and The Georgics
Translated by C. Day Lewis
Edited by R. O. A. M. Lyne

HORACE WALPOLE: The Castle of Otranto
Edited by W. S. Lewis

IZAAK WALTON and CHARLES COTTON:
The Compleat Angler
Edited by John Buxton
With an introduction by John Buchan